KV-240-296

- NOV 2024

WITHDRAWN FOR SALE

Autoimmune Liver Disease

FALK SYMPOSIUM 142

Autoimmune Liver Disease

Edited by

H.-P. Dienes
Institute for Pathology
University Clinic
Cologne
Germany

A.W. Lohse
Department of Medicine
University Medical Centre Hamburg-
Eppendorf
Germany

U. Leuschner
Medical Clinic
Johann Wolfgang Goethe University
Frankfurt
Germany

M.P. Manns
Department of Gastroenterology,
Hepatology and Endocrinology
Medical School Hannover
Hannover
Germany

Proceedings of the Falk Symposium 142 held in Freiburg, Germany,
October 12–13, 2004

Library of Congress Cataloging-in-Publication Data is available.

ISBN-10 1-4020-2894-6 HB
ISBN-13 9781-4020-2894-6 HB

Published by Springer,
PO Box 17, 3300 AA Dordrecht, The Netherlands

Sold and distributed in North, Central and South America
by Springer,
101 Philip Drive, Norwell, MA 02061 USA

In all other countries, sold and distributed
by Springer,
PO Box 322, 3300 AH Dordrecht, The Netherlands

Printed on acid-free paper

All Rights Reserved
© 2005 Springer and Falk Foundation e.V.

No part of the material protected by this copyright notice may be reproduced or utilized in any form or by any means, electronic, mechanical, including photocopying, recording or by any information storage and retrieval system, without written permission from the copyright owners.

Printed and bound in Great Britain by MPG Books Limited, Bodmin, Cornwall.

Contents

List of principal contributors

F Alvarez
Division of Gastroenterology,
  Hepatology and Nutrition
Sainte-Justine Hospital
3175 chemin de la Côte Sainte-
  Catherine
Montreal, Quebec, H3T 1C5
Canada

B Arnold
Deutsches Krebsforschungszentrum
  DKFZ
Dept of Molecular Immunology
Im Neuenheimer Feld 280
D-69120 Heidelberg
Germany

MF Bassendine
Liver Research Group
School of Clinical Medical Sciences
The Medical School, University of
  Newcastle
4th Floor, William Leech Building
Framlington Place
Newcastle upon Tyne, NE2 4HH
UK

PA Berg
Innere Medizin II
Klinikum der Universität Tübingen
Otfried-Müller-Str. 10
D-72076 Tübingen
Germany

P Bertolino
AW Morrow Gastroenterology and
  Liver Centre
Centenary Institute of Cancer
  Medicine and Cell Biology
Locked Bag No. 6
Newtown, NSW 2042
Australia

FB Bianchi
Università di Bologna
Policlinico S. Orsola
Istituto di Clinica Medica
Generale e Terapia Medica
Via Massarenti 9
I-40138 Bologna
Italy

KM Boberg
Medical Department
Rikshospitalet
N-0027 Oslo
Norway

M Burdelski
Pädiatrische Gastroenterologie
Universitätsklinikum Eppendorf
Martinistr. 52
D-20246 Hamburg
Germany

RW Chapman
Department of Hepatology and
  Gastroenterology
Oxford Radcliffe Hospital
Oxford, OX3 9DU
UK

IR Cohen
Department of Immunology
The Weizmann Institute of Science
IL-76100 Rehovot
Israel

AJ Czaja
Division of Gastroenterology and
  Hepatology
Mayo Clinic, 200 First Street SW
Rochester
Minnesota 55905
USA

VJ Desmet
Catholic University of Leuven
University Hospital St. Rafael
Department of Pathology
Minderbroederstraat 12
B-3000 Leuven
Belgium

HP Dienes
Institute for Pathology
University Clinic
Joseph-Stelzmann-Str. 9
D-50931 Köln
Germany

PT Donaldson
Complex Genetics Research Group
School of Clinical Medical Sciences
The Medical School, University of
  Newcastle
Framlington Place
Newcastle upon Tyne, NE2 4HH
UK

RA Flavell
Yale University School of Medicine
Howard Hughes Medical Institute
333 Cedar Street
New Haven, CT 06520
USA

PR Galle
Innere Medizin I
Klinikum der Universität
Langenbeckstr. 1
D-55131 Mainz
Germany

ME Gershwin
Division of Rheumatology
Allergy and Clinical Immunology
University of California at Davis
  School of Medicine, TB 192
Davis, CA 95616
USA

JL Heathcote
University of Toronto
Toronto Western Hospital
Fell Pavilion 6-170
399 Bathurst Street
Toronto, ON M5T 2S8
Canada

J Herkel
Department of Medicine
University Medical Clinic Eppendorf
Martinistr. 52
D-20246 Hamburg
Germany

MG von Herrath
Department of Developmental
  Immunology
La Jolla Institute for Allergy and
  Immunology
10355 Science Center Drive
San Diego, CA 92121
USA

DEJ Jones
School of Clinical Medical Sciences
4th Floor, William Leech Building
Medical School
Framlington Place
Newcastle upon Tyne, NE2 4HH
UK

D Kelly
The Liver Unit
Birmingham Children's Hospital
Steelhouse Lane
Birmingham, B4 6NH
UK

PH Krammer
Immungenetik D030
Deutsches Krebsforschungszentrum
Im Neuenheimer Feld 280
D-69120 Heidelberg
Germany

EL Krawitt
University of Vermont
College of Medicine
Given Medical Building
Burlington, VA 05405-0068
USA

U Leuschner
Medizinische Klinik
Johann Wolfgang Goethe Universität
D-60596 Frankfurt
Germany

AW Lohse
Department of Medicine
University Medical Clinic Eppendorf
Martinistr. 52
D-20246 Hamburg
Germany

IR Mackay
Monash University
Centre for Molecular Biology and
  Medicine
Clayton, Victoria 3168
Australia

MP Manns
Department of Gastroenterology,
  Hepatology and Endocrinology
Medical School Hannover
Carl-Neuberg-Str. 1
D-30625 Hannover
Germany

IG McFarlane
King's College Hospital
Liver Unit
Bessemer Road
London, SE5 9PJ
UK

K-H Meyer zum Büschenfelde
Trabener Str. 8
D-14193 Berlin
Germany

G Mieli-Vergani
Paediatric Liver Service
Institute of Liver Studies
King's College Hospital
London, SE5 9RS
UK

J Neuberger
The Liver and Hepatobiliary Unit
Queen Elizabeth Hospital
Birmingham, B15 2TH
UK

R Poupon
Hôpital Saint-Antoine
Service d'Hépato-Gastro-
  Entérologie
184, rue du Fauburg St. Antoine
F-75571 Paris
France

E Schrumpf
Medical Department
Rikshospitalet
N-0027 Oslo
Norway

A Stiehl
Innere Medizin IV
Klinikum der Universität
Bergheimer Str. 58
D-69115 Heidelberg
Germany

CP Strassburg
Department of Gastroenterology,
  Hepatology and Endocrinology
Hannover Medical School
Carl-Neuberg-Str. 1
D-30625 Hannover
Germany

D Vergani
Institute of Liver Studies
King's College Hospital
Denmark Hill
London, SE5 9RS
UK

H Wekerle
Neuroimmunologie
Max-Planck-Institut für
  Neurobiologie
Am Klopferspitz 18a
D-82152 Planegg
Germany

R Williams
University College Hospital Medical
  School
Institute of Hepatology
69–75 Chenies Mews
London, WC1E 6HX
UK

Preface

Autoimmune liver diseases, namely autoimmune hepatitis (AIH), primary biliary cirrhosis (PBC), primary sclerosing cholangitis (PSC) and their variants, occur world-wide, and in all age groups. The clinical presentation and course of the diseases can vary enormously, making diagnosis and management difficult. In the absence of a well-understood aetiology of autoimmune liver diseases, disease classification is based on consensus rather than hard facts. The clinical observation that many patients present features of more than one autoimmune liver disease makes definitions difficult and opens up controversies in diagnostic criteria as well as in treatment strategies. The variable natural history adds further difficulties in assessing the effectiveness of therapeutic interventions. On the other hand, enormous progress has been made in diagnosis and management of autoimmune liver diseases. In view of these difficulties and controversies it is important to bring together experts in the field in order to discuss the latest developments in basic and clinical research related to autoimmune liver diseases. This book, based on lectures given at the Falk Symposium 142, brings together all the relevant aspects of these interesting and clinically most relevant diseases. Chapters on basic mechanisms of immune regulation and dysregulation illuminate possible factors in the pathogenesis of autoimmune liver diseases. Diagnostic criteria and the best approach to patients are discussed. State-of-the-art recommendations on the management of patients with autoimmune liver diseases are combined with discussions of novel therapeutic approaches. An additional focus is placed on paediatric autoimmune liver disease, as morbidity in children is often the most serious, requiring liver transplantation in a much higher proportion of patients than in adults.

On behalf of my co-editors I wish to thank all the authors for their up-to-date reviews and their cooperation. Special thanks go to Dr. Christoph Schramm for his advice and help in organizing the meeting. Thanks also to Springer and Lancaster Publishing Services for their expert editing and flexible and pleasant cooperation. Most of all, thanks to the Falk family and the Falk Foundation, who for so many years have greatly helped to advance the field of hepatology, and the team at the Falk Foundation, who were perfect in all aspects of the management of the meeting and this book, and always a great pleasure to work with.

This book is both an excellent introduction to the field for the non-expert and a rich resource of the latest advances and discussion for the specialist. It will hopefully help all readers in their understanding of these interesting diseases, and thus help our present and future patients through better diagnosis and treatment of their condition.

Ansgar W. Lohse

Section I
Immune regulation I

Chair: K.-H. MEYER ZUM BÜSCHENFELDE and H. WEKERLE

1
Transforming growth factor-β in T cell tolerance

M. O. LI, Y. LAOUAR, Y. PENG, L. GORELIK and R. A. FLAVELL

INTRODUCTION

Understanding the regulatory mechanisms of immunity and tolerance remains a major challenge in immunology. As an immunosuppressive cytokine, tumour growth factor beta (TGF-β) plays a key role in dampening excessive pathological immune responses *in vivo*. Mice deficient in TGF-β developed severe multifocal inflammatory disease and succumbed to death shortly after birth[1]. Since TGF-β can be produced by – and act on – virtually every cell type, the regulatory network invoked by TGF-β remains incompletely understood. To investigate the role of TGF-β signalling in T cells we engineered a strain of transgenic mice expressing a dominant negative mutant of TGF-β receptor II[2]. In these mice the T cells were found to spontaneously differentiate into effector T cells, which critically influence the development of autoimmunity, tumour immunity and immune responses to pathogens. In addition, we took a gain-of-function approach to study TGF-β regulation of autoimmune diabetes by overexpressing this cytokine in the islets of the pancreas[3]. We demonstrated that a short pulse of TGF-β expression could expand the regulatory T cell population and inhibit diabetes in NOD mice. These studies have thus identified multiple mechanisms by which TGF-β regulates T cell tolerance.

MATERIALS AND METHODS

Transgenic mice

Mice expressing a dominant negative mutant of TGF-β receptor II under the control of CD4 promoter (DNR mice) were previously described[2]. All mice were kept under SPF conditions at Yale animal facility according to the approved protocols.

3

Tumour model

B16-F10 melanoma tumour cell line that is syngeneic to the C57BL/6 background was provided by A. Garen. EL4 cells were obtained from the American Type Culture Collection (Manassas, Virginia). Wild-type and DNR mice on C57BL/6 background were challenged with either B16-F10 cells intravenously or EL4 cells intraperitoneally and monitored for tumour growth.

Infection model

DNR mice were backcrossed onto BALB/c genetic background as described[4]. Wild-type and DNR mice were infected in the right hind foot with *Leishmania major* promastigoes of the WR309 substrain and monitored for disease progression.

Diabetes model

Under the control of a rat insulin promoter (RIP), tetracycline-controlled transactivator (TTA) is expressed specifically in insulin-producing cells to ensure a regulated TGF-β expression in TTA/TGF-β NOD mice as previously described[3]. NOD transgenic mice were fed with normal or doxycycline-supplemented food and monitored for disease development.

RESULTS

TGF-β and T effector cells

Mice with a T cell-specific blockade of TGF-β signalling developed immuno-pathology in multiple organs including lung, colon and kidney. Both CD4$^+$ and CD8$^+$ T cells from the DNR mice readily differentiated into effector/memory T cells and secreted cytokines upon *in-vitro* stimulation. Consistently, levels of T-dependent classes of immunoglobulins in the sera were found increased. These observations revealed an essential function for TGF-β in T cell tolerance in mice[2]. To investigate how TGF-β regulates tumour immunity we challenged mice with syngeneic tumour cells B16-F10 or EL4. In both cases the DNR mice were found resistant to the tumour challenge. TGF-β signalling in CD8$^+$ T cells was further shown to be essential for this protection, which is associated with the expansion of tumour-specific CD8$^+$ T cells[5]. CD4$^+$ T cells play a critical role in orchestrating both the humoral and cellular arms of immune responses. CD4$^+$ T cells from BALB/c mice are known to differentiate to a skewed Th2 phenotype which renders these mice susceptible to infection by intracellular pathogens such as *Leishmania*. To study the function of TGF-β signalling in pathogen infection, DNR mice were backcrossed to the BALB/c background and their response to *Leishmania major* infection investigated. Significantly, DNR mice were found to be resistant to the infection, which is associated with enhanced Th1-type cytokine production. Interestingly, Th2 cytokines were also found to be elevated in the DNR mice[6]. These observations suggested that

TGF-β inhibits both types of effector T cell differentiation. T helper cell differentiation is regulated by environmental cues and associated with epigenetic changes in the cytokine loci. Lineage determination factors, including GATA-3 for Th2 and T-bet for Th1 differentiation, were identified. Significantly, TGF-β inhibited the expression of both factors which contributed to its regulation of T helper cell differentiation[4,6].

TGF-β AND T REGULATORY CELLS

Several studies have demonstrated that CD4⁺CD25⁺ regulatory T cells produce elevated levels of TGF-β[7]. The fact that enhanced TGF-β signalling receptors reside on the membrane of CD4⁺CD25⁺ regulatory T cells underscores the potential for autocrine and/or paracrine receptor–ligand interactions in these cells. In this study we provide direct evidence that TGF-β is a positive regulator of CD4⁺CD25⁺ regulatory T cell expansion *in vivo*. We generated mice in which TGF-β expression can be induced temporally by the tetracycline regulatory system[3]. Data from RT-PCR and histochemistry studies showed that TGF-β gene and protein expression can be turned on and off in the islets efficiently within 1 week after changing the diet to a doxycycline-containing food source. Using this system, we were able to control and target the expression of the transgene at specific stages of diabetes development. Transgenic mice from all groups were monitored for diabetes development in comparison to transgene-negative control littermates for 60 weeks. We found

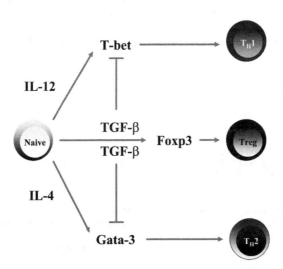

Figure 1   Regulation of effector T cell and Treg cell differentiation by TGF-β. TGF-β prevents Th1 and Th2 cell differentiation through inhibition of T-bed and GATA-3 expression. TGF-β induces the expression of Foxp3 gene and promotes Treg cell differentiation

that TGF-β expression significantly inhibited the development of diabetes in NOD transgenic mice, indicating that a short pulse of TGF-β in the islets during the priming phase of the disease was sufficient to provide protection by promoting the expansion of the intra-islet CD4$^+$CD25$^+$ T cell pool. Approximately 40–50% of intra-islet CD4$^+$ T cells expressed the CD25 marker and exhibited characteristics of regulatory T cells, including small size, high level of intracellular CTLA-4, expression of Foxp3, and transfer of protection against diabetes. Results from *in-vivo* incorporation of BrdUrd revealed that the generation of a high frequency of regulatory T cells in the islets is due to *in-situ* expansion upon TGF-β expression. These findings demonstrated a previously uncharacterized mechanism by which TGF-β inhibits autoimmune diseases via regulation of the size of the CD4$^+$CD25$^+$ regulatory T cell pool *in vivo*.

CONCLUSIONS

Understanding the role of TGF-β signalling in T cell regulation has come a long way since the discovery of its ability to regulate T cell proliferation. Our recent work demonstrated that TGF-β critically down-regulates effector T cell differentiation via inhibition of the expression of transcription factors T-bet and Gata-3. The role of TGF-β in regulatory T cell biology needs further investigation. Insights into the biology of regulatory T cells might provide important clues as to whether the different roles of TGF-β in regulatory T cell function pertain to heterogeneity of the regulatory T cell population or another regulatory T cell property. The complex role of TGF-β in the regulation of immune homeostasis is likely to keep many investigators fascinated and busy in deciphering the many mysteries of this molecular family for many years to come.

References

1. Shull MM, Ormsby I, Kier AB et al. Targeted disruption of the mouse transforming growth factor-beta 1 gene results in multifocal inflammatory disease. Nature. 1992;359:693–9.
2. Gorelik L, Flavell RA. Abrogation of TGFbeta signaling in T cells leads to spontaneous T cell differentiation and autoimmune disease. Immunity. 2000;12:171–81.
3. Peng Y, Laouar Y, Li MO, Green EA, Flavell RA. TGF-beta regulates *in vivo* expansion of Foxp3-expressing CD4$^+$CD25$^+$ regulatory T cells responsible for protection against diabetes. Proc Natl Acad Sci USA. 2004;101:4572–7.
4. Gorelik L, Constant S, Flavell RA. Mechanism of transforming growth factor beta-induced inhibition of T helper type 1 differentiation. J Exp Med. 2002;195:1499–505.
5. Gorelik L, Flavell RA. Immune-mediated eradication of tumors through the blockade of transforming growth factor-beta signaling in T cells. Nat Med. 2001;7:1118–22.
6. Gorelik L, Fields PE, Flavell RA. Cutting edge: TGF-beta inhibits Th type 2 development through inhibition of GATA-3 expression. J Immunol. 2000;165:4773–7.
7. Green EA, Gorelik L, McGregor CM, Tran EH, Flavell RA. CD4$^+$CD25$^+$ T regulatory cells control anti-islet CD8$^+$ T cells through TGF-beta–TGF-beta receptor interactions in type 1 diabetes. Proc Natl Acad Sci USA. 2003;100:10878–83.

2
Viruses as initiators, accelerators or terminators of autoimmunity? – a question of time and location of infection

U. CHRISTEN and M. G. VON HERRATH

INTRODUCTION

The development of autoimmunity is influenced by two major factors: genetic predisposition and environmental triggers or modulators. The susceptibility to develop a given autoimmune disease is largely dependent on genetic polymorphisms in the human population. In particular the inter-individual heterogeneity in the MHC haplotype can be a strong risk factor or, alternatively, a protective factor for autoimmunity. In addition, recently polymorphic expression of several other genes involved in the establishment of self-tolerance and immune regulation, for example the autoimmune regulator (AIRE) gene[1], the T cell immunoglobulin mucin (TIM) gene family[2] or CTLA-4 have been identified as potential factors that influence the susceptibility for autoimmune disease[3].

In theory viruses are prime candidates for environmental triggers of autoimmunity, since they induce a massive inflammation of the infected organ or tissue. Such an inflammatory insult is well suited to activate the host's antiviral defence systems. Components of the innate immune system, such as chemokines and cytokines, are major players early after virus infections. Chemokines are the factors that orchestrate the migration of leucocytes to the site of virus infection, where both virus-specific and non-specific lymphocytes are activated by a cocktail of cytokines, activated APC and other inflammatory factors. Such an environment can be an ideal 'breeding ground' for lymphocytes with reactivity for self-components that might be present because of an imperfect thymic or peripheral selection process. Such autoaggressive cells might include low-affinity T cells that have escaped negative selection in the thymus or functionally silenced (anergic) cells that have been reactivated in the periphery upon virus infection. Besides antigen non-specific (bystander) activation of autoaggressive lymphocytes, antigen-specific mechanisms are critically involved in the development of autoimmunity, such as the release of normally sequestered antigens by viral infection or the occurrence of structural similarities between the infecting pathogen and self-components (molecular mimicry).

VIRUSES AS INITIATORS

Historically, viral and bacterial infections have been associated with the occurrence of various autoimmune diseases, including type 1 diabetes (T1D), multiple sclerosis and ankylosing spondylitis (for a detailed review see ref. 4). In T1D possibly the strongest association has been demonstrated between rotavirus infections in young infants and the first occurrence of islet auto-antibodies[5]. However, this connection has been recently questioned[6]. Attempts to establish a direct epidemiological association between microbial infections and various autoimmune disorders have been unsuccessful so far because of several difficulties. First, and most importantly, not only patients suffering from autoimmune diseases but virtually all human individuals undergo a multitude of infections during their lifetime. Any infection could be a potential trigger on a susceptible MHC background, but many infections might be cleared by the time autoimmunity is diagnosed and precise footprints documenting the patient's history of viral and bacterial infections are therefore frequently difficult to find ('hit-and-run' triggers). Second, rather than initiate autoimmunity directly, virus infections might accelerate a preexisting autoimmune condition to progress to clinical disease. Multiple sequential events might be necessary to precipitate the disease and this would further complicate attempts to establish firm proof for the involvement of any environmental factor. Third, the precise timing and magnitude of inflammation and, for example, viral strain, might all play an important role based on studies in animal models.

Nevertheless, virus infections are an instrumental factor for breaking of self-tolerance and initiating autoimmunity in many experimental animal models for human autoimmune diseases, including T1D[7,8]. For example, in the RIP-LCMV model, the nucleoprotein (NP) or glycoprotein (GP) of lymphocytic choriomeningitis virus (LCMV) is expressed as a transgene under control of the rat insulin promoter (RIP) specifically by β cells in the pancreatic islets of Langerhans. These mice do not spontaneously develop diabetes, hyperglycaemia or insulitis[8]. However, upon infection with LCMV >95% of RIP-LCMV mice develop overt T1D[8]. Depending on the individual founder lines of transgenic mice the onset of diabetes is between 2 weeks (fast-onset) and several months (slow-onset)[9]. Further studies revealed the mechanism involved in the rapid compared to the slow-onset diabetes: transgenic lines expressing the LCMV-GP transgene exclusively in the β cells of the islets manifested rapid-onset T1D[9]. In these lines the high systemic numbers of autoaggressive CD8 T cells were sufficient to induce diabetes and did not require help from CD4 cells. In contrast, in lines expressing the LCMV-NP transgene in both the β cells and in the thymus, T1D took longer to occur after subsequent LCMV challenge. Several lines of evidence indicated that in RIP-NP mice the anti-self (viral)

CTL were of lower affinity and that CD4 T cells were essential to help anti-self (viral) CD8 lymphocytes to mediate T1D in adult transgenic mice[9].

VIRUSES AS ACCELERATORS

We used the RIP-LCMV system to evaluate how virus infections could influence an autoimmune process that is already ongoing. In a first approach we hypothesized that sequential infections with heterologous viruses would accelerate disease, if critical populations of autoaggressive lymphocytes were expanded. We chose to administer a secondary infection with Pichinde virus (PV) to LCMV-immune RIP-LCMV-NP mice (slow-onset line). It was previously demonstrated by studies from Ray Welsh's group that sequential infection with LCMV followed by PV expands a population of CD8 T cells that recognizes a subdominant epitope (NP_{205}), which confers molecular mimicry between LCMV and PV[10]. The frequency of LCMV-NP_{205} or PV-NP_{205}-specific T cells accounts for around 0.5–1% of all CD8 T cells after infection of wild-type C57BL/6 or RIP-LCMV-NP mice with either LCMV or PV, respectively[10,11]. This frequency of autoaggressive CD8 T cells is not sufficient to induce autoimmune disease, since naive RIP-LCMV-NP mice infected with PV alone do not develop T1D[11]. In contrast, sequential infection of RIP-LCMV-NP mice with LCMV followed by PV, which expands NP_{205}-specific CD8 T cells to a frequency of up to 10% of all CD8 T cells, strongly accelerates the development of clinically overt T1D[11]. Interestingly, the reverse approach (PV infection followed by LCMV) did not accelerate disease. Possibly only LCMV, which contains all the immunodominant and subdominant T cell epitopes, is capable of initially breaking self-tolerance to the islet antigen. Only after tolerance has been broken by LCMV can acceleration by PV-induced expansion of NP_{205}-specific T cells occur. The data from our study suggest that: (a) molecular mimicry of a subdominant epitope is not sufficient to initiate autoimmune disease, (b) sequential infection with two heterologous viruses that share a similar epitope (molecular mimicry) can boost an ongoing autoimmune process resulting in accelerated disease onset, and (c) the exact time and sequence of multiple virus infections is crucial for the progress of autoimmunity.

Similar observations have been done in experimental animal models for multiple sclerosis (MS)[12] and herpes stromal keratitis (HSK)[13], where, in the presence of a persistent infection, a chronic inflammatory milieu drives the activation and expansion of cross-reactive T cells. In the recombinant Theiler's murine encephalomyelitis virus (TMEV)-induced demyelination model for MS[12] a molecular mimic of an encephalitogenic myelin proteolipid epitope expressed by an engineered TMEV can exacerbate a previously established disease[14]. Further, in the murine herpes virus 1 (HSV-1)-induced HSK model, infection of mice with a HSV-1 mutant virus with a single amino acid change in the UL6 protein of HSV-1 affecting the putative mimicry epitope fails to induce HSK. However, the mutant HSV-1 was able to induce disease in predisposed mice that received CD4 cells from wild-type virus-infected donors[13].

9

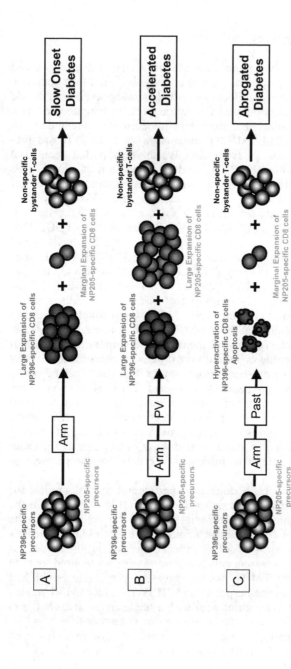

Figure 1 Dynamics of antigen-specific lymphocytes after heterologous virus infection. **A**: Infection of RIP-LCMV-NP mice with LCMV strain Armstrong (Arm) results in a large expansion of CD8 T cells that specifically recognize the immunodominant NP epitope NP_{396}. In contrast, only a marginal fraction of T cells are generated that react to the subdominant NP epitope NP_{205}. For the initiation of autoimmune disease, NP_{396}-specific, but not NP_{205}-specific, T cells have to be present at a sufficient magnitude. **B**: When LCMV-Arm-immune RIP-LCMV-NP mice are reinfected with PV, which shares a similar epitope (NP_{205}) with LCMV, NP_{205}-specific CD8 T cells expand to a critical frequency that is high enough to further boost the ongoing autoimmune destruction of the β cells in the islets of Langerhans and subsequently accelerates T1D. **C**: On the contrary, secondary infection of LCMV-Arm-immune RIP-LCMV-NP mice with LCMV strain Pasteur (Past) that grows to exceptionally high virus titres outside of the pancreas (in the PDLN) abrogates the development of T1D. The extensive LCMV-Past-induced inflammation of the PDLN recruits autoaggressive CD8 T cells from the pancreas along an IP-10 concentration gradient to the PDLN, where most of the NP_{396}-specific T cells die by apoptosis

VIRUSES AS TERMINATORS

Although viruses have been associated with the occurrence of autoimmune diseases and several animal models successfully use viruses for the initiation or acceleration of experimental autoimmune diseases, there are many studies that document a protective effect of viruses. For example the progress from benign to overt diabetes is aborted in NOD mice infected with LCMV[15]. Similarly, infection of NOD mice with several strains of group B Coxsackie viruses (CVB) significantly lowers the incidence of diabetes[16]. However, it has to be noted here that, when NOD mice are infected at a critical time when islet-specific T cells have already developed, CVB was demonstrated to accelerate rather than abrogate T1D[17,18]. In our models we observed that LCMV-immune RIP-LCMV-NP mice that received a secondary infection with the LCMV strain Pasteur (LCMV-Past) or NOD mice that were infected with LCMV did not develop T1D[19]. LCMV-Past is closely related to LCMV strain Armstrong (LCMV-Arm), which is normally used to induce T1D in the RIP-LCMV model[8], and shares all the immunodominant T cell epitopes with LCMV-Arm[20]. However, in contrast to LCMV-Past, secondary infection with LCMV-Arm does not abrogate T1D in the RIP-LCMV mouse[19]. The difference between the two viruses lies in their potential to grow to high virus titres. In LCMV-Arm-immune RIP-LCMV mice, LCMV-Past grows to more than three orders of magnitude higher titres in pancreas, spleen, and pancreatic draining lymph node (PDLN) than LCMV-Arm. Most importantly, the highest virus titre is found in the PDLN and not in the pancreas[19]. Associated with this extensive viral growth in the PDLN is a massively enhanced inflammation. In particular an IP-10 (CXCL10) chemokine gradient is observed if one compares IP-10 levels in the pancreas and the PDLN[19]. As a result lymphocytes disappear from the islets of Langerhans and recirculate to the PDLN, where the local inflammatory milieu causes an enhancement of apoptosis of auto-aggressive CD8 T cells[19]. Interestingly, the same mechanism operates in NOD mice that have been infected with LCMV-Past at 9 weeks of age: infection with LCMV-Past establishes an inflammatory gradient, characterized by high IP-10 expression in the PDLN, that results in a decrease of autoaggressive lympho-cytes within the islets of Langerhans and abrogates the development of T1D[19]. Our data indicate that: (a) virus infections can indeed have a beneficial effect, (b) a strong inflammatory insult at the right time and at the right place can abrogate an ongoing autoimmune process, (c) IP-10 is a key factor in orchestrating the trafficking of autoaggressive T cells. In another study we could demonstrate that IP-10 is essential for attracting autoaggressive T cells to the pancreas. When LCMV-Arm was used to initiate T1D in the RIP-LCMV model, blockade of IP-10 prevented insulitis and diabetes[21]. The observation that viruses can abrogate autoimmunity might provide mechanistic evidence for the 'hygiene hypothesis' that implicates the environment as a protective factor in the development of autoimmune diseases, including T1D or multiple sclerosis[22]. In that context it is important to consider the appearance of a north–south gradient for many autoimmune diseases[23].

CONCLUSIONS

Our review explains some of the difficulties in finding clear proof for certain viruses to act as initiators, accelerators, or terminators of autoimmunity. Even in our idealized model system for T1D we can demonstrate a whole spectrum of these outcomes. Infection with virus 'A' accelerates the ongoing autoimmune destruction and infection with virus 'B' at a similar state of disease progress and via the same route of administration abrogates disease (see Figure 1)[11,19]. Thus it is important to differentiate on a mechanistic level as to why infection by virus 'A' results in a different outcome than virus 'B'. One has to realize that it is rarely the case that a single environmental factor, such as infection with a pathogen, initiates, accelerates or terminates autoimmunity by itself. It is rather a combination of several factors that act in a possibly sequential fashion. It is the balance between disease progressive and abrogative factors that decides whether susceptible individuals develop autoimmune disease or not.

In the future the identification of additional environmental factors will become more and more important towards designing strategies that target the prevention of autoimmunity before its initiation. Such triggers can be, as discussed here, heterologous virus infections or alternatively exposure to drugs and chemicals that can either cause direct damage to tissues and cells or can have the potential to modify self-components that could be considered as 'foreign' neoantigens. In addition, novel treatments that specifically target the autoaggressive components or boost antigen-specific regulatory functions need to be developed to successfully abrogate ongoing autoimmune processes. However, one has to carefully design such regimens in terms of time and duration of treatment in order to achieve the desired effect.

References

1. Pitkanen J, Peterson P. Autoimmune regulator: from loss of function to autoimmunity. Genes Immun. 2003;4:12–21.
2. Kuchroo VK, Umetsu DT, DeKruyff RH, Freeman GJ. The TIM gene family: emerging roles in immunity and disease. Nat Rev Immunol. 2003;3:454–62.
3. Ueda H, Howson JM, Esposito L et al. Association of the T-cell regulatory gene CTLA4 with susceptibility to autoimmune disease. Nature. 2003;423:506–11.
4. von Herrath MG, Fujinami RS, Whitton JL. Microorganisms and autoimmunity: making the barren field fertile. Nat Rev Microbiol. 2003;1:151–7.
5. Honeyman MC, Coulson BS, Stone NL et al. Association between rotavirus infection and pancreatic islet autoimmunity in children at risk of developing type 1 diabetes. Diabetes. 2000;49:1319–24.
6. Blomqvist M, Juhela S, Erkkila S et al. Rotavirus infections and development of diabetes-associated autoantibodies during the first 2 years of life. Clin Exp Immunol. 2002;128:511–15.
7. Ohashi P, Oehen S, Buerki K et al. Ablation of tolerance and induction of diabetes by virus infection in viral antigen transgenic mice. Cell. 1991;65:305–17.
8. Oldstone MBA, Nerenberg M, Southern P, Price J, Lewicki H. Virus infection triggers insulin-dependent diabetes mellitus in a transgenic model: role of anti-self (virus) immune response. Cell. 1991;65:319–31.
9. von Herrath MG, Dockter J, Oldstone MBA. How virus induces a rapid or slow onset insulin-dependent diabetes mellitus in a transgenic model. Immunity. 1994;1:231–42.
10. Brehm MA, Pinto AK, Daniels KA, Schneck JP, Welsh RM, Selin LK. T cell immunodominance and maintenance of memory regulated by unexpectedly cross-reactive pathogens. Nat Immunol. 2002;3:627–34.

11. Christen U, Edelmann KH, McGavern DB et al. A viral epitope that mimics a self antigen can accelerate but not initiate autoimmune diabetes. J Clin Invest. 2004;114:1290–8.

12. Olson JK, Croxford JL, Calenoff MA, Dal Canto MC, Miller SD. A virus-induced molecular mimicry model of multiple sclerosis. J Clin Invest. 2001;108:311–18.

13. Panoutsakopoulou V, Sanchirico ME, Huster KM et al. Analysis of the relationship between viral infection and autoimmune disease. Immunity. 2001;15:137–47.

14. Croxford JL, Olson JK, Anger HA, Miller SD. Molecular minicry-induced initiation and exacerbation of CNS autoimmune disease: implications for MS pathogenesis. 2004 (Submitted for publication).

15. Oldstone MB. Prevention of type I diabetes in nonobese diabetic mice by virus infection. Science. 1988;239:500–2.

16. Tracy S, Drescher KM, Chapman NM et al. Toward testing the hypothesis that group B coxsackieviruses (CVB) trigger insulin-dependent diabetes: inoculating nonobese diabetic mice with CVB markedly lowers diabetes incidence. J Virol. 2002;76:12097–111.

17. Horwitz MS, Bradley LM, Harbertson J, Krahl T, Lee J, Sarvetnick N. Diabetes induced by Coxsackie virus: initiation by bystander damage and not molecular mimicry. Nat Med. 1998;4:781–5.

18. Horwitz MS, Ilic A, Fine C, Rodriguez E, Sarvetnick N. Coxsackievirus-mediated hyperglycemia is enhanced by reinfection and this occurs independent of T cells. Virology. 2003;314:510–20.

19. Christen U, Benke D, Wolfe T et al. Cure of prediabetic mice by viral infections involves lymphocyte recruitment along an IP-10 gradient. J Clin Invest. 2004;113:74–84.

20. Dutko FJ, Oldstone MB. Genomic and biological variation among commonly used lymphocytic choriomeningitis virus strains. J Gen Virol. 1983;64:1689–98.

21. Christen U, McGavern DB, Luster AD, von Herrath MG, Oldstone MB. Among CXCR3 chemokines, IFN-gamma-inducible protein of 10 kDa (CXC chemokine ligand (CXCL) 10) but not monokine induced by IFN-gamma (CXCL9) imprints a pattern for the subsequent development of autoimmune disease. J Immunol. 2003;171:6838–45.

22. Bach JF. Protective role of infections and vaccinations on autoimmune diseases. J Autoimmun. 2001;16:347–53.

23. Rosati G. The prevalence of multiple sclerosis in the world: an update. Neurol Sci. 2001;22:117–39.

3
Antigen-chip technology for accessing global information concerning the state of the body

I. R. COHEN, F. J. QUINTANA and Y. MERBL

INTRODUCTION

Until now immunologists have been able to focus on one or a few antibodies, cytokines or T cell clones at a time; immunology has lacked the comprehensive view of the immune system needed to optimize diagnosis, prognosis, drug development, treatment choice, patient stratification, and monitoring. Microarray technology combined with advanced informatic technologies has opened new opportunities for approaching the vast information stored in the immune system. As a first step in developing an antigen chip we have studied arrays of antigens using standard ELISA microtitre plates, and have applied a clustering algorithm (originally developed for DNA chips) to detect patterns of antibody reactivity within the sera of healthy persons and those with various diseases[1]. We have used this algorithm to analyse global antibody patterns and can successfully discriminate various conditions. For example, we have successfully discriminated between type 1 diabetic persons and healthy controls with 95% sensitivity and 90% specificity. The global patterns of antibodies are clearly more efficient in stratifying patients than any one-to-one antigen–antibody reaction. We have now developed an antigen microarray (antigen chip) that makes it possible to profile the global state of the immune system by analysing the patterns of autoantibodies to hundreds of different self antigens[2]. The present architecture of the chip can accommodate thousands of different antigens.

IMMUNE SYSTEM COMPUTER

The immune system is the key player in body maintenance. The immune system expresses both the genetic endowment of the individual and the life experience of the individual; the immune system deals with post-genomic adaptation to life. Like the central nervous system, the immune system is *self-organizing*: it

14

begins with genetically coded, primary instructions, to which it adds information culled from the individual's experience with the environment in health and disease. Just as each person develops a unique brain, each person develops an individualized immune system. If we need to know about the past immune history and the future susceptibility of the individual, it would be very useful to be able to consult the immune system.

The output of the immune system is a complex series of processes termed *inflammation*. Inflammation is initiated, regulated and terminated by cytokines, chemokines, adhesion molecules, antibodies and other immune molecules produced by macrophages, dendritic cells, B cells, T cells and other immune system cells. These cells and molecules orchestrate wound healing, blood vessel growth, connective tissue formation, apoptosis, tissue regeneration and much else that is needed for body maintenance and defence against invaders. The task of the immune system is to append the right type of inflammatory response dynamically over time according to the shifting needs of the tissues. Immune system cells patrol the body continuously and sense the defence and maintenance needs of the individual. The immune system integrates enormous amounts of information, stores the information in its antibody and lymphocyte repertoires and uses this information to express the type and grade of inflammation needed at each site and at each moment. The immune system records and knows the body's most critical secrets; the immune system uses what has been called the immunological homunculusn[3]. In short, the immune system functions as the *bio-informatic computer, defence force* and *public works department* of the body (Figure 1).

BIOMEDICAL APPLICATION

Theoretically, the immune system computer should be able to provide us with both the history and potential of its activity, which in practical terms translates into vital diagnostic and prognostic information. To learn about the state of an individual's body we only need consult the immune system computer. In practice, however, it has been feasible up until now to view only a very limited part of the computational output of the immune system. Traditionally, immunological diagnosis has been based on an attempt to correlate each disease with a specific immune reactivity, such as an antibody or a T cell response to a single antigen specific for the disease entity. This approach has been largely unsuccessful for three main reasons. First, a specific antigen or antigens may not have been identified in the disease (for example, Behçet's disease, rheumatoid arthritis, and others). Secondly, immunity to multiple self-antigens, and not to a single self-antigen, is manifest in various patients suffering from a single disease (for example, a dozen different antigens are associated with type 1 diabetes). Thirdly, a significant number of healthy persons may manifest antibodies or T cell reactivities to self-antigens targeted in autoimmune diseases, such as insulin, DNA, myelin basic protein, thyroglobulin and others. For this reason false-positive tests are not uncommon. Hence, there is a real danger of making a false diagnosis based on the determination of a given immune reactivity. Novel approaches, therefore, are needed to support

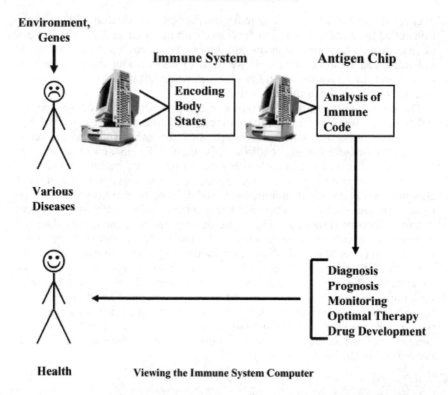

Figure 1 Viewing the immune system computer. The states of the immune system encode the various states of the body in disease. The antigen chip by viewing, as it were, the 'computer screen' system, can report to us on the state of the body

the diagnoses of specific diseases in a way that would justify specific therapeutic interventions. Because immunologists till now have been able to focus on one or a few antibodies, cytokines or T-cell clones at a time, we have lacked the comprehensive view of the immune system needed to facilitate diagnosis, prognosis, drug development, treatment optimization, patient stratification, prediction of response to treatment, and monitoring. Our approach is unique in that it gathers data from the global repertoire of antibodies reactive with key molecules of the body. Our antigen chip will provide a unique tool for post-genomic monitoring and management of human health and disease.

DISEASES

Antigen chips would appear to provide access to the immune system's bio-informatic computer, and thus constitute a platform for the development of numerous chip applications.

Autoimmune diseases are an obvious target for the use of antigen chips. We have used a global antibody assessment guided by bio-informatics to successfully separate various healthy and autoimmune disease populations[2]. *Infectious diseases*, too, require better diagnostic discrimination between persons who will be susceptible to a particular vaccination and persons who will not respond. Certain infections can trigger autoimmune responses, and it is important to be able to diagnose persons who are destined to develop autoimmune diseases. Hepatitis viruses are notorious for inducing chronic inflammation that may lead to hepatocellular carcinoma. The antigen chip could be used to screen for potentially dangerous changes in the autoimmune antibody repertoire resulting from such infections.

Immunotherapy of cancer is another situation in which it would be advantageous to classify persons with different types of immune reactivities to self-antigens; many, if not most tumour-associated antigens are self-antigens. Thus, it could be important in the design of therapeutic tumour vaccines to know what kind of immune potential is present in the patient.

Monitoring: assays for monitoring the state of the immune system are needed. Various immunologic therapies are now being developed. There is a critical need to develop markers that will enable the physician to monitor the response of the immune system to various treatments designed to modulate or arrest chronic inflammation and autoimmune diseases, vaccinate against infectious agents, or effect the immunotherapy of cancer. Furthermore, the monitoring of the overall breadth of the *recovering* immune system becomes crucially important in individuals who have received chemotherapy and stem cell transplants for leukaemias and other cancers. An immune system with a broader repertoire reflects one with more potential to combat infections.

Medicine badly needs predictive markers to stratify subjects and design trials based on 'inside information' provided by the immune system regarding the response of the test subjects to treatments. Immunomodulatory and anti-inflammatory drugs have focused on the disease as the only endpoint, and have failed to monitor the cause of the disease. The immune system can provide us with the inside information needed to optimize effectiveness and save time in arriving at dosing and other variables. Indeed, the patterns of reactivity detected using microarray chips will make it possible to profile subjects and identify those who are more likely or less likely to respond to a particular treatment. The new informatics can be expected to provide a new outlook on important medical problems; the antigen chip provides a voice for the immunological homunculus to tell us about the individual's body state[3].

Acknowledgements

I.R.C. is the Mauerberger Professor of Immunology at the Weizmann Institute of Science, Rehovot, Israel; the Director of the Center for the Study of Emerging Diseases, Jerusalem, Israel; and the Director of the National Institute for Biotechnology in the Negev, at Ben-Gurion University of the Negev, Beer Sheva, Israel.

References

1. Quintana FJ, Getz G, Hed G, Domany E, Cohen IR. Cluster analysis of human autoantibody reactivities in health and in type 1 diabetes mellitus: a bio-informatic approach to immune complexity. J Autoimmun. 2003;21:65–75.
2. Quintana FJ, Hagedorn PH, Elizur G, Merbl Y, Domany E, Cohen IR. Functional immunomics: microarray analysis of IgG autoantibody repertoires predicts the future response of mice to induced diabetes. Proc Natl Acad Sci USA. 2004;101(Suppl. 2):14615–21.
3. Cohen IR. The cognitive paradigm and the immunological homunculus. Immunol Today. 1992;13:490–4.

Section II
Immune regulation II

Chair: P.A. BERG and P.R. GALLE

4
CD95-mediated apoptosis: immune mission to death

P. H. KRAMMER

INTRODUCTION: THE IMMUNE SYSTEM

The immune system is a society of interacting cells consisting of T and B lymphocytes, natural killer (NK) cells, macrophages and professional antigen-presenting cells (APC) and their various subclasses. Most cellular components of the immune system are born in the bone marrow. B lymphocytes, NK cells and macrophages mature in the bone marrow and in the fetal liver. T lymphocytes mature in the bone marrow and in the thymus. T and B cell developments share many features, yet differ in others (Figure 1).

B cells

B cells express cell membrane receptors (antibodies) with a single antigen specificity. Millions of different B cells produce antibodies and can potentially capture millions of antigens. The sum of all antibody specificities is called 'the antibody repertoire'[1]. Based on the affinity of their antibodies B cells are selected in the bone marrow. The ones with high affinity for proteins derived from 'self' tissues are eliminated. Mature B lymphocytes exit the bone marrow and populate the secondary lymphoid organs, spleen and lymph nodes and the gut-associated lymphoid tissue. Once activated by an antigen, B cells undergo a second round of selection in the follicles of secondary lymphoid organs, after which they mature into plasma cells that produce and secrete antigen-specific antibodies, and recirculate to the bone marrow[2].

T cells

Pre-T lymphocytes emigrate from the bone marrow into the thymus. In the thymus they mature and are positively or negatively selected depending on the affinity of their T cell receptors (TCR) for self major histocompatibility antigens (MHC). MHC class I and II are molecules that sample peptide fragments from foreign and self proteins, respectively. They display these peptides at the cell's surface for scrutiny by T cells – a process called 'antigen

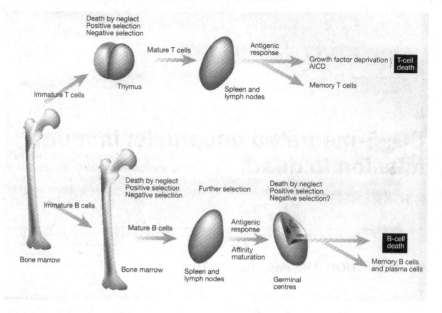

Figure 1 T and B lymphocyte development (see text). AICD, activation-induced cell death. (Reproduced with permission from: Krammer PH. CD95's deadly mission in the immune system. Nature. 2000;407:789–95)

presentation'. Each MHC class I or II protein presents a different fragment and thousands of MHC molecules protrude from every cell. The majority of the peptides presented in the thymus is derived from 'self' proteins. T cells with high affinity for self-MHC molecules and peptide are eliminated to assure tolerance to normal tissues and to prevent autoimmunity. T cells that interact with MHC class II molecules develop into cells that express the CD4 molecule on their surface, and those with affinity for MHC class I molecules turn into CD8[+] T lymphocytes. Only mature T cells that produce a functional TCR exit the thymus and populate the secondary lymphoid organs. Mature CD4[+] T cells function as helper T cells and secrete cytokines that regulate either cellular immune responses (T helper 1 cells) or antibody responses (T helper 2 cells). Mature CD8[+] T cells function as cytotoxic effector (killer) cells[3] (Figure 2).

Cells of the immune system work as a team. Following an attack by infectious agents, professional APC, dendritic cells in particular, present antigenic peptides of these infectious agents to T cells[4]. Antigen/MHC complexes, costimulating cell surface molecules on APC and cytokines, drive T cells into clonal expansion. These T cells, in turn, then communicate with other T or B cells to regulate their responses. Following a peak phase with highest clonal expansion of reactive cells, the immune response undergoes a down-phase. Most lymphocytes are eliminated; only a few survive and constitute the pool of memory cells[5].

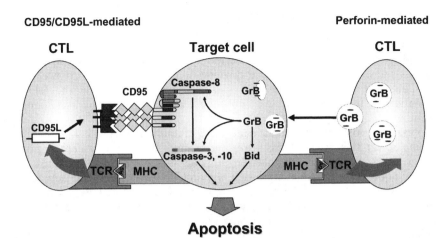

Figure 2 Cytotoxic T lymphocytes (CTL) can kill target cells by the CD95/CD95L system (left) or by the perforin/granzyme B (GrB) system (right). CD95 signalling involves a caspase cascade (see text). GrB entry into target cells and release involve perforin. GrB cleaves and activates caspases in the target cell. The target cell dies by apoptosis through either a CD95 or GrB signalling event

Life and death in the immune system

Several features of the immune system are unique: (1) its specificity: the repertoire of T and B lymphocytes, initially built from randomly selected antibody and TCR variable region genes, is shaped by selection to cope on the one hand with the vast universe of antigens and on the other hand with the danger of autoimmunity[3], and (2) its homeostatic control: following a clonal expansion phase antigen-reactive lymphocytes must be titrated back until the pool of lymphoid cells reaches baseline level again[6]. This is achieved by balanced fine-tuning between growth/expansion and death by apoptosis; generally, the immune system produces more cells than finally needed, and extra cells are eliminated by apoptosis.

Apoptosis is the most common form of death in cells of the immune system. It is astounding how many different pathways immune cells can choose to die. In principle, death can be by neglect when the antigen-specific receptors of the lymphoid cells are not stimulated or the lymphocytes are deprived of trophic cytokines. In a more active form, death can involve the death receptor/death ligand systems[7-9]. Apoptosis is such a central regulatory feature of the immune system that it is not surprising that too little or too much apoptosis results in severe diseases.

THE CD95/CD95L DEATH SYSTEM

A subset of TNF-receptor (TNF-R) family members is involved in death-transducing signals, and hence are referred to as the 'death receptors'. Members of this family contain one to five cysteine-rich repeats in their extracellular domain, and a death domain (DD) in their cytoplasmic tail. The DD is essential for transduction of the apoptotic signal. CD95 is one such family member, which plays a significant role in the immune system. It is a widely expressed glycosylated cell surface molecule of approximately 45–52 kDa (335 amino acids). It is a type I transmembrane receptor but alternative splicing can result in a soluble form, the function of which is unclear[10]. CD95 expression can be boosted by cytokines such as IFN-γ and TNF but also by the activation of lymphocytes[11,12]. CD95-mediated apoptosis is triggered by its natural ligand, CD95L, which is a TNF-related type II transmembrane molecule[7] and expressed in a far more restricted way than the receptor. Killer cells, or so-called cytotoxic T cells (CTL) remove, e.g. virus-infected cells, and those that express CD95L can do so by interacting with the CD95 receptor[13,14] on their targets (Figure 2). CD95L is seen on killer cell-derived vesicles[15,16], but can also be cleaved off the membrane by a metalloprotease[17–19]; whereas soluble human CD95L can induce apoptosis[20], soluble mouse CD95L cannot[21].

The CD95 DISC

Oligomerization, most likely trimerization, of CD95 is required for transduction of the apoptotic signal. A complex of proteins associates with activated CD95[8,22]. This death-inducing signalling complex, or DISC, forms within seconds after receptor engagement. First, the adaptor FADD(Mort1) binds via its own death domain to the death domain in CD95. FADD also carries a so-called death effector domain (DED) and, again by homologous interaction, recruits the DED containing procaspase-8 (also known as FLICE) into the DISC. Next, procaspase-8 is proteolytically activated and active caspase-8 released from the DISC into the cytoplasm in the form of a heterotetramer of two small and two large subunits[23]. Active caspase-8 cleaves various proteins in the cell including procaspase-3, which results in its activation and completion of the cell death programme. Various other proteins were described to bind to activated CD95 and the DISC, but their precise role and importance in the regulation of apoptosis remain to be defined.

Recently, using fluorescence resonance energy transfer, another model of CD95 signalling has been deduced. Extracellular pre-ligand-binding assembly domains (PLAD) were described for CD95 and TNF-R which are supposed to aggregate the receptors before ligand binding. To prevent premature signalling of preassociated receptors, a dangerous situation, intracellular receptor-associated apoptosis blockers were postulated[24,25]. On the basis of the 'PLAD-model' it is not entirely clear how ligand binding interferes with the PLAD association and leads to receptor association that initiates apoptosis. More structural work is needed to resolve these issues. It is also unclear whether the 'DISC-model' and the 'PLAD-model' complement each other to describe initial signalling events *in vivo*.

While some cytotoxic T lymphocytes kill their target cells by turning on the death receptor, others use perforin and granzyme B (GrB) to eliminate infected cells. With the help of perforin, GrB finds its way into the target cell and can kill it by directly cleaving and activating caspase-8[14] (Figure 2).

Two different pathways downstream of CD95

In so-called type I cells[26], the death signal is propagated by a caspase cascade initiated by activation of large amounts of caspase-8 at the DISC, followed by rapid cleavage of caspase-3 and other caspases, which in turn cleave vital substrates in the cell. In type II cells[26], however, hardly any DISC is formed; hence the caspase cascade cannot be propagated directly but has to be amplified via the mitochondria. Caspase-8 cuts the Bcl-2 family member BID and truncated BID 'activates' the mitochondria[27,28]. Mitochondria are 'activated' in both type I and type II cells, but are not strictly necessary for type I cells to die. Upon 'activation' mitochondria release pro-apoptotic molecules, such as cytochrome C[29] and Smac/DIABLO[30,31]. Together with APAF-1 and procaspase-9 in the cytoplasm these molecules form the so-called apoptosome, the second initiator complex of apoptosis. Caspase-9 activates further downstream caspases and the end-result, again, is apoptosis (Figure 3). The reason for the differences between type I and type II cells is currently unclear, but perhaps biochemical differences at the receptor level hold the answer. Recently, Bid-deficient mice were generated[32], which are resistant to CD95-induced hepatocellular apoptosis. In Bid$^{-/-}$ cells from these mice mitochondrial dysfunction was delayed, cytochrome C not released, effector caspase activity reduced and the cleavage of apoptosis substrates altered. Taken together, these data support the type I/type II concept of CD95 signalling.

FLIP (FLICE-inhibitory proteins)

Additional DED-containing proteins have been found in a certain class of herpes viruses. These proteins contain two DED and also bind to the CD95/FADD complex. This inhibits recruitment and activation of caspase-8, formerly known as FLICE; hence their name FLIP, for FLICE-inhibitory proteins. In transfected cells, v-FLIP inhibited apoptosis induced by several apoptosis-inducing receptors (CD95, TNF-R1, TRAMP/DR3 and TRAIL-R1), suggesting that these receptors use similar signalling pathways[33–35]. Two human homologues of v-FLIP have been identified by several groups at the same time, and are known under a variety of names. Although it is widely assumed that the cellular FLIP block apoptosis, the data are ambiguous and c-FLIP might be pro- or anti-apoptotic depending on the cellular context. Recent results with cells from cFLIP-deficient mice support the role of c-FLIP as an anti-apoptotic molecule[36].

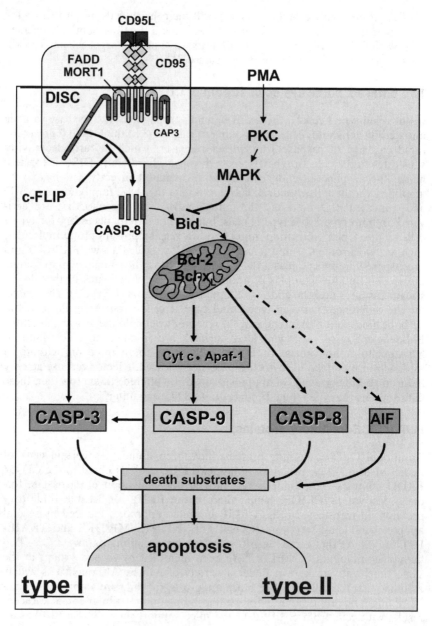

Figure 3 Signalling pathways induced by CD95. CD95 signadling pathway (including DISC formation) used in type I and type II cells (see text)

SIGNALLING BY OTHER DEATH RECEPTORS

Signalling of apoptosis by other members of the death receptor subfamily seems to follow the same basic rules and is initiated by the same sequential steps[37-39]: (1) ligand binding, (2) receptor trimerization and DISC formation, (3) attraction of the adapter molecule FADD into the DISC, (4) association of procaspase-8, (5) autocatalytic cleavage of the procaspase and formation of active caspase-8 (a heterotetramer). Active caspase-8 then serves as the initiator caspase activating other further downstream executioner caspases that cleave cellular death substrates resulting in the morphological and biochemical catastrophe termed apoptosis. The following paragraphs briefly discuss how these signalling pathways are used in the immune system.

T LYMPHOCYTE DEATH IN THE THYMUS

The T cell repertoire is shaped in the thymus by apoptosis and survival signals. A young adult mouse with $1-2 \times 10^8$ thymocytes generates between 20 and 40×10^6 new T cells per day[40]. The number of T cells that exit the thymus and enter the peripheral T cell pool, however, is only about 2–3% of the ones initially generated. Despite the high death rate of T cells in the thymus, only a limited number of apoptotic cells can be observed in histological sections. Thus, apoptotic thymocytes are removed efficiently and, most importantly, this is achieved without signs of inflammation[41-43].

Pre-T lymphocytes after entry into the thymus differentiate and rearrange their TCR genes. Those T cells that fail to productively rearrange their TCR gene, and thus cannot be stimulated by self-MHC-peptide complexes, die by neglect. In T lymphocytes of FADD-dominant negative transgenic mice the requirement for pre-TCR signals is bypassed[44]. In these mice T cell survival and differentiation is promoted. Since FADD is an essential adapter of several death receptor DISC, these data suggest a role for death receptors at this early stage of T cell development. Thymocytes that successfully passed pre-TCR selection mature further, develop into $CD4^+/CD8^+$ (double-positive, DP) T cells and undergo further TCR-affinity-driven positive and negative selection on thymic stromal cells. Following these selection processes mature single positive (SP) $CD4^+$ MHC class II- and $CD8^+$ MHC class I-restricted T cells exit the thymus and generate the peripheral T cell pool. Like crossing several borders the T cell crosses several checkpoints to assure self-MHC restriction and self-tolerance.

Initially, most investigators agreed that the CD95 system is not involved in negative selection since the TCR repertoire in mice with a defect in this system (lpr, lpr[eg] and gld mice) was not altered[45]. Upon closer inspection, however, it was found that negative selection may involve the CD95 system when T cells encounter high antigen concentrations[46]. The role of other members of the TNF-R superfamily, TNF-RI and II, CD30 and CD40, remains controversial. Likewise, survival signals of thymocytes at different maturation stages remain ill-defined. Numerous data suggest that members of the Bcl-2 family influence survival of immature T lymphocytes, i.e. positive selection, but not negative selection[3].

Finally, a modulating role in thymocyte survival and apoptosis has been ascribed to quite a number of different molecules such as glucocorticoid hormones, cytokines, costimulating cell surface receptors, signalling molecules, transcription factors[3] and NO[47]. Considering the available data, our understanding of the molecular basis of apoptosis and selection of T lymphocytes in the thymus still remains fragmentary.

DELETION OF PERIPHERAL T CELLS BY APOPTOSIS

Deletion by apoptosis is also observed in mature peripheral T cells. It occurs by neglect in those T cells not sufficiently stimulated by growth signals[48] and, importantly, it occurs at the peak or the down-phase of the immune response (Figure 4)[49–54] to down-regulate the number of reactive cells and to terminate the immune response. This so-called activation-induced cell death (AICD) may also serve as a second line of defence against autoimmunity by deleting autoreactive cells in the periphery.

Following activation T cells go through several phases (Figure 4): (1) an IL-2-dependent clonal expansion and effector phase following antigen challenge; (2) a down-phase in which most antigen-specific T cells are eventually eliminated; and (3) a phase in which certain T cells that survive the down-phase enter the memory T cell pool. In the first phase T cells are apoptosis-resistant and memory cells are also thought to be relatively resistant to apoptosis[55]. In the down-phase, however, T cells become progressively apoptosis-sensitive in the presence of IL-2. Thus, IL-2 serves a dual role: it is initially mandatory for clonal expansion and later to sensitize T cells towards apoptosis[56]. *In vitro*, T

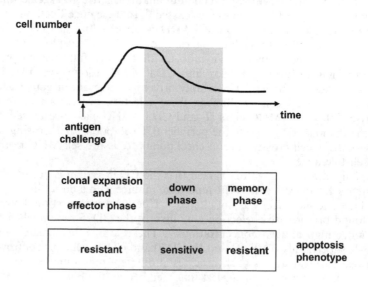

Figure 4   The course of a T cell immune response and the apoptosis phenotype of the T cells (see text)

cell activation leads to expression of CD95L and allows T cells to eliminate CD95[+] neighbouring cells and those cells that secrete CD95L can commit suicide[11,54]. However, while the CD95/CD95L system seems to be used at the initiation of AICD the TNF-RII/TNF-α system is important at a later phase[57,58]. Furthermore, ROS and perforin/granzyme B are also involved in AICD[59,60] that can also occur in a caspase-independent fashion[61]. Apparently, activated T cells have various ways to die. The molecular basis for their choice and the identity of the pathways used *in vivo* remain to be determined.

Sensitive T cells behave like type I cells, form a CD95-DISC and initiate a caspase cascade that results in apoptosis. Resistant T cells, however, show reduced DISC formation, and amplification via the mitochondria is blocked by up-regulation of the anti-apoptotic molecule Bcl-X_L. The role of c-FLIP in this scenario is controversial[62]. Taken together, T cells can die via different routes and the shift from apoptosis resistance to sensitivity coincides with a shift from a type II to a type I pathway of apoptosis.

COSTIMULATION AND T CELL SURVIVAL

T cells activated by one signal via the TCR can be saved from AICD by a second signal from costimulatory molecules, adhesion molecules or cytokine receptors. CD28 is a major costimulating coreceptor expressed on T cells, is stimulated by CD80 and CD86 expressed on APC and functions to increase cytokine production and cytokine receptor induction. Under certain conditions CD28 can sensitize T cells towards apoptosis, but generally CD28 enhances cell proliferation and viability of T cells.

The effect of costimulation was observed on three levels: (1) a strong up-regulation of c-FLIP[63,64], (2) up-regulation of Bcl-X_L[65], and (3) down-regulation of CD95L mRNA and protein at a defined time (8–12 h after stimulation). Thus, costimulation blocks both the type I and II pathways in T cells and, at least temporarily, also blocks CD95L expression[63]. At present it is not clear how antigen activated T cells with an apoptosis-resistant phenotype transit into memory T cells, and whether memory T cells are locked in the apoptosis-resistant state, but molecules such as c-FLIP and Bcl-X_L may hold some of the answers.

B LYMPHOCYTE DEATH

Three cell surface molecules represent key elements of B cell life and death regulation: BCR, CD40 and CD95. It depends greatly on the stage of maturation and activation of the B cell, on the quantity and quality of the signal provided, and on the context of cytokines and other components of the cellular environment whether triggering the BCR, e.g. by antigens, induces survival or death[66,67]. Evidence from studies of normal and malignant B cells suggests that BCR activation induces apoptosis via the mitochondrial pathway. Many components of the signalling pathway, however, are still elusive. Thus, it is unclear which signals link BCR stimulation to mitochondrial activation[68,69].

Like in T cells costimulated by CD28, BCR-activated B cells can be rescued from apoptosis by costimulation via CD40 activated by CD40L expressed on T cells and macrophages. This stimulus may represent the most important survival signal for B cells notwithstanding that such signals at a different maturation stage may also prepare B cells for death[2,70]. Although it has been described that transgenic bcl-2 prevents death and impairs affinity maturation in germinal centres, it is unclear, for example, in which other situations BCl-2 and other family members and the IAP block apoptosis[71], and when IL-4 and other cytokines act as survival signals[2]. In addition, it is unclear how plasma cells die and which anti-apoptotic signals counterregulate their death[2,72].

Thus, the principles of B and T cell development, repertoire selection and involvement of apoptosis in death by neglect and negative selection are similar. However, there are some fundamental B cell-specific characteristics (Figure 1).

First, autoreactive B cells are eliminated in the bone marrow but in response to antigenic stimulation B cells undergo a second diversification and affinity maturation step in the germinal centres of the secondary lymphoid organs by a process called somatic hypermutation; low-affinity or autoreactive B cell mutants are eliminated by apoptosis, the rest mature into memory B cells and long-lived plasma cells[2]. Plasma cells may constitute an important component of B cell memory, in particular the ones that recirculate to the bone marrow where they are kept alive by as-yet-undefined stromal signals[73].

While T cells can use the CD95L to commit activation-induced suicide[7], B cells generally do not express CD95L and die from a direct BCR-mediated signal. This opens the possibility for a murder case in which the T cells kill CD95[+] B cells. This may apply to susceptible tolerant B cells or to B cells insufficiently stimulated by survival signals or those whose BCR are unoccupied by antigen[74].

Recently, the discovery of new receptor/ligand pairs within the TNF-R/TNF superfamilies has shed further light on regulation of B cell life and death[9,72,75–77]. BLyS (Tall-1, Thank, BAFF, zTNF4) and APRIL expressed on T cells and dendritic cells were found to bind to the receptors TACI and BCMA expressed on B cells up-regulating NF-κB, B cell proliferation and immunoglobulin production. The receptor/ligand systems seem to act in concert to regulate B cell function. Overstimulation of these systems can lead to autoimmunity and autoantibody formation as seen in systemic lupus erythematosus (SLE). Blocking these systems may be used as a new treatment modality in such diseases.

APC, T AND B CELL INTERACTIONS

T and B cells influence each other and influence persistence, clonal expansion and apoptosis of other cells; but it is the APC that prime the T cells and initiate T cell-dependent immunity[4]. APC are able to engulf apoptotic and necrotic cells, and present their antigens to T cells, but at present the jury is out as to whether material from apoptotic or necrotic cells activates or soothes the T cells[78,79]. APC are not passive bystander cells. Activated APC synthesize CD95L, TRAIL, TNF and other factors that modulate the activity and function of T cells[80]. In turn, activated T cells influence APC function and thus the course of the immune response. At the initiation of the immune response APC must be resistant to apoptosis to exert their function[81]. Thus, switching these cells off to down-regulate the response becomes an important issue. Two members of the TNF-R superfamily, CD40 and CD95, play adverse roles in this context; the CD40/CD40L system allows APC survival and the CD95/CD95L induces their death[82]. The plasticity of the immune system might require that the cells can give and receive life and death signals at the same time, and that it is the cellular context which determines which signal dictates the cellular response.

DISEASES OF THE IMMUNE SYSTEM INVOLVING APOPTOSIS

Apoptosis is a fundamental process of regulation of the immune system and its derangement leads to severe diseases. Several examples of such diseases with either too little or too much apoptosis are discussed below.

GENETIC DEFECTS IN THE CD95/CD95L SYSTEM

Several mouse mutations have been identified that cause complex disorders of the immune system, manifested as lymphadenopathy and autoimmunity. One is the recessive lpr (lymphoproliferation) mutation. The symptoms of the lpr disease are similar to those in SLE. The mutations lpr[cg] (allelic to lpr) and gld (generalized lymphoproliferative disease) cause a very similar disease. In all three cases aberrant T cells accumulate; in lpr mice a splicing defect results in greatly reduced expression of CD95. In lpr[cg] mice a point mutation in the intracellular DD of CD95 abolishes transmission of the apoptotic signal. In gld mice a point mutation in the C terminus of CD95L impairs its ability to interact successfully with its receptor. Thus, a failure of apoptosis accounts for the complex immune disorder in lpr and gld mutant mice[83].

In humans a similar disease with a dysfunction of the CD95 (type Ia ALPS)/CD95L (type Ib ALPS) system has been reported. Children with this 'auto-immune lymphoproliferative syndrome' (ALPs or Canale Smith syndrome) show massive, non-malignant lymphadenopathy, an altered and enlarged T cell population, and a severe autoimmunity. Many of these children show a crippling mutation in the death domain (DD) of CD95 but are heterozygous for this defect. Because the parents are not sick, whereas the children are, a secondary as-yet-unknown defect must exist that is responsible for the symptoms[24,25,84,85].

In some cases (type II ALPS) defective CD95-mediated apoptosis is observed without mutations in CD95 or CD95L[86]. This suggests that other defects that affect CD95 signalling exist, and it has recently been suggested that mutations in caspase-10 may cause such a complementing defect.

Thus, the inability of the immune system to eliminate self-reactive lympho-cytes by apoptosis can cause autoimmunity. It could be possible that autoanti-gen-driven prolonged T cell activation locks the cells into an AICD-resistant

phenotype. Alternatively, apoptosis might produce changes in cellular constituents which affect antigen processing and self-tolerance. Increased resistance
to apoptosis and persistence of autoreactive activated T cells have been found
in models of experimental autoimmunity. In these systems a clinically beneficial
effect of drugs was observed which resensitized such T cells towards apoptotic
deletion[87].

APOPTOSIS AND LYMPHOID TUMOURS

Apoptosis, or rather the lack thereof, plays an important role in the generation
of tumours. Follicular lymphomas result from a translocation of bcl-2 into the
immunoglobulin heavy chain locus and deregulated bcl-2 expression under the
influence of the immunoglobulin enhancer. Overexpression of bcl-2 suppresses
apoptosis and favours tumour cell proliferation[88]. This is supported by the
increased tumour incidence in, for example, bcl-2 transgenic animals[89].
Similarly, c-FLIP and most other apoptosis inhibitors may have oncogenic
potential. In contrast, pro-apoptotic molecules could, in principle, act as
tumour suppressors. Furthermore, resistance to chemotherapy may result from
anti-apoptotic mechanisms similar to those known to block apoptosis in
normal cells, and future therapeutic strategies will be aimed at sensitizing
tumour cells to apoptosis while sparing normal tissue from drug damage[90].

Tumours develop multiple mechanisms to evade elimination by the immune
system. These mechanisms comprise lack of expression of costimulatory or
MHC molecules and active strategies such as the production of immunosuppressive cytokines. CD95L may also have an immunosuppressive function. A
number of tumours, including lymphoid tumours, are resistant to apoptosis
and express functional CD95L constitutively or after chemotherapy. This
situation may enable tumour cells to delete anti-tumour lymphocytes and to
suppress anti-tumour immune responses, a phenomenon called 'tumour
counterattack'.

CD95L is expressed constitutively in immune-privileged sites such as the
testis and the eye, and might contribute to the immune-privileged status by
inducing apoptosis in infiltrating lymphocytes, and this may be exploited in
trying to delay rejection of allografts. However, it has also been reported that
overexpression of CD95L in grafts does not simply confer immune privilege
but, instead, induces a granulocytic response that accelerates rejection. Thus,
the *in-vivo* consequences of CD95L expression are far from clear, and whether
the counterattack by tumours underlies their escape from the immune system *in
vivo* needs to be validated by further experiments. Once established, 'counterattack' could possibly be exploited for therapeutic use by either inhibiting it for
cancer therapy or setting it up in organ transplantation[91].

APOPTOSIS IN AIDS

AIDS, characterized by a depletion of CD4$^+$ T helper cells, is a disease with too much apoptosis. The number of CD4$^+$ T cells, for example, in the peripheral blood of individuals productively infected with HIV is low (in the range of one in several thousand). This implies that T cell depletion in this disease also affects non-infected CD4$^+$ T cells. How do they die? As discussed earlier, T cells can choose between several death-signalling pathways. These different signalling pathways might all be affected in AIDS. Further experiments are urgently needed to determine the contribution of such pathways to CD4$^+$ T cell depletion in this disease. The experiments described below discuss initial attempts to elucidate this question.

The regulatory viral gene products (e.g. HIV-1 Tat) produced by HIV-infected cells penetrate non-infected cells and render these cells hypersensitive to TCR-induced CD95-mediated apoptosis. HIV Tat induces a pro-oxidative state in the affected cells, increases CD95L expression, and facilitates TCR-triggered CD95-mediated suicide. Further sensitization of the CD4$^+$ T cells results from the binding of HIV gp120 to CD4 and from crosslinking of bound gp120 by gp120 antibodies[92–98]. In addition to CD95-mediated apoptosis a novel and rapid type of apoptosis induced by both HIV-binding cell surface receptors, CD4 and CXCR4, in T cell lines, human peripheral blood lymphocytes, and CD4/CXCR4 transfectants was found. The potency of this phenomenon, and its specificity for CD4$^+$ T cells, suggest that it might play a significant role in T helper cell depletion in AIDS. On the basis of these data the use of antichemokine receptor antibodies meant to prevent HIV-1 infection might be dangerous. The use of the natural ligand of CXCR4, SDF-1α, or its derivates, however, could be considered for therapy, as it inhibits infection as well as CXCR4-mediated apoptosis. Studying the apoptotic signalling cascade triggered by CD4 and CXCR4 might prove to be useful in the identification of therapeutic strategies aimed at intervening with the progressive loss of CD4$^+$ T cells in HIV-1-infected individuals[61].

CONCLUSIONS

Without death by apoptosis life of cells of the immune system and their precise and specific function would not be possible. We have begun to understand the signalling pathways of apoptosis in lymphocytes and the rules which determine lymphocyte interactions. We have also begun to understand the survival signals which counteract apoptosis, and we may be capable, eventually, to manipulate the cells therapeutically. Finally, apoptosis adds new insight into the pathogenesis of diseases which might be of therapeutic benefit in the future.

Acknowledgements

I thank all my previous and present collaborators for help with the chapter and the figures, particularly S. Baumann, K.-M. Debatin, T. Defrance, C.S. Falk, S. Kirchhoff, A. Krueger, B. Kyewski, M. Peter, I. Schmitz, A. Strecker and H.

Walczak. I also thank H. Sauter and B. Pétillon for expert secretarial assistance. This work was funded by Deutsche Krebshilfe Dr Mildred Scheel Stiftung; German Israeli Cooperation in Cancer Research; AIDS grant German Federal Health Agency; Tumor Centre Heidelberg/Mannheim; BMBF Förderschwerpunkte 'Clinical-biomedical research' and 'Apoptosis'; AIDS Verbund Heidelberg; Wilhelm-Sander Stiftung; Ernst-Jung-Stiftung; Förderschwerpunkt Transplantation; and SFB 405 and 601 (DFG). I apologize to all my colleagues who have done excellent work in the field and whose papers have not been quoted comprehensively. It was not possible to be encyclopaedic in this exponentially growing field. I dedicate this chapter to the Basel Institute for Immunology that was lately announced to be closed by Roche after having prospered in full academic freedom for almost 30 years.

References

1. Tonegawa S. Somatic generation of antibody diversity. Nature. 1983;302:575–81.
2. Craxton A, Otipoby KL, Jiang A, Clark E. A. Signal transduction pathways that regulate the fate of B lymphocytes. Adv Immunol. 1999;73:79–152.
3. Sebzda E, Mariathasan S, Ohteki T, Jones R, Bachmann MF, Ohashi PS. Selection of the T cell repertoire. Annu Rev Immunol. 1999;17:829–74.
4. Drakesmith H, Chain B, Beverley P. How can dendritic cells cause autoimmune disease? Immunol Today. 2000;21:214–17.
5. DosReis GA, Shevach EM. Peripheral T-cell self-reactivity and immunological memory. Immunol Today. 1998;19:587–8.
6. Berzins SP, Godfrey DI, Miller JF, Boyd RL. A central role for thymic emigrants in peripheral T cell homeostasis. Proc Natl Acad Sci USA. 1999;96:9787–91.
7. Krammer PH. CD95(APO-1/Fas)-mediated apoptosis: live and let die. Adv Immunol. 1999;71:163–210.
8. Peter ME, Scaffidi C, Medema JP, Kischkel FC, Krammer PH. The death receptors. In: Kumar S, editor. Apoptosis, Problems and Diseases. Heidelberg: Springer, 1998:25–63:
9. Marsters SA, Yan M, Pitti RM, Haas PE, Dixit VM, Ashkenazi A. Interaction of the TNF homologues BLyS and APRIL with the TNF receptor homologues BCMA and TACI. Curr Biol. 2000;10:785–8.
10. Cascino I, Fiucci G, Papof G, Ruberti G. Three functional soluble forms of the human apoptosis-inducing Fas molecule are produced by alternative splicing. J Immunol. 1995;54: 2706–13.
11. Klas C, Debatin KM, Jonker RR, Krammer PH. Activation interferes with the APO-1 pathway in mature human T cells. Int Immunol. 1993;5:625–30.
12. Leithäuser F, Dhein J, Mechtersheimer G et al. Constitutive and induced expression of Apo-1, a new member of the nerve growth factor/tumor necrosis factor receptor superfamily, in normal and neoplastic cells. Lab Invest. 1993;69:415.
13. Golstein P. Controlling cell death. Science. 1997;275:1081–2.
14. Medema JP, Toes RE, Scaffidi C et al. Cleavage of FLICE (caspase-8) by granzyme B during cytotoxic T lymphocyte-induced apoptosis. Eur J Immunol. 1997;27:3492–8.
15. Martinez-Lorenzo MJ, Anel A, Gamen S et al. Activated human T cells release bioactive Fas ligand and APO2 ligand in microvesicles. J Immunol. 1999;163:1274–81.
16. Li JH, Rosen D, Ronen D et al. The regulation of CD95 ligand expression and function in CTL. J Immunol. 1998;161:3943–9.
17. Mariani SM, Matiba B, Bäumler C, Krammer PH. Regulation of cell surface APO-1/Fas (CD95) ligand expression by metalloproteases. Eur J Immunol. 1995;25:2303–7.
18. Tanaka M, Suda T, Haze K et al. Fas ligand in human serum. Nat Med. 1996;2:317–22.
19. Yagita H, Hanabuchi S, Asano Y, Tamura T, Nariuchi H, Okumura K. Fas-mediated cytotoxicity – a new immunoregulatory and pathogenic function of Th1 CD4+ T cells. Immunol Rev. 1995;146:223–39.

20. Tanaka M, Suda T, Takahashi T, Nagata S. Expression of the functional soluble form of human fas ligand in activated lymphocytes. EMBO J. 1995;14:1129–35.
21. Suda T, Hashimoto H, Tanaka M, Ochi T, Nagata S. Membrane Fas ligand kills human peripheral blood T lymphocytes, and soluble Fas ligand blocks the killing. J Exp Med. 1997;186:2045–50.
22. Kischkel FC, Hellbardt S, Behrmann I et al. Cytotoxicity-dependent APO-1 (Fas/CD95)-associated proteins form a death-inducing signaling complex (DISC) with the receptor. EMBO J. 1995;14:5579–88.
23. Muzio M, Chinnaiyan AM, Kischkel FC et al. FLICE, a novel FADD-homologous ICE/CED-3-like protease, is recruited to the CD95 (Fas/APO-1) death-inducing signaling complex. Cell. 1996;85:817–27.
24. Siegel RM, Frederiksen JK, Zacharias DA et al. Fas preassociation required for apoptosis signaling and dominant inhibition by pathogenic mutations. Science. 2000;288:2354–7.
25. Chan FK, Chun HJ, Zheng L, Siegel RM, Bui KL, Lenardo MJ. A domain in TNF receptors that mediates ligand-independent receptor assembly and signaling. Science. 2000; 288:2351–4.
26. Scaffidi C, Fulda S, Srinivasan A et al. Two CD95 (APO-1/Fas) signaling pathways. EMBO J. 1998;17:1675–87.
27. Luo X, Budihardjo I, Zou H, Slaughter C, Wang X. Bid, a Bcl2 interacting protein, mediates cytochrome c release from mitochondria in response to activation of cell surface death receptors. Cell. 1998;94:481–90.
28. Li H, Zhu H, Xu C, Yuan J. Cleavage if BID by caspase 8 mediates the mitochondrial damage in the Fas pathway of apoptosis. Cell. 1998;94:491–501.
29. Kroemer G. The pharmacology of T cell apoptosis. Adv Immunol. 1995;58:211–96.
30. Du C, Fang M, Li Y, Li L, Wang X. Smac, a mitochondrial protein that promotes cytochrome c-dependent caspase activation by eliminating IAP inhibition. Cell. 2000;102: 33–42.
31. Verhagen AM, Ekert PG, Pakusch M et al. Identification of DIABLO, a mammalian protein that promotes apoptosis by binding to and antagonizing IAP proteins. Cell. 2000; 102:43–53.
32. Yin XM, Wang K, Gross A et al. Bid-deficient mice are resistant to Fas-induced hepatocellular apoptosis. Nature. 1999;400:886–91.
33. Thome M, Schneider P, Hofmann K et al. Viral Flice-inhibitory proteins (FLIPs) prevent apoptosis induced by death receptors. Nature 1997;386:517–21.
34. Hu S, Vincenz C, Buller M, Dixit VM. A novel family of viral death effector domain-containing molecules that inhibit both CD-95- and tumor necrosis factor receptor-1-induced apoptosis. J Biol Chem. 1997;272:9621–4.
35. Bertin J, Armstrong RC, Ottilie S et al. Death effector domain-containing herpesvirus and poxvirus proteins inhibit both Fas- and TNFR1-induced apoptosis. Proc Natl Acad Sci USA. 1997;94:1172–6.
36. Yeh WC, Itie A, Elia AJ et al. Requirement for Casper (c-FLIP) in regulation of death receptor-induced apoptosis and embryonic development. Immunity. 2000;12:633–42.
37. Bodmer JL, Holler N, Reynard S et al. TRAIL receptor-2 signals apoptosis through FADD and caspase-8. Nat Cell Biol. 2000;2:241–3.
38. Sprick MR, Weigand MA, Rieser E et al. FADD/MORT1 and caspase-8 are recruited to TRAIL receptors 1 and 2 and are essential for apoptosis mediated by TRAIL receptor 2. Immunity. 2000;12:599–609.
39. Kischkel FC, Lawrence DA, Chuntharapai A, Schow P, Kim KJ, Ashkenazi A. Apo2L/TRAIL-dependent recruitment of endogenous FADD and caspase-8 to death receptors 4 and 5. Immunity. 2000;12:611 20.
40. Chen WF, Scollay R, Clark-Lewis I, Shortman K. The size of functional T-lymphocyte pools within thymic medullary and cortical cell subsets. Thymus. 1983;5:179–95.
41. Surh CD, Sprent J. T-cell apoptosis detected in situ during positive and negative selection in the thymus. Nature. 1994;372:100–3.
42. Amsen D, Kruisbeek AM. Thymocyte selection: not by TCR alone. Immunol Rev. 1998; 165:209–29.
43. Kishimoto H, Sprent J. The thymus and central tolerance. Clin Immunol. 2000;95:S3–7.
44. Newton K, Harris AW, Strasser A. FADD/MORT1 regulates the pre-TCR checkpoint and can function as a tumour suppressor. EMBO J. 2000;19:931–41.

45. Krammer PH, Debatin KM. When apoptosis fails. Curr Biol. 1992;2:383–5.
46. Kishimoto H, Surh CD, Sprent J. A role for Fas in negative selection of thymocytes *in vivo*. J Exp Med. 1998;187:1427–38.
47. Aiello S, Noris M, Piccinini G et al. Thymic dendritic cells express inducible nitric oxide synthase and generate nitric oxide in response to self- and alloantigens. J Immunol. 2000; 164:4649–58.
48. Nelson BH, Willerford DM. Biology of the interleukin-2 receptor. Adv Immunol. 1998;70: 1–81.
49. Dhein J, Walczak H, Bäumler C, Debatin KM, Krammer PH. Autocrine T-cell suicide mediated by APO-1/(Fas/CD95). Nature. 1995;73:438–41.
50. Alderson MR, Tough TW, Davis-Smith T et al. Fas ligand mediates activation-induced cell death in human T lymphocytes. J Exp Med. 1995;181:71–7.
51. Brunner T, Mogil RJ, LaFace D et al. Cell-autonomous Fas (CD95)/Fas-ligand interaction mediates activation-induced apoptosis in T-cell hybridomas. Nature. 1995;373:441–44.
52. Ju ST, Panka DJ, Cui H et al. Fas(CD95)/FasL interactions required for programmed cell death after T-cell activation. Nature. 1995;373:444–8.
53. Van Parijs L, Ibraghimov A, Abbas AK. The roles of costimulation and Fas in T cell apoptosis and peripheral tolerance. Immunity. 1996;4:321–8.
54. Singer GG, Abbas AK. The fas antigen is involved in peripheral but not thymic deletion of T lymphocytes in T cell receptor transgenic mice. Immunity. 1994;1:365–71.
55. Inaba M, Kurasawa K, Mamura M, Kumano K, Saito Y, Iwamoto I. Primed T cells are more resistant to Fas-mediated activation-induced cell death than naive T cells. J Immunol. 1999;163:1315–20.
56. Van Parijs L, Refaeli Y, Lord JD, Nelson BH, Abbas AK, Baltimore D. Uncoupling IL-2 signals that regulate T cell proliferation, survival, and Fas-mediated activation-induced cell death. Immunity. 1999;11:281–8.
57. Zheng L, Fisher G, Miller RE, Peschon J, Lynch DH, Lenardo MJ. Induction of apoptosis in mature T cells by tumour necrosis factor. Nature. 1995;377:348–51.
58. Zimmerman C, Brduscha-Riem K, Blaser C, Zinkernagel RM, Pircher H. Visualization, characterization, and turnover of CD8$^+$ memory T cells in virus-infected hosts. J Exp Med. 1996;183:1367–75.
59. Hildeman DA, Mitchell T, Teague TK et al. Reactive oxygen species regulate activation-induced T cell apoptosis. Immunity. 1999;10:735–44.
60. Spaner D, Raju K, Rabinovich B, Miller RG. A role for perforin in activation-induced T cell death *in vivo*: increased expansion of allogeneic perforin-deficient T cells in SCID mice. J Immunol. 1999;162:1192–9.
61. Berndt C, Möpps B, Angermüller S, Gierschik P, Krammer PH. CXCR4 and CD4 mediate a rapid CD95-independent cell death in CD4$^+$ T cells. Proc Natl Acad Sci USA. 1998;95: 12556–61.
62. Kirchhoff S, Muller WW, Krueger A, Schmitz I, Krammer PH. TCR-mediated up-regulation of c-FLIPshort correlates with resistance toward CD95-mediated apoptosis by blocking death-inducing signaling complex activity. J Immunol. 2000;165:6293–300.
63. Kirchhoff S, Muller WW, Li-Weber M, Krammer PH. Up-regulation of c-FLIPshort and reduction of activation-induced cell death in CD28-costimulated human T cells. Eur J Immunol. 2000;30:2765–74
64. Van Parijs L, Refaeli Y, Abbas AK, Baltimore D. Autoimmunity as a consequence of retrovirus-mediated expression of C-FLIP in lymphocytes. Immunity. 1999;11:763–70.
65. Boise LH, Noel PJ, Thompson CB. CD28 and apoptosis. Curr Opin Immunol. 1995;7:620–5.
66. Lam KP, Rajewsky K. Rapid elimination of mature autoreactive B cells demonstrated by Cre-induced change in B cell antigen receptor specificity *in vivo*. Proc Natl Acad Sci USA. 1998;95:13171–5.
67. Lam KP, Kuhn R, Rajewsky K. *In vivo* ablation of surface immunoglobulin on mature B cells by inducible gene targeting results in rapid cell death. Cell. 1997;90:1073–83.
68. Bouchon A, Krammer PH, Walczak H. Critical role for mitochondria in B cell receptor-mediated apoptosis. Eur J Immunol. 2000;30:69–77.
69. Berard M, Mondiere P, Casamayor-Palleja M, Hennino A, Bella C, Defrance T. Mitochondria connects the antigen receptor to effector caspases during B cell receptor-induced apoptosis in normal human B cells. J Immunol. 1999;163:4655–62.

70. Lagresle C, Mondiere P, Bella C, Krammer PH, Defrance T. Concurrent engagement of CD40 and the antigen receptor protects naive and memory human B cells from APO-1/Fas-mediated apoptosis. J Exp Med. 1996;183:1377–88.
71. Smith KG, Light A, O'Reilly LA, Ang SM, Strasser A, Tarlinton D. bcl-2 transgene expression inhibits apoptosis in the germinal center and reveals differences in the selection of memory B cells and bone marrow antibody-forming cells. J Exp Med. 2000;191:475–84.
72. Knodel M, Kuss AW, Lindemann D, Berberich I, Schimpl A. Reversal of Blimp-1-mediated apoptosis by A1, a member of the Bcl-2 family. Eur J Immunol. 1999;29:2988–98.
73. Merville P, Dechanet J, Desmouliere A et al. Bcl-2$^+$ tonsillar plasma cells are rescued from apoptosis by bone marrow fibroblasts. J Exp Med. 1996;183:227–36.
74. Foote LC, Marshak-Rothstein A, Rothstein TL. Tolerant B lymphocytes acquire resistance to Fas-mediated apoptosis after treatment with interleukin 4 but not after treatment with specific antigen unless a surface immunoglobulin threshold is exceeded. J Exp Med. 1998;187:847–53.
75. Xia XZ, Treanor J, Senaldi G et al. TACI is a TRAF-interacting receptor for TALL-1, a tumor necrosis factor family member involved in B cell regulation. J Exp Med. 2000;192:137–44.
76. Gross JA, Johnston J, Mudri S et al. TACI and BCMA are receptors for a TNF homologue implicated in B-cell autoimmune disease. Nature. 2000;404:995–9.
77. Shu HB, Johnson H. B cell maturation protein is a receptor for the tumor necrosis factor family member TALL-1. Proc Natl Acad Sci USA. 2000;97:9156–61.
78. Steinman RM, Turley S, Mellman I, Inaba K. The induction of tolerance by dendritic cells that have captured apoptotic cells. J Exp Med. 2000;191:411–16.
79. Albert ML, Sauter B, Bhardwaj N. Dendritic cells acquire antigen from apoptotic cells and induce class I-restricted CTLs. Nature. 1998;392:86–9.
80. Fanger NA, Maliszewski CR, Schooley K, Griffith TS. Human dendritic cells mediate cellular apoptosis via tumor necrosis factor-related apoptosis-inducing ligand (TRAIL). J Exp Med. 1999;190:1155–64.
81. Ashany D, Savir A, Bhardwaj N, Elkon KB. Dendritic cells are resistant to apoptosis through the Fas (CD95/APO-1) pathway. J Immunol. 1999;63:5303–11.
82. Bjorck P, Banchereau J, Flores-Romo L. CD40 ligation counteracts Fas-induced apoptosis of human dendritic cells. Int Immunol. 1997;9:365–72.
83. Nagata S. Human autoimmune lymphoproliferative syndrome, a defect in the apoptosis-inducing Fas receptor: a lesson from the mouse model. J Hum Genet. 1998;43:2–8.
84. Fisher GH, Rosenberg FJ, Straus SE et al. Dominant interfering Fas gene mutations impair apoptosis in a human autoimmune lymphoproliferative syndrome. Cell. 1995;81:935–46.
85. Rieux-Laucat F, Le Deist F, Hivroz C et al. Mutations in Fas associated with human lymphoproliferative syndrome and autoimmunity. Science. 1995;268:1347–9.
86. Wang J, Zheng L, Lobito A et al. Inherited human caspase 10 mutations underlie defective lymphocyte and dendritic cell apoptosis in autoimmune lymphoproliferative syndrome type II. Cell. 1999;98:47–58.
87. Zhou T, Song L, Yang P, Wang Z, Lui D, Jope RS. Bisindolylmaleimide VIII facilitates Fas-mediated apoptosis and inhibits T cell-mediated autoimmune diseases. Nat Med. 1999;5:42–8.
88. Chao DT, Korsmeyer SJ. BCL-2 family: regulators of cell death. Annu Rev Immunol. 1998;16:395–419.
89. Adams JM, Harris AW, Strasser A, Ogilvy S, Cory S. Transgenic models of lymphoid neoplasia and development of a pan-hematopoietic vector. Oncogene. 1999;18:5268–77.
90. Debatin KM. Activation of apoptosis pathways by anticancer drugs. Adv Exp Med Biol. 1999;457:237–44.
91. Igney FH, Behrens CK, Krammer PH. Tumor counterattack – concept and reality. Eur J Immunol. 2000;30:725–31.
92. Westendorp MO, Frank R, Ochsenbauer C et al. Sensitization of T cells to CD95-mediated apoptosis by HIV-1 Tat and gp120. Nature. 1995;375:497–500.
93. Krammer PH, Dhein J, Walczak H et al. The role of APO-1-mediated apoptosis in the immune system. Immunol Rev. 1994;142:175–91.
94. Debatin KM, Fahrig-Faissner A, Enenkel-Stoodt S, Kreuz W, Benner A, Krammer PH. High expression of APO-1 (CD95) on T lymphocytes from human immunodeficiency virus-1-infected children. Blood. 1994;83:3101–3.

95. Baumler CB, Bohler T, Herr I, Benner A, Krammer PH, Debatin KM. Activation of the CD95 (APO-1/Fas) system in T cells from human immunodeficiency virus type-1-infected children. Blood. 1996;88:1741–6.
96. Li CJ, Friedman DJ, Wang C, Metelev V, Pardee AB. Induction of apoptosis in uninfected lymphocytes by HIV-1 Tat protein. Science. 1995;268:429–31.
97. Finkel TH, Tudor-Williams G, Banda NK et al. Apoptosis occurs predominantly in bystander cells and not in productively infected cells of HIV- and SIV-infected lymph nodes. Nat Med. 1995;1:129–34.
98. Gougeon ML, Montagnier L. Programmed cell death as a mechanism of CD4 and CD8 T cell deletion in AIDS. Molecular control and effect of highly active anti-retroviral therapy. Ann NY Acad Sci. 1999;887:199–212.

5
Regulation of autoimmunity – lessons from liver and skin

B. ARNOLD

INTRODUCTION

The immune system is an adaptive defence system capable of specifically recognizing and eliminating an apparently limitless variety of foreign invaders. The enormous diversity of the antigen-specific receptors on B and T lymphocytes, the key players in the adaptive immune response, is generated by random rearrangement of the respective genes. The main challenge of such a defence system is to have as broad a T and B cell repertoire as possible in the absence of autoreactivity. Several tolerance mechanisms are operating in parallel under physiological conditions to silence T cells either during their development in the thymus or in the periphery.

Thymocytes are triggered to undergo apoptosis when their T cell receptor (TCR) binds to the respective peptide/MHC complex with sufficiently high avidity. The basis for the tolerogenic response of the immature thymocytes in comparison to the activation of the mature T cells seems to be a result of differences in the signals transmitted by the TCR and in the set of genes triggered by the second messengers[1]. The thymus harbours not only ubiquitously expressed proteins and blood-borne antigens but also antigens which were formerly thought to be tissue-specific.

Medullary thymic epithelial cells have been identified as a specialized cell type with promiscuous expression of a broad range of tissue-specific gene products[2]. Nevertheless, T lymphocytes with self-destructive capacity are often found in healthy individuals, suggesting that thymic control of self-reactive T cells is incomplete and that, in addition, peripheral mechanisms are operating to avoid autoimmune diseases.

There are several possibilities relating to how tissue-specific, self-reactive T cells leaving the thymus could be silenced in the periphery. Such T cells remain naive if the tissue antigen is expressed at an insufficient level to transmit a signal, or if the respective tissue is inaccessible for these T cells. This phenomenon has been termed ignorance[3]. When an inflammatory process leads to up-regulation of antigen expression and tissue accessibility the autoreactive T cells might be controlled by regulatory T cells. Such regulatory

T cells can be generated in the thymus and need for their survival in the periphery the presence of the respective self-antigen as shown in studies of autoimmune thyroiditis[4]. Rats whose thyroids had been ablated by ^{131}I had a deficit in peripheral regulatory T cell activity specific for thyroid antigens. However, CD4[+] thymocytes from these athyroid rats were still able to prevent thyroiditis upon adoptive transfer. This result implies that the intra-thymic development of regulatory T cell precursors is normal in these rats, and that the failure to demonstrate specific functional regulatory T cells in the periphery is due to the absence of the specific peripheral autoantigen.

Alternatively, self-antigens which are sufficiently highly expressed in intact peripheral organs of adult mice could be presented to CD8 T cells in the draining lymph nodes by professional antigen-presenting cells (APC) via an exogenous class I-restricted pathway . This cross-presentation could lead to tolerance based on deletion of the respective T cells in the absence of T cell help[5]. The APC performing the cross-presentation have been identified as dendritic cells[6]. There is general support for the concept that dendritic cells can either stimulate or turn off T cell reactivity in lymphoid compartments. It is, however, debated whether or not non-professional APC can also induce tolerance directly in the respective tissues. The aim of this contribution is, therefore, to show that parenchymal cells such as keratinocytes, hepatocytes and liver sinusoidal endothelial cells do contribute to peripheral T cell tolerance towards tissue-specific antigens.

PERIPHERAL T CELL TOLERANCE TO A TRULY SESSILE TISSUE ANTIGEN EXPRESSED ON PARENCHYMAL CELLS

Thymus-independent peripheral T cell tolerance induction has been demonstrated in transgenic mice expressing the MHC class I molecule K^b only on keratinocytes (2.4KerIV-K^b)[7,8]. These transgenic animals accepted K^b-positive skin grafts and syngeneic tumours expressing K^b, enabling us to analyse the observed tolerance in these animals in detail. Corresponding TCR transgenic mice were employed with a TCR recognizing only the intact form, but not a processed form, of the K^b antigen[9]. The fate of these T cells could be followed using an anti-clonotypic antibody (Desire-1).

It was important to determine that the tolerance was induced in the periphery by K^b on keratinocytes and not by minute amounts of K^b in the thymus that may have escaped detection. To this end, adult thymectomized 2.4KerIV-K^b transgenic mice were grafted with the thymi of non-transgenic newborn mice of the same H-2 haplotype, irradiated, and reconstituted with bone marrow cells from Des-TCR mice[10]. These chimeric mice accepted K^b-positive skin grafts. Although no K^b expression was detectable on bone-marrow cells of the transgenic mice, the possibility was excluded that in the chimeras potentially K^b-positive bone-marrow cells from the 2.4KerIV-K^b mice could have repopulated the non-transgenic thymus, thereby leading to thymic tolerance induction. Non-transgenic (H-2^{dxk}) mice were lethally irradiated, reconstituted with bone marrow of different origin, and skin grafted 6–8 weeks later. Mice reconstituted with bone marrow of Des-TCR (H-2^{bxk}) origin accepted K^b-bearing skin grafts, because bone marrow-derived K^b-positive cells in the thymus deleted the K^b-reactive T cells. In contrast, reconstitution with bone marrow of either Des-TCR (H-2^{dxk}) or 2.4KerIV-K^b × Des-TCR (H-2^{dxk}) mice led to skin graft rejection. These results demonstrate the absence of K^b-positive cells in the bone marrow of 2.4KerIV-K^b × Des-TCR mice.

Similar experiments were performed with CRP-K^b mice expressing K^b exclusively in hepatocytes. These studies excluded any antigen-specific contribution by the thymus to the observed tolerance[11]. Intact MHC class I molecules cannot be transferred from parenchymal cells to professional APC in a way that allows activation of T cells[12]. Therefore, T cell tolerance seen in these model systems is due to direct encounter of K^b on keratinocytes in the skin or hepatocytes in the liver.

So far, hepatocytes and keratinocytes were taken as examples for tolerance induction to a truly sessile antigen by parenchymal cells in the respective tissue in the absence of cross-presentation. This tolerance mechanism could not have been demonstrated with a nominal antigen being also cross-presented in the regional lymph nodes by dendritic cells. In the physiological situation both mechanisms, direct contact with the antigen on tissue cells and recognition of cross-presented peptides on dendritic cells, may operate in parallel to ensure tolerance induction in the respective T cells.

MIGRATION OF NAIVE T CELLS AND ITS CONSEQUENCE FOR TOLERANCE INDUCTION TO TISSUE-RESTRICTED ANTIGENS

Studies on lymphocyte migration in the adult sheep have shown that in general naive T cells do not gain access to peripheral non-lymphoid tissues, in contrast to activated/memory T cells which can migrate into tissues[13]. Therefore, it has been argued that tissue antigens are ignored by naive T cells unless they are picked up by dendritic cells and presented in the draining lymph nodes[5]. Unfortunately, most of the studies mentioned above used adult animals either by confronting mature T cells with exogenous antigen or by transferring T cells specific for constitutively expressed endogenous antigen. Large-scale trafficking of virgin T cells through extra-lymphoid tissues in fetal sheep[14], as well as in newborn rats and mice and, therefore, the possibility of a direct contact between naive T cells with tissue-specific self-antigens on parenchymal cells, is often neglected.

To test the hypothesis that such an interaction during neonatal life could lead to T cell tolerance the above-mentioned 2.4Ker IV-K^b-mice were crossed onto the Rag-2-deficient background and reconstituted as neonates or adults with bone marrow from Des-TCR transgenic mice[10]. Tolerance could only be observed in the neonatally reconstituted mice and was not detectable in the adult recipients. Blockage of T cell migration neonatally with antibodies directed against P- and E-selectin prevented tolerance induction. Thus, the accessibility of neonatal skin for naive T cells, and the interaction with self-antigens on keratinocytes, contribute to the control of the respective auto-reactive T cells.

The difference between neonatal and adult tolerance induction may only have been detected because the experiments were performed with mice not suffering infections. Under physiological conditions tolerance to tissue-specific antigens might also be established in the adult as a consequence of the inflammatory processes induced by pathogens. Naive T cells have been reported to enter inflammatory sites. Up-regulation of MHC class II molecules on keratinocytes or of co-stimulatory molecules such as COS-ligand on fibroblasts seen at inflammatory sites may allow activation of autoreactive T cells by these parenchymal cells. This activation could finally lead to tolerance induction.

A DOMINANT FORM OF CD8 T CELL TOLERANCE

The lack of tolerance in the Des-TCR 2.4KerIV-K^b.Rag-2$^{-/-}$ adult chimeras raises the following questions: why are normal double-transgenic Des-TCR × 2.4KerIV-K^b mice tolerant and which mechanism prevents naive T cells which continuously leave the thymus in adult life from rejecting K^b-positive grafts? To consider a dominant tolerance process in our system would require that T cells which had been rendered tolerant to K^b in the neonatal phase persist indefinitely and interfere with the activation of naive, K^b-specific T cells when confronted with a K^b-positive graft.

To test this hypothesis we asked whether these tolerant T cells in day 15 thymectomized, adult double-transgenic mice could influence the reactivity of transferred naive T cells. T cells from Des-TCR.RAG-2$^{-/-}$ mice could reject a syngeneic tumour transfected with the K^b and B7 genes (P815-K^b.B7) after transfer into day 15 thymectomized 2.4KerIV-K^b.Rag2$^{-/-}$ animals transgenic for an anti-*Leishmania* TCR (not crossreacting with K^b). In contrast, day 15 thymectomized Des-TCR × 2.4KerIV-K^b.RAG-2$^{-/-}$ mice, which had received the same number of naive CD8$^+$, Des-TCR cells, accepted the respective tumour graft (R. Reibke et al., unpublished data). Thus, T cells with a given TCR can either develop into destructive effector cells, or into regulator cells, depending on the conditions of antigen encounter.

To understand the molecular basis of the observed tolerance, gene expression of tolerant and activated/naive CD8 T cells was studied using microarray technology and quantitative RT-PCR. Gene products up-regulated in CD4, CD25 regulatory T cells such as CTLA4, GITR, FoxP3 and CD25 were not differentially expressed. On the other hand, such as in CD4, CD25 T cells, increased expression of gene products involved in the Notch signalling pathway was found in these regulatory CD8 T cells (T. Oelert et al., unpublished data).

Further studies should reveal whether there are common pathways in CD4 and CD8 regulatory T cells used.

TOLERANCE INDUCTION BY LIVER SINUSOIDAL ENDOTHELIAL CELLS

Similar to tolerance induction after intravenous injection of antigen, intraportal antigen application results in systemic antigen-specific tolerance, indicating that the liver is particularly capable of actively inducing peripheral tolerance. The capacity of hepatocytes to induce T cell tolerance was mentioned above. The narrow diameter of the sinusoids promotes contact between passing leucocytes and the liver sinusoidal endothelial cells (LSEC) that line the sinusoids. A series of surface molecules associated with professional APC are constitutively expressed by LSEC, indicating an immunological role of this resident hepatic cell population[15]. Indeed, LSEC can very efficiently cross-present antigen to CD8$^+$ T cells[16]. Upon stimulation *in vitro*, CD8$^+$ T cells secrete normal amounts of IL-2 and interferon-γ. This capacity is lost after 3 days of co-culture with LSEC, in contrast to the continuous production of these lymphokines by CD8$^+$ T cell populations activated with antigen-presenting splenocytes. Furthermore, Des-TCR, CD8$^+$ T cells primed by antigen-presenting LSEC did not show K^b-specific cytotoxicity, whereas CD8$^+$ T cells primed by splenocytes displayed specific cytotoxicity. Participation of LSEC in tolerance induction to circulating soluble antigens was verified by injecting C57BL/6 mice intravenously with ovalbumin. LSEC were isolated 24 h later and adoptively transferred into C57BL/6 mice. After 1 week the procedure was repeated and the mice were challenged with a syngeneic tumour expressing ovalbumin. Only mice which had received ovalbumin-pulsed LSEC accepted the tumour, whereas all untreated mice showed tumour rejection. This indicates that T cell tolerance can be induced by cross-presenting LSEC in animals with a normal T cell repertoire, and that tolerance to intravenously applied antigen is mediated by LSEC.

In addition, these findings point to mechanisms which can prevent a destructive T cell response after initial T cell activation depending on the microenvironment.

DEFINITION OF TOLERANCE

Some of the present confusion in the field of tolerance is due to views regarding tolerance. Burnet's clonal selection theory postulated that self-reactive T cells must be eliminated to ensure self-protection. This view was supported by demonstrating reactivity to alloantigens either by *in-vivo* graft rejection, or *in-vitro* induction of cytotoxic T lymphocytes (CTL) and the absence of any responses to self-major histocompatibility complex (MHC) antigens in both assay systems. Since the self-reactive T cells were deleted one could use any assay system to demonstrate tolerance.

However, non-deletional mechanisms of tolerance were observed. Expression of an alloantigen on thymic epithelium achieved by grafting during neonatal life resulted in *in-vivo* tolerance to the 'foreign' alloantigen present during ontogeny as judged by skin graft acceptance[17]. CTL responses *in vitro* against this alloantigen were normal. Thus, reactivity or non-reactivity to a given self-antigen was dependent upon the test system, clearly showing that CTL assays might not reflect the *in-vivo* situation.

Furthermore, it was shown in several transgenic mouse models that activated self-reactive T cells capable of causing tissue damage could be present in an animal without causing an autoimmune disease in the respective organ expressing the target antigen. Transgenic mice expressing the MHC class-I molecule H-2K^b (K^b) in the insulin-producing β cells of the pancreas were crossed with Des-TCR transgenic mice. Although T cells expressing the highest level of Des-TCR were deleted intrathymically in such double-transgenic mice, K^b-specific T cells were detected in the periphery. These cells caused the rejection of K^b-expressing skin grafts, but ignored islet K^b antigens even after priming[18]. Thus, in the presence of *in-vivo* T cell reactivity one could not observe any attack of the intact organ expressing the respective target antigen.

Finally, K^b-specific T cells were still present in the tolerant Des-TCR × CRP-K^b mice, expressing K^b only on hepatocytes under the regulatory sequences of the gene encoding the C-reactive protein (CRP). It was, therefore, of interest to ask whether the observed tolerance could be reversed *in vivo*, and whether this would result in autoaggression[19]. Repeated injections of bacterial CpG oligonucleotides led to liver damage, as seen by the rise in liver enzyme activity in the blood. Thus, the CD8 Des-TCR T cells could be activated by the K^b expressing hepatocytes under inflammatory conditions. However, this destructive response was only transient. As soon as CpG application was stopped the autoimmune response disappeared. Even repeated injections for 45 days could not induce a chronic autoimmune disease.

These examples demonstrate that the presence of activated T cells and the respective target structure in an intact organ do not necessarily lead to tissue damage. Following the present terminology we call these animals non-tolerant because we detect T cell reactivity *in vitro* or *in vivo*. However, if we are interested in the physiological mechanisms preventing autoimmune diseases we may follow Ron Schwartz's definition[20] that tolerance represents a 'physiological state in which the immune system does not react destructively against the organism that harbours it'. In this case the above mentioned animals are clearly tolerant, as they do not suffer from chronic autoimmune diseases (Table 1).

Table 1 How do we define tolerance?

	T cell reactivity		Self destruction	
	In vitro	In vivo	Transient	Chronic
Tolerance based on clonal deletion	No	No	No	No
Tolerance II	Yes	No	No	No
Tolerance III	Yes	Yes	No	No
Tolerance IV	Yes	Yes	Yes	No
Autoimmune diseases	Yes			

References

1. Kruisbeck AM, Amsen D. Mechanisms underlying T cell tolerance. Curr Opin Immunol. 1996;8:233–44.
2. Derbinski J, Schulte A, Kyewski B, Klein L. Promiscuous gene expression in medullary thymic epithelial cells mirrors the peripheral self. Nature Immunol. 2001;2:1032–9.
3. Zinkernagel RM, Ehl S, Aichele P, Oehen S, Kündig T, Hengartner H. Antigen localization regulates immune responses in a dose- and time-dependent fashion: a geographical view of immune reactivity. Immunol Rev. 1997;156:199–209.
4. Seddon B, Mason D. The third function of the thymus. Immunol Today. 2000;21:95–9.
5. Heath WR, Kurts C, Miller JFAP, Carbone FR. Cross-tolerance: a pathway for inducing tolerance to peripheral tissue antigens. J Exp Med. 1998;187:1549–53.
6. Kurts C, Cannarile M, Klebba I, Brocker T. Cutting edge: dendritic cells are sufficient to cross-present self-antigens to CD8 T cells *in vivo*. J Immunol. 2001;166:1439–42.
7. Schönrich G, Momburg F, Malissen M et al. Distinct mechanisms of extrathymic T cell tolerance due to differential expression of self antigen. Int Immunol. 1992;4:581–90.
8. Schönrich G, Kalinke U, Momburg F et al. Downregulation of T cell receptors on self-reactive T cells as a novel mechanism for extrathymic tolerance induction. Cell. 1991;65: 293–304.
9. Guimezanes A, Barrett-Wilt GA, Gulden-Thompson P et al. Identification of endogenous peptides recognized by *in vivo* or *in vitro* generated alloreactive cytotoxic T lymphocytes: distinct characteristics correlated with CD8 dependence. Eur J Immunol. 2001;31:421–32.
10. Alferink J, Tafuri A, Vestweber D, Hallmann R, Hämmerling GJ, Arnold B. Peripheral T cell trafficking controls neonatal tolerance to tissue antigens. Science. 1998;282:1338–41.
11. Ferber I, Schönrich G, Schenkel J, Mellor A, Hämmerling GJ, Arnold B. Levels of peripheral T cell tolerance induced by different doses of tolerogen. Science. 1994;263:674–6.
12. Kurts C, Heath WR, Carbone FR, Allison J, Miller JF, Kosaka H. Constitutive class I-restricted exogenous presentation of self antigens *in vivo*. J Exp Med. 1996;184:923.
13. Mackay CR. Homing of naive, memory and effector lymphocytes. Curr Opin Immunol. 1993;5:423–7.
14. Kimpton WG, Washington RNP, Cahill RN. Virgin ab and gd T cells recirculate extensively through peripheral tissues and skin during normal development of the fetal immune system. Int Immunol. 1995;7:1567–77.
15. Knolle PA, Gerken G. Local control of the immune response in the liver. Immunol Rev. 2000;174:21–34.
16. Limmer A, Ohl J, Kurts C et al. Efficient presentation of exogenous antigen by liver endothelial cells to CD8+ T cells results in antigen-specific T cell tolerance. Nature Med. 2000;6:1348–54.
17. Corbel C, Martin C, Ohki H, Coltey M, Hlozanek I, Le Douarin NM. Evidence for peripheral mechanisms inducing tissue tolerance during ontogeny. Int Immunol. 1990;2: 33–40.
18. Heath WR, Allison J, Hoffmann MW et al. Autoimmune diabetes as a consequence of locally produced interleukin-2. Nature. 1992;359:547–9.
19. Limmer A, Sacher T, Alferink J et al. Failure to induce organ-specific autoimmunity by breaking of tolerance: importance of the microenvironment. Eur J Immunol. 1998;28: 2395–406.
20. Schwartz RH. Immunological tolerance. In: Paul WE, editor. Fundamental Immunology, 3rd edn. New York: Raven Press, 1993:677–731.

Section III
Pathogenesis I

Chair: I.G. McFARLANE and F.B. BIANCHI

6
Autoreactive T cells: any evidence in autoimmune liver disease?

Y. MA, M. S. LONGHI, D. P. BOGDANOS, G. MIELI-VERGANI and
D. VERGANI

THE LYMPHOID LIVER

Classically, lymphoid organs were considered to be those tissues with exclusively immunological functions; for example, the thymus, lymph nodes and spleen[1]. This definition has recently been expanded to include organs whose primary function is not immunological but which clearly require dedicated and often elaborate immunological mechanisms to mediate their functions. The gut and the uterus, for example, fall into this category; the liver is also regarded as a lymphoid organ with unique immunological properties[2–4]. Because of its location and function the liver is continuously exposed to a large antigenic load that includes pathogens, toxins, tumour cells, dietary and self-antigens. The range of local immune mechanisms required to cope with this diverse immunological challenge is the focus of ongoing studies. The liver has an 'epithelial constitution' and contains large numbers of phagocytic cells, antigen-presenting cells (APC) and lymphocytes, and is a site for the production of cytokines, complement components and acute-phase proteins[4].

In the normal liver the hepatocytes comprise 70% of the total number of cells. The remaining 30% cells are mostly concerned with immunological functions and include lymphocytes, sinusoidal endothelial cells, Kupffer cells, biliary epithelial cells and hepatic stellate cells[4–8].

The normal human liver has an intrinsic lymphocyte population mainly resident in the portal tract but also scattered throughout the parenchyma. One to two million of these lymphoid cells are retrievable from 200 mg of normal liver tissue, indicating that an average human liver of 1.5 kg contains approximately 1×10^{10} lymphocytes. Intrahepatic lymphocytes (IHL) consist of natural killer T (NKT) lymphocytes (50% of the IHL), T cells (25% of IHL), natural killer (NK) cells (20%) and B lymphocytes (5%). The relative proportions among IHL of different classes of lymphocytes differ widely from those in the peripheral blood where up to 80% are T cells, 20% B cells, 13% NK cells, and 0.2–2% are NKT cells[4–9].

The relative frequencies of lymphocyte subpopulations vary from individual to individual. This is likely to reflect differences in each individual's immunological status that would be influenced by factors such as genetic background, history of infections, and current antigenic exposure. Ongoing immune challenges to the liver would presumably affect the relative proportions of resident hepatic lymphocytes and infiltrating circulating lymphocytes[10,11].

The liver is an organ in which blood from the intestines, rich in bacterial and dietary antigens, intermingles with the circulating lymphocytes. The constitutive presence of antigenic material imposes constraints on immune responses generated in the liver, where distinctive control mechanisms have to determine whether antigen encounter will result in immunological unresponsiveness (tolerance)[10,11]. Liver autoimmunity implies loss of self-tolerance, and the question arises as to how this happens.

BREAKDOWN OF LIVER TOLERANCE

Various mechanisms have been proposed to account for the initiation of an autoimmune liver response with no single initiating event being able to account for all instances of autoimmunity[11]. Two general conditions, however, must prevail: self-reactive B and T lymphocytes must exist in the immunological repertoire and autoantigens must be presented in conjunction with MHC class II molecules by APC[11,12]. The presence of lymphocytes in the liver is commonly considered to be associated with disease pathogenesis[4,5,12]. Lymphocyte infiltration is well documented in inflammatory conditions such as autoimmune and viral liver diseases[13]. The liver harbours lymphoid cells that participate in innate and adaptive immune responses and, like peripheral lymphoid tissues, is a site where naive T cells can be activated by local antigen[4-9]. Hepatocytes or cholangiocytes can carry antigens perceived as foreign, and for which neither of the appropriate options, elimination or complete tolerance, has been achieved[11,14]. Such antigens may be derivatives of hepatotropic viruses, drugs or xenobiotics, or the host itself, as is the case in autoimmune liver diseases[11,14].

Infections and liver tolerance

Viral or other microbial infections of the liver induce an immune response to clear the pathogen, the competence of the immune response determining whether chronic disease will develop or, as usually happens, clearance will occur[13]. Hepatotropic viruses are generally non-cytopathic, and the host immune response is responsible for both viral clearance and the associated inflammatory cell injury[15-17]. A misdirected immune response to hepatocellular or biliary ductular constituents released in the course of such injury is a postulated pathway for liver autoimmunity[11,14].

Drugs and liver tolerance

The liver is the major site of drug metabolism. Drugs and their metabolites can be toxic if their degradation pathways are defective[18,19]; thus, intracellular antigens may be released from the damaged cells[20]. Presentation of protein adducts may eventually lead to an immune response, including the production of autoantibodies, inflammation and hepatocyte necrosis[19,20]. In addition, the covalent binding of a drug metabolite to its enzyme may result in the modification of this enzyme, and in the appearance of antibodies that recognize the alkylated form as a neoantigen[19,20].

Apoptosis and liver tolerance

Apoptosis, as the basis for programmed cell death, is responsible for the normal regular turnover of hepatocytes and the elimination of liver cells in inflammatory pathologies[10,11,21]. Although less obvious in biliary epithelium, apoptosis is also present in autoimmune cholangiopathies[14]. Aside from hepatocytes and their disposal, apoptosis is relevant to the breakdown and/or maintenance of liver tolerance[10,11,21]. First, death by apoptosis allows for non-inflammatory elimination of cell components in contrast to necrosis which is proinflammatory and potentially autoantigenic[12]. However, if apoptosis is overwhelming, apoptotic fragments can induce an autoimmune response and indeed autoimmune disease, as has been seen in the case of systemic lupus erythematosus[22]. Second, apoptosis is a mechanism used for the removal of autoreactive T and B lymphocytes, as is illustrated by lymphoproliferative diseases due to deficiency of apoptotic genes[23].

Autoimmune liver diseases

Autoimmune liver diseases represent a broad spectrum of disorders that can affect one or the other of the two anatomical components, in particular the hepatocytes in autoimmune hepatitis (AIH) types 1 and 2 and *de-novo* AIH following transplantation and the cholangiocytes and bile ducts in primary biliary cirrhosis (PBC), and sclerosing cholangitis (SC)[12]. AIH-1 is character-ized by seropositivity for antibodies to nuclear (ANA) and/or smooth muscle antibodies (SMA) and AIH type 2 for anti-liver kidney microsomal type 1 antibodies (LKM1)[24]. The serology of SC is similar to that of AIH-1[25,26]. The serological hallmark of PBC is the presence of high-titre antimitochondrial antibodies (AMA)[27]. This chapter will focus on the role of autoreactive T cells in patients with AIH.

AIH

Examination of the liver tissue from patients with AIH reveals a dramatic picture. A multitude of mononuclear cells, lymphocytes, plasma cells, and macrophages infiltrates the portal tract, invades the adjacent parenchyma, and surrounds dying hepatocytes. This picture, defined by histologists as interface hepatitis, first suggested that autoaggressive cellular immunity was involved in the pathogenesis of the disease[28,29].

Immunocytochemical studies have identified that the T lymphocytes mounting α/β T cell receptor predominate. Among the T cells a majority are positive for the CD4 helper/inducer phenotype, and a sizeable minority for the CD8 cytotoxic/suppressor phenotype. Lymphocytes of non-T cell lineage are fewer and include (in decreasing order of frequency) killer/natural killer (K/NK) cells (CD16/CD56 positive), macrophages, and B cells[30]. The presence of NKT (cells that express both NK and T cell markers) is the focus of ongoing studies[31,32].

A powerful stimulus must promote the massive inflammatory cell infiltration that is present at diagnosis. Regardless of the initial trigger, it is probable that such a high number of activated inflammatory cells cause liver damage. There are different possible pathways that an immune attack can follow to inflict damage on hepatocytes (Figure 1).

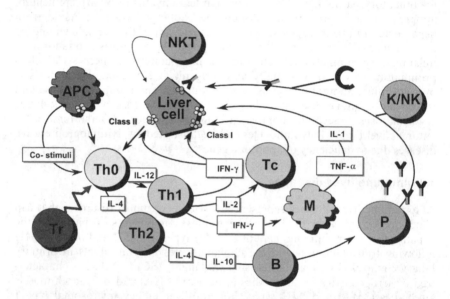

Figure 1 Scenario of an autoimmune attack to the liver cell. A specific autoantigenic peptide is presented to an uncommitted T helper (Th0) lymphocyte within the HLA class II molecule of an antigen-presenting cell (APC). Th0 cells become activated and, according to the presence in the microenvironment of interleukin (IL)-12 or IL-4 and the nature of the antigen, differentiate into Th1 or Th2 and initiate a series of immune reactions determined by the cytokines they produce: Th2 secrete mainly IL-4 and IL-10 and direct autoantibody production by B lymphocytes (B) and plasma cell (P); Th1 secrete IL-2 and interferon-γ (IFN-γ), which stimulate T cytotoxic (Tc) lymphocytes, enhance expression of class I and induce expression of class II HLA molecules on hepatocytes and activate macrophages (M); activated macrophages release IL-1 and tumour necrosis factor α (TNF-α). If T regulatory cells (Tr) do not oppose, a variety of effector mechanisms are triggered: liver cell destruction could derive from the action of Tc lymphocytes; cytokines released by Th1 and recruited macrophages; complement activation or engagement of killer/natural killer (K/NK) by the autoantibody bound to the hepatocyte surface. The role of the recently described natural killer T cells (NKT) is currently being explored

IMPAIRMENT OF T-REGULATORY CELLS

An impairment of immunomodulatory mechanisms, which would enable the autoimmune response to develop, has been suggested in several reports over the years. Thus, children and young adults with AIH have low levels of T cells expressing the CD8 marker[33], and impaired suppressor cell function[34]. These old studies have shown that: (1) the impairment of immunoregulation segregates with the possession of the HLA haplotype B8/DR3, which predisposes to AIH[35]; and (2) this immunoregulatory defect can be corrected by therapeutic doses of corticosteroids[36]. Furthermore, patients with AIH have been reported to have a specific defect in a subpopulation of T cells controlling the immune response to liver-specific membrane antigens[37].

We have recently obtained further evidence for an impairment of immunoregulatory function in AIH[38]. Among those T cell subsets with potential immunosuppression function, such as NKT cells, T helper 3 (Th3), T regulatory 1 (Tr1), CD8$^+$CD28$^-$ and $\gamma\delta$ T cells, CD4$^+$ T cells constitutively expressing the interleukin 2 receptor (IL-2R) α chain (CD25) (T-regs) have emerged as the dominant immunoregulatory population[39]. These cells, which represent 5–10% of the total population of peripheral CD4$^+$ T lymphocytes in healthy mice and humans, control the innate and the adaptive immune response by preventing the proliferation and effector function of autoreactive T cells. Their precise mechanism of action, possibly involving a contact with the target cells, as supported by *in-vitro* data[40,41], or implicating the release of cytokines with regulatory properties (IL-10, TGF-β), as proved by *in-vivo* studies[42–44], is still under investigation. In addition to CD25, which is also present on T cells undergoing activation, T-regs express a number of additional markers such as the glucocorticoid-induced tumour necrosis factor receptor (GITR), CD62L, the cytotoxic T-lymphocyte associated protein 4 (CTLA-4) and the forkhead/winged helix transcription factor FOXP3, whose expression has been associated with the acquisition of regulatory properties[45–47].

We have shown that, in patients with AIH, T-regs are defective in number in comparison to normal controls, and this numerical reduction relates to the stage of disease, being more evident at diagnosis than during drug-induced remission[38]. The percentage of T-regs inversely correlates with markers of disease severity, such as anti-soluble liver antigen (SLA) and LKM1 autoantibody titres, suggesting that a reduction in regulatory T cells favours the serological manifestations of autoimmune liver disease. In addition, T-regs ability to expand is significantly lower in patients than in controls, but their ability to suppress IFN-γ production by CD4$^+$CD25$^-$ T cells is maintained, as demonstrated by experiments showing that T-regs from AIH patients can reduce the number of IFN-γ-producing CD4$^+$CD25$^-$ T cells in a manner equally efficient to T-regs from normal controls[38]. If loss of immunoregulation is central to the pathogenesis of autoimmune liver disease, our findings suggest that treatment should concentrate on restoring T-regs ability to expand, with consequent increase in their number. This is at least partially achieved by standard immunosuppression, since we have observed an increase in T-regs numbers during remission. As mentioned above, well before the first description of CD4$^+$CD25$^+$ regulatory T cells, impaired 'suppressor' T cell function was described in autoimmune hepatitis and *in vitro* restoration of 'suppressor' T cell activity was reported following incubation of patients' lymphocytes with therapeutic doses of corticosteroids, while the same effect was not observed in chronic hepatitis B[36]. An effect of immunosuppressive drugs on 'suppressor' T cell activity was reported in AIH after incubation of patients' lymphocytes with therapeutic doses of corticosteroids and more recently in a murine model of diabetes, where tolerance achieved by a short treatment with $1\alpha,25$-dihydroxy-vitamin D3 and mycophenolate mofetil, was associated with induction of $CD4^+CD2^+$ regulatory T cells[48].

Ongoing studies are focused on defining the functional characteristics of T-regs and their relationship with other subsets of immunoregulatory lymphocytes in order to gain a better insight into the mechanisms involved in the loss of self-tolerance.

CD4 AUTOREACTIVE T CELLS

To trigger an autoimmune response, a peptide must be embraced by a human leucocyte antigen (HLA) class II molecule and presented to uncommitted T helper (Th0) cells by professional APC, with the costimulation of ligand–ligand (CD28 on Th0, CD80 on APC) interaction between the cells (Figure 1)[49]. Once the autoimmune response has been initiated, and in the absence of effective immunosuppressive treatment, the tissue-damaging process would accelerate and persist[49]. Hepatocytes from patients with AIH, in contrast to normal hepatocytes, express HLA class II molecules[50]. Although lacking the antigen-processing machinery typical of APC, these hepatocytes may present peptides through a bystander mechanism[51]. In the presence of impaired immunoregulation and inappropriate expression of HLA class II antigens on the hepatocytes, an autoantigenic peptide could be presented to the helper/inducer cells leading to their activation. Although no direct evidence exists as yet that an autoantigenic peptide is presented by hepatocytes and recognized by CD4 T helper cells, activation of these cells has been documented in AIH[30,52]. We have previously reported that paediatric patients with AIH have increased numbers of circulating activated T lymphocytes expressing CD4, IL-2 receptor (IL-2R) and HLA DR, which reflect disease activity, since their levels are highest when the disease is most active[52]. To identify the antigen specificity of the circulating T cells, including activated T cells, in these patients we studied the frequency of liver cell membrane reactive T cells in patients and healthy individuals by limiting dilution analysis (LDA). The higher the frequency of T cells specific for a given antigen, the more potent the immune response to that antigen. Although liver autoantigen-specific T cell precursors are also found in normal subjects, patients with autoimmune liver disease have an over 10-fold higher frequency of liver antigen-specific T cell precursors in their circulation[53]. This finding suggests that the pool of liver autoreactive T cells undergoes a significant expansion in patients with autoimmune liver disease, indicating that these cells may have a pathogenic role by perpetuating, and possibly initiating, the immune attack on the liver.

The presence of a CD3/CD4-positive dense mononuclear cell infiltration in the liver of patients with AIH indicates a key role for these cells in disease pathogenesis. Given that T cells recognize antigens in a precise fashion, studies in the early 1990s were conducted at a single T cell level in order to identify antigen-specific T cell recognition. In an attempt to investigate liver antigen-reactive T cells, we generated T cell clones from both peripheral blood and liver parenchyma of paediatric patients with AIH using IL-2 in our cloning protocol to ensure that only cells actively involved in immune responses, and therefore expressing the IL-2 receptor (IL2R), were expanded[54]. The high frequency of CD4$^+$ αβ T cells among clones generated from peripheral blood confirmed an earlier report of a high percentage of CD4$^+$ αβ T cells among the IL2R$^+$ population in the circulation of patients with autoimmune liver diseases[52]. However, in contrast to the clones derived from the peripheral blood, a large proportion of clones derived from liver biopsies consisted of either CD4$^-$CD8$^-$ γδ cells or CD8$^+$ αβ T cells[53–55]. When reactivity to putative liver autoantigens was studied using standard proliferation assays, both αβ and γδ T cell clones demonstrated specific expansion in the presence of liver membrane antigens, albeit αβ clones were more reactive than γδ clones. Some of the liver membrane-reactive clones also proliferated in the presence of liver-specific lipoprotein (LSP) and/or asialoglycoprotein receptor (ASGPR). Furthermore, the clones were restricted in their response by HLA class II molecules. Because CD4 is the phenotype of helper T cells, we investigated whether these clones were able to help autologous B cells in the production of immunoglobulin *in vitro*, and found that their co-culture with B cells dramatically increased autoantibody production[53,54].

The best candidate for the study of T cell ligands in AIH is LKM1 since its target in AIH-2 has been identified as cytochrome P4502D6[56]. Data on PBMC and liver-specific T cell responses in AIH-2 have been conducted in adult patients by Löhr et al.[57–59]. Using a cloning protocol identical to ours, Löhr et al. generated T cell clones from peripheral blood and liver biopsies of patients with AIH-2, which were characterized by CD4$^+$ phenotype, reactivity to CYP2D6, Th1 cytokine profile and restriction by HLA class II molecule[59]. However, this study was focused on a single B cell antigenic region on CYP2D6, namely the peptide sequence 262–285[59]. Using a systematic approach based on the construction of overlapping peptides covering the whole CYP2D6 molecule, we have documented the fine specificity of *ex-vivo* CYP2D6-specific T cell responses in patients with AIH-2 and provided information on their HLA restriction, clinical relevance, disease specificity and cytokine profile (unpublished data). Using *ex-vivo* T cells, seven broad regions of CYP2D6 are recognized in HLA DR7-positive patients, and four partially overlapping regions in DR7-negative patients, while medium term T cell lines (TCL) have a much narrower response, targeting only few of the peptides involved in the *ex-vivo* recognition. This is an important observation for the interpretation of published results. Thus, TCL and T cell clones (TCC) are frequently used to overcome the problem of studying the small number of antigen-specific T cells among peripheral blood mononuclear cells, an approach that, however, does not allow investigating the full spectrum of TCR specificities, because of the *in-vitro* selection process, in which lower affinity T cells or those requiring finely adjusted stimulation parameters, such as antigen concentration, incubation time and cytokines, are disadvantaged and outgrown. Epitope mapping based on TCL and TCC may therefore provide a skewed repertoire leading to an underestimation of the real number of immunogenic peptides.

CD8 AUTOREACTIVE T CELLS

In addition to the unfolding role of CYP2D6-specific CD4 T cells in auto-immune hepatitis type 2 (AIH-2), there is growing evidence implicating an HLA class I restricted CD8 response in the pathogenesis of autoimmune liver damage. In 1990 Wen et al. described CD8 T cell clones specific for asialoglycoprotein receptor (ASGPR) in patients with autoimmune hepatitis[54]. More recently an involvement of an HLA class I restricted CD8 T cell immune response has been demonstrated in patients with primary biliary cirrhosis[60]. Performing an HLA class I restricted epitope mapping of the E2 components of pyruvate dehydrogenase complex (PDC-E2) Matsumura and colleagues identified two epitopes on PDC-E2 able to induce a CD8 T cell immune response[61].

A study currently in progress in our laboratory is evaluating the presence of a CD8 T cell autoimmune response to CYP2D6 in AIH-2 with the aims of defining a hierarchy of epitope immunodominance on CYP2D6 and characterizing functionally CYP2D6-specific CD8 T cells.

CONCLUDING REMARKS

The data presented above suggest that a failure of immune homeostatic processes, normally keeping the response against self-antigens under control, is involved in the pathogenesis of AIH. The prime mechanism for tolerance breakdown remains to be elucidated. There is some experimental evidence that molecular mimicry mechanisms between viral and self-mimicking sequences may be involved, and such mechanisms are the focus of ongoing studies[62-65].

References

1. Miller JF, Davies JS. Embryological development of the immune mechanism. Annu Rev Med. 1964;15:23–36.
2. Head JR. Uterine natural killer cells during pregnancy in rodents. Nat Immun. 1996;15:7–21.
3. Abreu-Martin MT, Targan SR. Regulation of immune responses of the intestinal mucosa. Crit Rev Immunol. 1996;16:277–309.
4. Doherty DG, O'Farrelly C. Innate and adaptive lymphoid cells in the human liver. Immunol Rev. 2000;174:5–20.
5. Hata K, Zhang XR, Iwatsuki S, Van Thiel DH, Herberman RB, Whiteside TL. Isolation, phenotyping, and functional analysis of lymphocytes from human liver. Clin Immunol Immunopathol. 1990;56:401–19.
6. Norris S, Collins C, Doherty DG et al. Resident human hepatic lymphocytes are phenotypically different from circulating lymphocytes. J Hepatol. 1998;28:84–90.
7. Norris S, Doherty DG, Collins C et al. Natural T cells in the human liver: cytotoxic lymphocytes with dual T cell and natural killer cell phenotype and function are

phenotypically heterogeneous and include Valpha24-JalphaQ and gammadelta T cell receptor bearing cells. Hum Immunol. 1999;60:20–31.

8. Doherty DG, Norris S, Madrigal-Estebas L et al. The human liver contains multiple populations of NK cells, T cells, and CD3⁺CD56⁺ natural T cells with distinct cytotoxic activities and Th1, Th2, and Th0 cytokine secretion patterns. J Immunol. 1999;163:2314–21.

9. Mehal WZ, Azzaroli F, Crispe IN. Immunology of the healthy liver: old questions and new insights. Gastroenterology. 2001;120:250–60.

10. Crispe IN. Hepatic T cells and liver tolerance. Nat Rev Immunol. 2003;3:51–62.

11. Kita H, Mackay IR, Van De Water J, Gershwin ME. The lymphoid liver: considerations on pathways to autoimmune injury. Gastroenterology. 2001;120:1485–501.

12. Mackay IR. Hepatoimmunology: a perspective. Immunol Cell Biol. 2002;80:36–44.

13. Knolle PA, Gerken G. Local control of the immune response in the liver. Immunol Rev. 2000;174:21–34.

14. Reynoso-Paz S, Coppel RL, Mackay IR, Bass NM, Ansari AA, Gershwin ME. The immunobiology of bile and biliary epithelium. Hepatology. 1999;30:351–7.

15. Chisari FV, Ferrari C. Hepatitis B virus immunopathogenesis. Annu Rev Immunol. 1995; 13:29–60.

16. Cerny A, Chisari FV. Pathogenesis of chronic hepatitis C: immunological features of hepatic injury and viral persistence. Hepatology. 1999;30:595–601.

17. Bertoletti A, Ferrari C. Kinetics of the immune response during HBV and HCV infection. Hepatology. 2003;38:4–13.

18. Bissell DM, Gores GJ, Laskin DL, Hoofnagle JH. Drug-induced liver injury: mechanisms and test systems. Hepatology. 2001;33:1009–13.

19. Liu ZX, Kaplowitz N. Immune-mediated drug-induced liver disease. Clin Liver Dis. 2002; 6:467–86.

20. Robin MA, Le Roy M, Descatoire V, Pessayre D. Plasma membrane cytochromes P450 as neoantigens and autoimmune targets in drug-induced hepatitis. J Hepatol. 1997;26(Suppl. 1):23–30.

21. Patel T. Apoptosis in hepatic pathophysiology. Clin Liver Dis. 2000;4:295–317.

22. Cocca BA, Cline AM, Radic MZ. Blebs and apoptotic bodies are B cell autoantigens. J Immunol. 2002;169:159–66.

23. Vaishnaw AK, Toubi E, Ohsako S et al. The spectrum of apoptotic defects and clinical manifestations, including systemic lupus erythematosus, in humans with CD95 (Fas/APO-1) mutations. Arthritis Rheum. 1999;42:1833–42.

24. Alvarez F, Berg PA, Bianchi FB et al. International Autoimmune Hepatitis Group Report: review of criteria for diagnosis of autoimmune hepatitis. J Hepatol. 1999;31:929–38.

25. Gregorio GV, Portmann B, Reid F et al. Autoimmune hepatitis in childhood: a 20-year experience. Hepatology. 1997;25:541–7.

26. Gregorio GV, Portmann B, Karani J et al. Autoimmune hepatitis/sclerosing cholangitis overlap syndrome in childhood: a 16-year prospective study. Hepatology. 2001;33:544–53.

27. Neuberger J. Primary biliary cirrhosis. Lancet. 1997;350:875–9.

28. De Groote J, Desmet VJ, Gedigk P et al. A classification of chronic hepatitis. Lancet. 1968; 2:626–8.

29. Scheuer P. Chronic agressive hepatitis. In: Liver Biopsy Interpretation. Ballière Tindall, London; 1974:68–72.

30. Senaldi G, Portmann B, Mowat AP, Mieli-Vergani G, Vergani D. Immunohistochemical features of the portal tract mononuclear cell infiltrate in chronic aggressive hepatitis. Arch Dis Child. 1992;67:1447–53.

31. Abo T, Kawamura T, Watanabe H. Physiological responses of extrathymic T cells in the liver. Immunol Rev. 2000;174:135–49.

32. Takeda K, Hayakawa Y, Van Kaer L, Matsuda H, Yagita H, Okumura K. Critical contribution of liver natural killer T cells to a murine model of hepatitis. Proc Natl Acad Sci USA. 2000;97:5498–503.

33. Nouri-Aria KT, Donaldson PT, Hegarty JE, Eddleston AL, Williams R. HLA A1-B8-DR3 and suppressor cell function in first-degree relatives of patients with autoimmune chronic active hepatitis. J Hepatol. 1985;1:235–41.

34. Nouri-Aria KT, Lobo-Yeo A, Vergani D, Mieli-Vergani G, Eddleston AL, Mowat AP. T suppressor cell function and number in children with liver disease. Clin Exp Immunol. 1985;61:283-9.

35. Donaldson PT, Doherty DG, Hayllar KM, McFarlane IG, Johnson PJ, Williams R. Susceptibility to autoimmune chronic active hepatitis: human leukocyte antigens DR4 and A1-B8-DR3 are independent risk factors. Hepatology. 1991;13:701-6.

36. Nouri-Aria KT, Hegarty JE, Alexander GJ, Eddleston AL, Williams R. Effect of corticosteroids on suppressor-cell activity in 'autoimmune' and viral chronic active hepatitis. N Engl J Med. 1982;307:1301-4.

37. Vento S, Hegarty JE, Bottazzo G, Macchia E, Williams R, Eddleston AL. Antigen specific suppressor cell function in autoimmune chronic active hepatitis. Lancet. 1984;1:1200-4.

38. Longhi MS, Ma Y, Bogdanos DP, Cheeseman P, Mieli-Vergani G, Vergani D. Impairment of CD4$^+$CD25$^+$ regulatory T-cells in autoimmune liver disease. J Hepatol. 2004;41:31-7.

39. Shevach EM. CD4$^+$ CD25$^+$ suppressor T cells: more questions than answers. Nat Rev Immunol. 2002;2:389-400.

40. Thornton AM, Shevach EM. CD4$^+$CD25$^+$ immunoregulatory T cells suppress polyclonal T cell activation *in vitro* by inhibiting interleukin 2 production. J Exp Med. 1998;188:287-96.

41. Ng WF, Duggan PJ, Ponchel F et al. Human CD4(+)CD25(+) cells: a naturally occurring population of regulatory T cells. Blood. 2001;98:2736-44.

42. Green EA, Gorelik L, McGregor CM, Tran EH, Flavell RA. CD4$^+$CD25$^+$ T regulatory cells control anti-islet CD8$^+$ T cells through TGF-beta–TGF-beta receptor interactions in type 1 diabetes. Proc Natl Acad Sci USA. 2003;100:10878-83.

43. Nakamura K, Kitani A, Fuss I et al. TGF-beta 1 plays an important role in the mechanism of CD4$^+$CD25$^+$ regulatory T cell activity in both humans and mice. J Immunol. 2004;172: 834-42.

44. Annacker O, Pimenta-Araujo R, Burlen-Defranoux O, Barbosa TC, Cumano A, Bandeira A. CD25$^+$ CD4$^+$ T cells regulate the expansion of peripheral CD4 T cells through the production of IL-10. J Immunol. 2001;166:3008-18.

45. Piccirillo CA, Thornton AM. Cornerstone of peripheral tolerance: naturally occurring CD4$^+$CD25$^+$ regulatory T cells. Trends Immunol. 2004;25:374-80.

46. Fontenot JD, Gavin MA, Rudensky AY. Foxp3 programs the development and function of CD4$^+$CD25$^+$ regulatory T cells. Nat Immunol. 2003;4:330-6.

47. Khattri R, Cox T, Yasayko SA, Ramsdell F. An essential role for Scurfin in CD4$^+$CD25$^+$ T regulatory cells. Nat Immunol. 2003;4:337-42.

48. Gregori S, Casorati M, Amuchastegui S, Smiroldo S, Davalli AM, Adorini L. Regulatory T cells induced by 1 alpha,25-dihydroxyvitamin D3 and mycophenolate mofetil treatment mediate transplantation tolerance. J Immunol. 2001;167:1945-53.

49. Vergani D, Choudhuri K, Bogdanos DP, Mieli-Vergani G. Pathogenesis of autoimmune hepatitis. Clin Liver Dis. 2002;6:439-49.

50. Lobo-Yeo A, Senaldi G, Portmann B, Mowat AP, Mieli-Vergani G, Vergani D. Class I and class II major histocompatibility complex antigen expression on hepatocytes: a study in children with liver disease. Hepatology. 1990;12:224-32.

51. Chen M, Shirai M, Liu Z, Arichi T, Takahashi H, Nishioka M. Efficient class II major histocompatibility complex presentation of endogenously synthesized hepatitis C virus core protein by Epstein–Barr virus-transformed B-lymphoblastoid cell lines to CD4(+) T cells. J Virol. 1998;72:8301-8.

52. Lobo-Yeo A, Alviggi L, Mieli-Vergani G, Portmann B, Mowat AP, Vergani D. Preferential activation of helper/inducer T lymphocytes in autoimmune chronic active hepatitis. Clin Exp Immunol. 1987;67:95-104.

53. Wen L, Ma Y, Bogdanos DP et al. Pediatric autoimmune liver diseases: the molecular basis of humoral and cellular immunity. Curr Mol Med. 2001;1:379-89.

54. Wen L, Peakman M, Lobo-Yeo A et al. T-cell-directed hepatocyte damage in autoimmune chronic active hepatitis. Lancet. 1990;336:1527-30.

55. Wen L, Peakman M, Mieli-Vergani G, Vergani D. Elevation of activated gamma delta T cell receptor bearing T lymphocytes in patients with autoimmune chronic liver disease. Clin Exp Immunol. 1992;89:78-82.

56. Vergani D, Mieli-Vergani G. Autoimmune hepatitis. Autoimmun Rev. 2003;2:241-7.

57. Lohr H, Manns M, Kyriatsoulis A et al. Clonal analysis of liver-infiltrating T cells in patients with LKM-1 antibody-positive autoimmune chronic active hepatitis. Clin Exp Immunol. 1991;84:297–302.

58. Lohr H, Treichel U, Poralla T, Manns M, Meyer zum Buschenfelde KH. Liver-infiltrating T helper cells in autoimmune chronic active hepatitis stimulate the production of autoantibodies against the human asialoglycoprotein receptor *in vitro*. Clin Exp Immunol. 1992;88: 45–9.

59. Lohr HF, Schlaak JF, Lohse AW et al. Autoreactive CD4[+] LKM-specific and anticlonotypic T-cell responses in LKM-1 antibody-positive autoimmune hepatitis. Hepatology. 1996;24:1416–21.

60. Kita H, Lian ZX, Van de Water J et al. Identification of HLA-A2-restricted CD8(+) cytotoxic T cell responses in primary biliary cirrhosis: T cell activation is augmented by immune complexes cross-presented by dendritic cells. J Exp Med. 2002;195:113–23.

61. Matsumura S, Kita H, He XS et al. Comprehensive mapping of HLA-A0201-restricted CD8 T-cell epitopes on PDC-E2 in primary biliary cirrhosis. Hepatology. 2002;36:1125–34.

62. Bogdanos DP, Mieli-Vergani G, Vergani D. Virus, liver and autoimmunity. Dig Liver Dis. 2000;32:440–6.

63. Bogdanos DP, Lenzi M, Okamoto M et al. Multiple viral/self immunological cross-reactivity in liver kidney microsomal antibody positive hepatitis C virus infected patients is associated with the possession of HLA B51. Int J Immunopathol Pharmacol. 2004;17: 83–92.

64. Kerkar N, Choudhuri K, Ma Y et al. Cytochrome P4502D6(193-212): a new immunodominant epitope and target of virus/self cross-reactivity in liver kidney microsomal autoantibody type 1-positive liver disease. J Immunol. 2003;170:1481–9.

65. Bogdanos DP, Choudhuri K, Vergani D. Molecular mimicry and autoimmune liver disease: virtuous intentions, malign consequences. Liver. 2001;21:225–32.

7
Autoimmune hepatitis in transgenic mouse models: what lessons can be learnt?

P. BERTOLINO, D. G. BOWEN, M. ZEN and G. W. McCAUGHAN

INTRODUCTION

The liver is continuously exposed to bacteria, pathogens, toxins, and antigens absorbed via the gut. Current evidence suggests that immune responses within the liver are biased towards tolerance rather than immunity[1-3]. Intrahepatic tolerance can, however, sometimes be broken and autoimmune hepatitis (AIH) may result; this condition occurs with an annual incidence of 1.9 in 100 000 in northern Europeans[4]. The mechanisms underlying the development of AIH have not yet been elucidated, but environmental factors coupled to genetic predisposition are thought to be key parameters in the disease process[4]. This disease is characterized by destruction of the liver parenchyma, in association with the generation of autoantibodies and autoimmune T cell responses. Although it is clear that AIH is immune-mediated, the mechanisms leading to the onset of this immune process and the resultant hepatocyte destruction remain ill-defined. In some cases self-reactive T cells are specific for autoantigen expressed by hepatocytes (such as cytochrome P450 2D6)[5-7], suggesting that parenchymal liver cells may be directly killed by cytotoxic T lymphocytes (CTL). In other cases, however, AIH is observed in association with extra-hepatic autoimmune diseases in the absence of demonstrated reactivity to hepatocyte autoantigens[8,9].

The parameters that determine the balance between tolerance and auto-immunity in the liver also remain unclear. It is generally accepted that the default immune response pathways within the liver lead to tolerance, and that these are active processes requiring antigen-specific T cell activation[10-12]. It has been hypothesized that pre-existing inflammation is a key parameter required for the induction of immunity within the liver[13-15]. The presence of inflammation might change the tolerogenic intrahepatic environment and the nature of the immune response generated: inflammation induces the differentiation and maturation of professional antigen-presenting cells (APC), mostly dendritic cells (DC), through up-regulation of costimulatory molecules including CD80,

CD86 and CD40 and export of MHC class II molecule/peptide complexes to the cell surface[16]. These mature DC would recirculate from the liver to the draining lymph nodes (LN) where they would allow effective T cell activation[16]. In the case of the liver, it has also been suggested that not only is inflammation required for the initiation of an effective immune response within the liver, but it is also important to allow T cell access to hepatocytes, and thus would be a pre-requisite for autoimmunity within this organ[13,15]. However, this hypothesis is contradictory to some of our results, which indicate that T cells can induce hepatitis in the absence of pre-existing inflammation[17]. To explain these divergent findings we have proposed that, in addition to inflammation, the balance between tolerance and immunity within the liver is determined by the site at which T cells undergo initial priming[18]. We have recently shown that alone amongst the solid organs, and probably due to its unique architecture and vasculature, the liver can act as a site of activation for naive CD8[+] T cells, and competes with lymph nodes for primary T cell activation[17–19]. To investigate the role of the site of primary activation in immune-mediated hepatitis, we have developed a variety of transgenic mouse models in which antigen is expressed at different sites. These lines exhibit differential susceptibility to the development of experimental autoimmune hepatitis upon adoptive transfer of syngeneic donor T cells expressing a transgenic antigen-specific TCR. In this chapter we will summarize results obtained using these models, and hypothesize how the site of primary T cell activation could influence the onset and maintenance of AIH.

UNIQUE STRUCTURE OF THE HEPATIC SINUSOID

We and others have suggested that the ability of the liver to induce tolerance is associated with the unique vasculature of the liver sinusoids, which would favour T cell trapping and activation[1,17,20,21]. Homing and entry of naive and activated T cells into lymphoid tissues and organs are determined by complex molecular interactions involving adhesion molecules that recognize their ligands on endothelial cells[22]. Naive T lymphocytes do not normally have access to the parenchyma of most organs, as they do not express adhesion molecules and chemokine receptors required for adhesion to endothelial cells or subsequent trans-endothelial migration[23,24]. However, naive T cells express L-selectin (CD62L) which binds to peripheral LN specific vascular addressins (PNAd) and has been shown to play a critical role in initial binding (tethering) and subsequent rolling of the lymphocytes through the high endothelial venules (HEV) of the LN under normal conditions of flow[22]. Chemokines and intercellular adhesion molecule 1 (ICAM-1)/leucocyte function-associated antigen-1 (LFA-1) interactions also play an important role in these adhesion steps[25]. In the LN, naive T cells interact with professional APC, in particular DC, expressing relevant peptide/MHC complexes. This contact induces T cell activation resulting in expression and up-regulation of adhesion molecules, which allows activated T cells to undergo trans-endothelial migration and to infiltrate the tissues.

The structure of the hepatic sinusoid, and the nature of blood flow within it, however, differ markedly from those observed in other organs. In contrast to LN and most solid organs, blood flow within hepatic sinusoids is 50–100 times slower than in capillaries[26]. Sinusoids are formed by monolayers of hepatocytes defining narrow channels of approximately 10 µm in diameter that are lined by a layer of specialized endothelial cells and liver-resident macrophage known as Kupffer cells (KC)[1]. Unlike other endothelial cells, liver sinusoidal endothelial cells (LSEC) are perforated by multiple 120 nm diameter holes (fenestrations), and do not form tight junctions with adjacent endothelial cells[26]. This unique structure, combined with the lack of a basement membrane, defines a perisinusoidal space between LSEC and hepatocytes, known as the space of Disse, which is in continuity with the sinusoidal lumen (Figure 1).

These unique conditions of slow blood flow, in combination with the narrow diameter of the hepatic sinusoids and their unusual structure, favours contact between lymphocytes and liver cells. We have proposed that, if hepatic cells express the relevant antigen, the unusual nature of these intrahepatic interactions might allow activation of naive T cells within the liver itself, independent of the usual requirements for expression of adhesion molecules for initial cellular contact and interaction. Such activation might be mediated by LSEC and KC that directly line the sinusoidal lumen, but could potentially also be mediated by hepatocytes, as contact between lymphocytes and hepatocytes

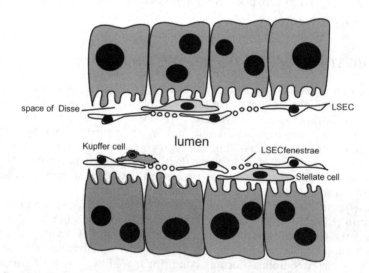

Figure 1 Schematic structure of the hepatic sinusoid. Liver sinusoids are lined by liver sinusoidal endothelial cells (LSEC), which separate the sinusoid lumen from hepatocytes. Kupffer cells patrol the sinusoids and bind to LSEC and occasionally hepatocytes through the gaps of adjacent LSEC. Stellate cells are located in the space of Disse. Naive T lymphocytes recirculating within the sinusoids may be activated by contact with KC or LSEC located within the lumen of the sinusoids, or by contact with hepatocytes. Interaction of T cells with hepatocytes may occur either through the gap between two adjacent LSEC or through LSEC fenestrations. This type of interaction may be facilitated by the slow blood flow through the narrow lumen of the sinusoids

might occur via LSEC fenestrations[26]. Indeed, recent electron microscopy studies in our group have demonstrated that some cytoplasmic extensions of intrahepatic T cells project through LSEC fenestrations and are in direct contact with hepatocytes (A. Warren, in preparation). In addition, previous electron microscopy has demonstrated that hepatocyte microvillae may project through LSEC fenestrations into the sinusoidal lumen[27,28], where they might also potentially contact lymphocytes.

EVIDENCE FOR PRIMARY INTRAHEPATIC T CELL ACTIVATION FROM TRANSGENIC MOUSE MODELS

T cells can enter the liver through two types of vascular endothelium: via portal vessels, resulting in infiltration of the portal tracts, or via sinusoidal endothelium, resulting in parenchymal infiltration[29]. Adhesion at these sites is differentially regulated. Most prior investigations regarding recruitment of lymphocytes to the liver have been focused on recruitment of activated T cells. These studies have established that the liver possesses distinctive capacity for the retention of activated CD8$^+$ cells[17,19–21] and CD4$^+$ T cells[30], some of which express markers of apoptosis, leading to the conclusion that the liver might act as a graveyard for these cells[31]. The capacity of the liver to retain activated T cells has been noted even in the absence of intrahepatic expression of cognate antigen[21].

In contrast to activated T cells, naive syngeneic T cells recirculate through the liver without being retained in the absence of intrahepatically expressed cognate antigen[17,19,21,30]. To examine whether antigen expression by hepatic cells would allow naive T cells to be antigen-specifically retained within the liver and undergo activation *in situ*, we have developed several transgenic models in which a small subset of TCR transgenic T cells may be identified within the recipient using a clonotypic Ab. All these models use Des-TCR transgenic mice in which all CD8$^+$ T cells express a transgenic TCR specific for self-peptides associated with the mouse class I molecule H-2K^b [10]. LN cells from these mice (composed of mostly naive T cells) were adoptively transferred into a number of syngeneic transgenic mouse lines on B10.BR (H-2^k) background, which expressed the allogeneic H-2K^b molecule under the control of different promoters (Figure 2):

1. *Met-K^b mice*, which express H-2K^b under the control of the sheep metallothionein promoter[32]. In this lineage, expression of H-2K^b was detected on hepatocytes but not on LSEC, KC or liver DC[17,18]. However, due to leaky expression of the transgene, indirect evidence suggests that H-2K^b is also expressed at very low levels on bone marrow derived APC in the thymus and LN[11,17].

2. *Alb-K^b mice*, which express H-2K^b under the control of the mouse albumin promoter[33]. In contrast to Met-K^b mice, H-2K^b expression in these animals is restricted to the liver on hepatocytes and has not been reported to occur in other cells[33].

3. *178.3 mice*, which ubiquitously express H-2K^b under the control of the MHC class I promoter[19,34].

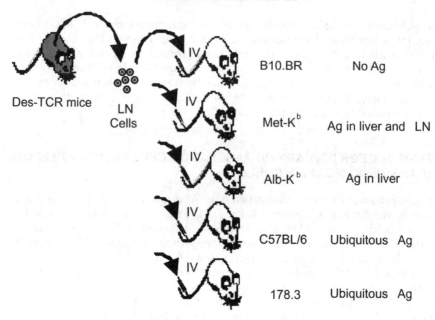

Figure 2 Transgenic mouse models: pooled LN cells from anti-H-2K^b-specific Des-TCR transgenic mice were adoptively transferred into syngeneic control non-transgenic B10.BR mice or a number of lineages expressing H-2K^b: syngeneic transgenic 178.3 mice, which express H-2K^b under the control of the MHC class I promoter; syngeneic transgenic Met-K^b mice, which express H-2K^b under the control of the sheep metallothionein promoter; syngeneic transgenic Alb-K^b mice, which express H-2K^b under the control of the mouse albumin promoter; and C57BL/6 mice, which express wild-type levels of H-2K^b

Initial experiments, aimed at investigating whether intrahepatic primary T cell activation could occur, were performed using Met-K^b mice. In these experiments we found that transgenic CD8$^+$ T cells were specifically retained within the liver of Met-K^b mice as soon as 15 min after transfer, and expressed the very early activation marker CD69 2 h post-transfer, suggesting that these cells have been activated *in situ*[17]. Careful examination of various organs and lymphoid tissues of Met-K^b mice has revealed that, consistent with leaky H-2K^b expression on bone marrow-derived APC in Met-K^b mice, CD8$^+$ Des-TCR T cells were also simultaneously activated in the LN, but not blood, spleen and other organs[17]. However, at this time point CD69$^+$ cells were not detected in the blood, suggesting that Des-TCR T cell activation in Met-K^b mice occurred independently in both liver and LN, and that under circumstances in which antigen is expressed both on hepatocytes and in the LN, both sites are competing for T cell migration and initial activation[18].

Intrahepatic primary CD8$^+$ T cell activation was also demonstrated when Des-TCR T cells were transferred into Alb-K^b mice[18]. In Alb-K^b mice, activated T cells expressing CD69 were also present within the liver at 2 h following adoptive transfer; however, in contrast to Met-K^b mice, no activated

$CD69^+$ T cell could be detected within the LN, reflecting lack of H-2K^b expression at this site[18]. The livers of Met-K^b and Alb-K^b mice contained a similar proportion of CD69$^+$ T cells (M. Zen, unpublished data), suggesting that there was a comparable recruitment of T cells at early time points.

More surprising results were obtained when Des-TCR T cells were adoptively transferred into 178.3 mice[19]. The majority of donor T cells were retained within the liver within 1 h after adoptive transfer, and expressed CD69 at 2 h[19]. This was not a transgenic artefact, as it was also reproduced using C57BL/6 recipient mice which express wild-type levels of H-2K^b. Intravital microscopy studies have confirmed that, in the presence of intrahepatic antigen, purified naive CD8$^+$ Des-TCR T cells were retained within the liver as early as 2 min post-adoptive transfer (Bertolino et al., submitted). These results were confirmed and quantified using ^{51}Cr radiolabelling (Bertolino et al., submitted), which yielded similar findings to those previously obtained by flow cytometry[19], and confirmed that 40–60% of naive T cells were retained within the livers of 178.3 recipient mice within 1 h of transfer, despite ubiquitous expression of antigen.

Our previous data indicated that donor T cells retained within the liver in the ubiquitous presence of antigen did not undergo immediate apoptosis, but became activated within 2 h after transfer, suggesting that they had been activated *in situ* by liver cells[19]. Donor T cells appeared to remain trapped within the liver for at least 24 h. Surprisingly despite the presence of cognate antigen within the secondary lymphoid organs, very few T cells migrated into the LN of recipient animals, suggesting that trapping and activation within the liver prevented the cells from migrating into lymphoid tissues[19], an observation consistent with competition between liver and LN for primary T cell activation. Using this model we have shown that early antigen-specific retention of naive CD8$^+$ T cells in 178.3 animals is predominantly ICAM-1/LFA-1 dependent (Bertolino et al., submitted).

Collectively, results obtained in these various mouse models indicate that intrahepatic accumulation of activated CD8$^+$ T cells can be an early active event, rather than exclusively a late passive event dependent on prior activation in peripheral lymphoid tissues, as has previously been believed. The degree of T cell retention is likely to be dependent upon the affinity of the TCR and the amount of MHC/peptide complexes expressed by intrahepatic cells, but also may be dependent upon the nature of the liver cells expressing the antigen. It is likely that T cell access to hepatocytes is more limited than access to KC and LSEC, which might potentially lead to a reduction in the numbers of cells accumulating within the liver in the presence of antigen expressed solely by hepatocytes.

The above data indicate that the liver is an exception to the general rule of T cell activation and recirculation, which predicts that naive T cells recirculate via the lymph and blood, but do not enter peripheral tissues prior to activation in secondary lymphoid organs.

TRANSGENIC MOUSE LINES EXHIBIT DIFFERENTIAL SUSCEPTIBILITY TO THE DEVELOPMENT OF HEPATITIS

Although the same number of transgenic T cells was adoptively transferred into the various transgenic recipient mice, this did not always result in the development of hepatitis. The adoptive transfer of Des-TCR T cells into Met-K[b] mice induced a severe but transient hepatitis peaking at days 5–6[11], suggesting that activated T cells acquired effective CTL function and could induce hepatitis in the absence of pre-existing inflammation[17]. Under the same experimental conditions, 178.3 mice also developed transient hepatitis[19]; however, this hepatitis occurred within a shorter period of time, peaking at day 2, and resolving by day 3–4. In contrast, consistent with previous reports[13], Alb-K[b] mice did not develop biochemical hepatitis following adoptive transfer of the same number of transgenic T cells, despite the expression of higher levels of H-2K[b] by hepatocytes[18].

The differential occurrence of hepatitis between these models suggested that antigen expression on hepatocytes was not the only requirement for the induction of hepatitis in the absence of prior hepatic inflammation. We have therefore investigated whether the site in which T cells undergo primary cell activation was a determinant of these differences. Of the three lineages, the Met-K[b] model was the only one in which a significant number of transgenic T cells were activated within LN. We have therefore proposed that the type of hepatitis occurring in these mice (characterized by a day 5–6 peak) is associated with primary T cell activation within LN. Our recent experimental data support this model. In Met-K[b] mice the blocking of LN entry by the majority of naive Des-TCR CD8[+] T cells via the administration of anti-CD62L antibody led to the abrogation of hepatitis, unlike control Met-K[b] mice treated with isotype control Ab, which developed hepatitis as previously described in this lineage[18]. Anti-CD62L antibody treatment did not affect intrahepatic retention and activation of adoptively transferred naive CD8[+] T cells[18]. In addition, donor T cells activated within the liver of anti-CD62L antibody-treated Met-K[b] mice proliferated at a similar rate to those activated within the LN and livers of isotype-control treated Met-K[b] animals[18], indicating that, despite undergoing apparently normal proliferation, T cells activated within the liver received a different signal which programmed them to an alternative fate.

The lack of hepatitis in Alb-K[b] mice, in which T cells are activated in the liver but not in the LN, is consistent with this model. However, results obtained in 178.3 mice, in which very few T cells were activated in lymphoid tissues, indicate that hepatitis can also occur when a high number of T cells was trapped and activated in the liver. To investigate the nature of hepatitis occurring in 178.3 mice, we have generated radiation-induced bone marrow chimeras in which H-2K[b] was expressed on bone marrow-derived APC; however, hepatocyte expression of this antigen was excluded. Adoptive transfer of transgenic T cells into these mice resulted in a hepatitis similar to that observed in intact 178.3 mice[19]. These experiments indicated that antigen expression by hepatocytes was not a requirement for hepatitis in this model, and that liver damage occurring in 178.3 and 178.3 bone marrow chimeras resulted from 'bystander' hepatocyte injury: T cells were activated in an

Met-K^b mice   178.3 mice

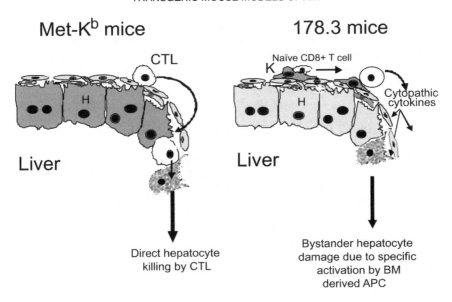

Figure 3 Transgenic mouse models indicate that liver damage might be mediated by two distinct mechanisms: in Met-K^b mice (left panel), T cells activated in the lymph nodes become CTL, undergo transendothelial migration and directly kill hepatocytes (mechanism 1). In 178.3 animals (right panel), hepatocytes are killed by cytopathic cytokines, such as TNF-α and IFN-γ, released following antigen-specific T cell activation in the liver (mechanism 2). In contrast to mechanism 1, mechanism 2-mediated liver damage does not require antigen expression by hepatocytes. H, hepatocytes, K, Kupffer cell

antigen-specific manner within the liver by antigen-expressing bone marrow-derived APC, but induced indirect damage to non-antigen-bearing hepatocytes (Figure 3). By investigating the mechanisms of hepatitis in this model, we have excluded a role for Fas in this phenomenon. However, hepatocyte damage was totally abrogated by treating recipient mice with antibodies specific for IFN-γ and TNF-α, suggesting that bystander damage was mediated by these two cytokines[19]. Bystander hepatitis did not occur under the same conditions in B10.BR mice reconstituted with Met-K^b bone marrow (D. Bowen, unpublished observations), confirming that hepatitis occurring in the Met-K^b model was not due to bystander injury, but rather was dependent upon antigen expression by hepatocytes, and thus likely to be mediated by direct CTL–hepatocyte interactions (Figure 3).

Why do T cells activated within the liver cause bystander damage in 178.3 hosts while this phenomenon is not observed in the Met-K^b or Alb-K^b models? It is likely that the occurrence of such 'bystander hepatitis' is dependent upon the higher frequency of Des-TCR T cells which are intrahepatically retained and activated in 178.3 mice, and possibly also on the type of liver cell which induces activation (bone marrow-derived APC versus hepatocytes). In addition, it is possible that such indirect hepatocyte injury requires relatively high levels of antigen expression by bone marrow-derived APC within the liver, a condition which is present within 178.3 mice, but not in the Met-K^b or Alb-K^b lineages. Regardless of the mechanisms responsible for the development of bystander hepatitis in the 178.3 model, it is likely that this type of intrahepatic activation does not generate effective CTL capable of directly mediating hepatocyte killing.

FATE OF CD8$^+$ T CELLS ACTIVATED WITHIN THE LIVER

We and others have demonstrated[11,33] that, following adoptive transfer of Des-TCR T cells, both Met-K^b and Alb-K^b recipients ultimately became tolerant to H-2K^b. However, controversy still existed as to which mechanism of tolerance was operating in these models. Early reports using Alb-K^b mice suggested that tolerance is achieved by anergy resulting in down-regulation of both CD8 and TCR[33], while our earlier reports using Met-K^b mice indicated that liver-activated T cells undergo peripheral deletion[11]. These studies were difficult to compare, as they were performed with mice bred on different genetic back-grounds, and did not take into account variations in the site of primary T cell activation. In an effort to reconcile these contradictory conclusions we have recently compared Alb-K^b and Met-K^b mice within the same background, and demonstrated that, in both situations, intrahepatically activated T cells exhibited reduced survival in comparison to intranodally activated T cells[18] (M. Zen, unpublished observations). In addition, we have shown that, before dying, activated T cells were not effective CTL and were unable to mediate severe liver damage[18]. We have hypothesized that primary activation within the liver may explain some of the tolerogenic properties of this organ[1]. The molecular mechanisms determining this programme remain to be defined, but *in-vitro* experiments have yielded some important clues. Transgenic CD8$^+$ T cells activated by hepatocytes in culture became CTL before dying prematurely[35,36]. Premature death was Fas- and TNFR-independent, but was contingent on low expression of *IL-2*, *IFN-γ* and *bcl-x$_L$* survival genes[36]. MAb crosslinking of CD28 molecules during T cell activation by hepatocytes *in vitro* resulted in enhanced cytokine and *bcl-x$_L$* expression, and prevented hepato-cyte-activated T cells from dying prematurely in culture[36]. Collectively, these experiments suggest that T cells activated by hepatocytes die by neglect resulting from the failure of hepatocytes to provide costimulation, which is necessary for sufficient expression of *IL-2* and *bcl-x$_L$*; however whether this is really the mechanism of liver-induced tolerance needs to be demonstrated.

FROM EXPERIMENTAL MODELS TO AUTOIMMUNE HEPATITIS: WHAT LESSONS CAN BE LEARNT?

Although the models described above are not intended to mimic a specific human autoimmune liver disease, they are important tools to dissect hepatic–immune interactions from which lessons can be learnt. Comparison between the Met-K^b and Alb-K^b models has revealed that, although immune responses can be primed within the liver, they are abortive and very inefficient in

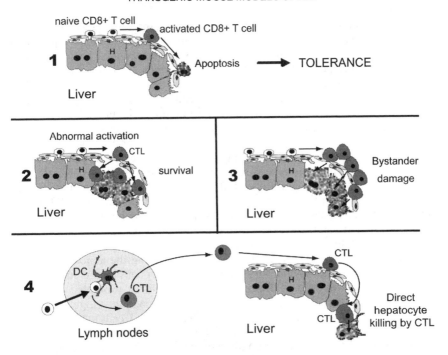

Figure 4 Models of AIH. In normal physiology, naive CD8$^+$ T cells activated within the liver die, possibly by neglect, without inducing hepatitis (mechanism 1). Under pathological conditions, T cell death following intrahepatic T cell activation may be impaired, resulting in the survival of autoreactive T cells that could accumulate and induce liver damage (mechanism 2). Alternatively, if present at high frequencies, T cells activated by bone marrow-derived APC in the liver or lymphoid tissues, could induce bystander hepatocyte damage by secreting cytopathic cytokines (mechanism 3). Finally, in a more classical view, AIH could be induced by autoreactive T cells activated within lymphoid tissues that accumulate within the liver (mechanism 4). H, hepatocytes

promoting effective CTL capable of inducing AIH. Under normal conditions the default pathway of activation within the liver would lead to tolerance (Figure 4, mechanism 1). According to our results the intrahepatic induction of priming might be associated with the development of AIH when there is a high frequency of autoreactive T cells trapped within this organ, or when the tolerogenic pathway of intrahepatic activation is defective. As suggested by the results in the 178.3 model, a high frequency of T cells, primed in the liver or lymphoid tissues, is likely to induce a 'cytokine storm' leading to bystander hepatocyte injury, even when hepatocytes do not express the autoantigen (Figure 4, mechanism 3). We have hypothesized that such mechanisms of indirect CD8$^+$ T cell-mediated hepatocyte injury may play a role in some forms of immune-mediated liver injury, including hepatitis associated with extra-hepatic autoimmune disease: in the presence of high-level autoantigen expression by bone marrow-derived APC within the liver, hepatocellular injury might be mediated in such an indirect fashion, even in the absence of hepatocyte autoantigen expression. Furthermore, bystander hepatocellular injury and death might lead to the 'unmasking' of previously intracellular proteins, which could lead to epitope spreading following cross-presentation of such previously sequestered antigens. AIH has been described as becoming clinically apparent following a variety of infections, and it is possible that such epitope spreading to previously sequestered antigens might be a factor in the precipitation of this disease following infection. It is of note that a number of viruses are associated with the development of hepatitis in the absence of demonstrable tropism for hepatocytes[37]; such hepatocellular injury could be caused by similar bystander mechanisms and, perhaps in the presence of immune regulatory defects, might precipitate immune recognition of hitherto strictly intracellular antigens. In addition, it is tempting to speculate that in some forms of AIH where autoantigens have been demonstrated to be expressed by hepatocytes, a failure of the intrahepatic pathways of activation to generate tolerance might in some way contribute to the development of disease (Figure 4, mechanism 2). A variety of factors might lead to this anomaly, including the abnormal expression of adhesion or co-stimulatory molecules on liver cells due to concomitant intrahepatic inflammation. Alternatively, AIH could be promoted by T cells activated within the LN, which then recognize their cognate antigen expressed within the liver (Figure 4, mechanism 4). In our model, T cells activated within the LN that induced hepatitis in Met-K^b mice died following liver injury. The mechanisms of this deletion remain unknown, but it is likely that, in the absence of such controlling mechanisms, liver damage would continue, and result in chronic AIH.

In conclusion, observations from these transgenic models would suggest that future strategies for the study of AIH should include the identification of parameters that limit immune responses within the liver, as defects within these pathways might contribute to the genesis and maintenance of intrahepatic autoimmunity.

Acknowledgements

This work was supported by the National Health and Medical Research Council of Australia (NHMRC). D.G.B. was supported by an NHMRC Medical Postgraduate Research Scholarship. The authors thank Drs Bernd Arnold, Grant Morahan, Jacques Miller and Matthias Hoffmann for providing us with the transgenic mice mentioned in this chapter.

References

1. Bertolino P, McCaughan GW, Bowen DG. Role of primary intrahepatic T-cell activation in the 'liver tolerance effect'. Immunol Cell Biol. 2002;80:84–92.
2. Wick MJ, Leithauser F, Reimann J. The hepatic immune system. Crit Rev Immunol. 2002; 22:47–103.
3. Crispe IN. Hepatic T cells and liver tolerance. Nat Rev Immunol. 2003;3:51–62.
4. Diamantis I, Boumpas DT. Autoimmune hepatitis: evolving concepts. Autoimmun Rev. 2004;3:207–14.

5. Lohr H, Manns M, Kyriatsoulis A et al. Clonal analysis of liver-infiltrating T cells in patients with LKM-1 antibody-positive autoimmune chronic active hepatitis. Clin Exp Immunol. 1991;84:297–302.
6. Lohr HF, Schlaak JF, Lohse AW et al. Autoreactive CD4$^+$ LKM-specific and anticlonotypic T-cell responses in LKM-1 antibody-positive autoimmune hepatitis. Hepatology. 1996;24:1416–21.
7. Arenz M, Pingel S, Schirmacher P, Meyer zum Büschenfelde KH, Lohr HF. T cell receptor Vbeta chain restriction and preferred CDR3 motifs of liver–kidney microsomal antigen (LKM-1)-reactive T cells from autoimmune hepatitis patients. Liver. 2001;21:18–25.
8. Runyon BA, LaBrecque DR, Anuras S. The spectrum of liver disease in systemic lupus erythematosus. Report of 33 histologically-proved cases and review of the literature. Am J Med. 1980;69:187–94.
9. Keshavarzian A, Rentsch R, Hodgson HJ. Clinical implications of liver biopsy findings in collagen-vascular disorders. J Clin Gastroenterol. 1993;17:219–26.
10. Schönrich G, Kalinke U, Momburg F et al. Down-regulation of T cell receptors on self-reactive T cells as a novel mechanism for extrathymic tolerance induction. Cell. 1991;65:293–304.
11. Bertolino P, Heath WR, Hardy CL, Morahan G, Miller JF. Peripheral deletion of autoreactive CD8$^+$ T cells in transgenic mice expressing H-2K^b in the liver. Eur J Immunol. 1995;25:1932–42.
12. Bishop GA, Sun J, Sheil R, McCaughan GW. High dose/activation-associated tolerance. A mechanism for allograft tolerance. Transplantation. 1997;64:1377–82.
13. Limmer A, Sacher T, Alferink J et al. Failure to induce organ-specific autoimmunity by breaking of tolerance: importance of the microenvironment. Eur J Immunol. 1998;28:2395–406.
14. Limmer A, Sacher T, Alferink J, Nichterlein T, Arnold B, Hammerling GJ. A two-step model for the induction of organ-specific autoimmunity. Novartis Foundation Symposium 1998;215:159–67; discussion 167–71.
15. Sacher T, Knolle P, Nichterlein T, Arnold B, Hammerling GJ, Limmer A. CpG-ODN-induced inflammation is sufficient to cause T-cell-mediated autoaggression against hepatocytes. Eur J Immunol. 2002;32:3628–37.
16. Banchereau J, Briere F, Caux C et al. Immunobiology of dendritic cells. Annu Rev Immunol. 2000;18:767–811.
17. Bertolino P, Bowen DG, McCaughan GW, Fazekas De St Groth B. Antigen-specific primary activation of CD8$^+$ T cells within the liver. J Immunol. 2001;166:5430–8.
18. Bowen DG, Zen M, Holz L, Davis T, McCaughan GW, Bertolino P. The site of primary T cell activation is a determinant of the balance between intrahepatic tolerance and immunity. J Clin Invest. 2004;114:701–12.
19. Bowen DG, Warren A, Davis T et al. Cytokine-dependent bystander hepatitis due to intrahepatic murine CD8$^+$ T-cell activation by bone marrow-derived cells. Gastroenterology. 2002;123:1252–64.
20. Ando K, Guidotti LG, Cerny A, Ishikawa T, Chisari FV. CTL access to tissue antigen is restricted *in vivo*. J Immunol. 1994;153:482–8.
21. Mehal WZ, Juedes AE, Crispe IN. Selective retention of activated CD8$^+$ T cells by the normal liver. J Immunol. 1999;163:3202–10.
22. Salmi M, Jalkanen S. How do lymphocytes know where to go? Current concepts and enigmas of lymphocyte homing. Adv Immunol. 1997;64:139–218.
23. Mackay CR, Marston WL, Dudler L. Naive and memory T cells show distinct pathways of lymphocyte recirculation. J Exp Med. 1990;171:801–17.
24. Mackay CR, Marston W, Dudler L. Altered patterns of T cell migration through lymph nodes and skin following antigen challenge. Eur J Immunol. 1992;22:2205–10.
25. Lehmann JC, Jablonski-Westrich D, Haubold U, Gutierrez-Ramos JC, Springer T, Hamann A. Overlapping and selective roles of endothelial intercellular adhesion molecule-1 (ICAM-1) and ICAM-2 in lymphocyte trafficking. J Immunol. 2003;171:2588–93.
26. MacSween RNM, Scothorne RJ. Developmental anatomy and normal structure. In: MacSween RNM, Anthony PP, Scheuer PJ, editors. Pathology of the Liver. New York: Churchill Livingstone, 1979:1–49.

27. Cogger VC, Mross PE, Hosie MJ, Ansselin AD, McLean AJ, Le Couteur DG. The effect of acute oxidative stress on the ultrastructure of the perfused rat liver. Pharmacol Toxicol. 2001;89:306–11.
28. Le Couteur DG, Cogger VC, Markus AM et al. Pseudocapillarization and associated energy limitation in the aged rat liver. Hepatology. 2001;33:537–43.
29. Salmi M, Adams D, Jalkanen S. Cell adhesion and migration. IV. Lymphocyte trafficking in the intestine and liver. Am J Physiol. 1998;274:G1–6.
30. Hamann A, Klugewitz K, Austrup F, Jablonski-Westrich D. Activation induces rapid and profound alterations in the trafficking of T cells. Eur J Immunol. 2000;30:3207–18.
31. Huang L, Soldevila G, Leeker M, Flavell R, Crispe IN. The liver eliminates T cells undergoing antigen-triggered apoptosis *in vivo*. Immunity. 1994;1:741–9.
32. Morahan G, Brennan FE, Bhathal PS, Allison J, Cox KO, Miller JFAP. Expression in transgenic mice of class I histocompatibility antigens controlled by the metallothionein promoter. Proc Natl Acad Sci USA. 1989;86:3782–6.
33. Schönrich G, Momburg F, Malissen M et al. Distinct mechanisms of extrathymic T cell tolerance due to differential expression of self antigen. Int Immunol. 1992;4:581–90.
34. Romermann D, Heath WR, Allison J et al. Ligand density determines the efficiency of negative selection in the thymus. Transplantation. 2001;72:305–11.
35. Bertolino P, Trescol-Biemont MC, Rabourdin-Combe C. Hepatocytes induce functional activation of naive CD8+ T lymphocytes but fail to promote survival. Eur J Immunol. 1998; 28:221–36.
36. Bertolino P, Trescol-Biemont MC, Thomas J et al. Death by neglect as a deletional mechanism of peripheral tolerance. Int Immunol. 1999;11:1225–38.
37. O'Farrelly C, Crispe IN. Prometheus through the looking glass: reflections on the hepatic immune system. Immunol Today. 1999;20:394–8.

8
Autoantibodies to SLA/LP: specificity and pathogenetic relevance?

U. CHERUTI, C. WANG, A. W. LOHSE and J. HERKEL

INTRODUCTION

The mechanisms that drive autoimmunity to hepatic antigens and the pathogenesis of autoimmune liver diseases are not clear. Autoimmunity to the SLA/LP molecule may serve as a prototype for understanding pathogenetic mechanisms of autoimmune liver disease, because the autoimmune response to SLA/LP is strictly disease-specific and highly uniform. This chapter aims at summarizing the current understanding of SLA/LP autoimmunity.

DIAGNOSTIC RELEVANCE OF SLA/LP AUTOANTIBODIES

The diagnosis of autoimmune hepatitis (AIH) is based on a combination of clinical, serological and histological findings[1]. The presence of autoantibodies is an important component of the diagnosis[2]; however, most of the commonly detected autoantibodies, e.g. antinuclear antibodies (ANA), smooth muscle antibodies (SMA), and antibodies to liver–kidney microsome type 1 (anti-LKM1) are not disease-specific[3–5].

In contrast, antibodies to soluble liver antigen (SLA)[6] and to liver/pancreas (LP)[7] have been described as specific for autoimmune liver disease[8]. Recently, identity of the SLA and LP antigens has been reported, and the target antigen of SLA/LP autoantibodies has been cloned[9]. Using the recombinant SLA/LP molecule the strict specificity of SLA/LP autoantibodies as markers for AIH has been confirmed with more than 2000 sera tested[10]. It should be noted, however, that SLA/LP autoantibodies may also be present in paediatric AIH patients[11] and in AIH-associated overlap syndromes, such as the overlap of AIH and primary biliary cirrhosis[12].

Worldwide, SLA/LP autoantibodies are present in about 20% of patients suffering from AIH; aside from Japan, where the frequency seems to be lower[10,13,14]. According to the presence of SLA/LP autoantibodies, AIH patients with these antibodies are being classified as suffering from AIH type 3[15]. However, the clinical differences between patients with or without SLA/LP autoantibodies are negligible(10); thus, the usefulness of classifying patients with SLA/LP autoantibody into a distinct clinical subgroup is questionable.

MOLECULAR AND FUNCTIONAL PROPERTIES OF THE SLA/LP PROTEIN

At least three variant forms of the SLA/LP molecule have been described: (1) the presumably major variant, a 1326 base pair open-reading frame sequence (gene bank accession numbers AJ238617 and AJ277541)[9,16], which encodes for a 441 amino acid protein sequence with a theoretical molecular weight of 48.8 kDa and a theoretical pI of 8.64; (2) an aminoterminal splice variant with a 1269 base pair open-reading frame sequence (AF146396)[9], which encodes for a 422 amino acid protein sequence with a theoretical molecular weight of 46.9 kDa and a theoretical pI of 8.52; (3) a 35 kDa carboxyterminal protein fragment[17]. The 48.8 kDa and the 46.9 kDa variants seem to account for the typical staining pattern produced by patient sera in Western blots of liver protein extract, which is a double band with apparent molecular weights of about 45–50 kDa[9].

The major 48.8 kDa variant has been identified independently by two different approaches: one was by screening with SLA autoantibody-positive serum of an hepatocyte-derived cDNA expression library[9]; the other was by immunoprecipitation with patient serum from HeLa cell extract of a 48 kDa protein, which was associated with a tRNA that facilitates the co-translational incorporation of selenocysteine into proteins[16,18]. On grounds of this association it has been speculated that the SLA/LP molecule might be involved in selenoprotein metabolism. A model structure of the SLA/LP molecule, which was based on fold recognition, is compatible with a pyridoxal phosphate-dependent transferase[19], suggesting that the SLA/LP protein could function as serine hydroxymethyltransferase within the poorly characterized mammalian selenocysteine pathway[20]. However, direct evidence for a role of SLA/LP in selenocysteine metabolism is missing.

The shorter variants of 46.9 kDa, which has been cloned from a Jurkat cell line derived cDNA expression library[9], and of 35 kDa, which has been cloned from a liver-derived cDNA expression library[17], are both incompatible with a role as transferase, and their putative function remains obscure.

The SLA/LP molecule was highly conserved in evolution, and sequences from various species, including human, mouse, rat, fruit fly, nematode and archaebacterium, display high degrees of similarity or homology (HomoloGene:15031; Figure 1). SLA/LP is expressed in a variety of mammalian tissues; expression in liver is among the most prominent.

FEATURES OF THE AUTOIMMUNE RESPONSE TO SLA/LP

The autoantibody response to SLA/LP is not random, but displays a remarkable degree of uniformity: analysis of AIH patient sera revealed that SLA/LP autoantibodies are of a preferred dominant subtype (IgG1) and recognize the

```
Man             ----MDSNNFLGNCGVGEREGRVASALVARRHYRFIHGIGRSGDISAVQPKAAGSSLLNK
Mouse           ----MDSNNFLGNCGVGEREGRVASALVARRHYRFIHGIGRSGDISAVQPKAAGSSLLNK
Fly             MLASLDSNNYPHKVGLGEREARIACKLVARRHYNFGHGIGRSGDLLEAQPKAAGSTLLAR
Worm            ----------MIPVGAGEREGRVLTPLVQRLHSNLTHGIGRSGNLLEIQPKALGSSMLAC
Archaebacterium ----MDTDKDPNVVQIGEREARVYTKLQRDGVFDFCHGVGRSGNLIDPQPKAPGASVMYK

Man             ITNSLVLDIIKLAGVHTVANCFVVPMATGMSLTLCFLTLRHKRPKAKYIIWPRIDQKSCF
Mouse           ITNSLVLNVIKLAGVHSVASCFVVPMATGMSLTLCFLTLRHKRPKAKYIIWPRIDQKSCF
Fly             LTNALILDLIRGIGLPSCAGCFLVPMCTGMTLTLCLQSLRKRRPGARYVLWSRIDQKSCF
Worm            LSNEFAKHALHLLGLHAVKSCIVVPLCTGMSLSLCMTSWRRRRPKAKYVVWLRIDQKSSL
Archaebacterium LTNKLLESFLKALGLK--VNAIATPVATGMSLALCLSA-ARKKYNSNVVIYPYAAHKSPI

Man             KSMITAGFEPVVIENVLEGDELRTDLKAVEAKVQ-ELGPDCILCIHSTTSCFAPRVPDRL
Mouse           KSMVTAGFEPVVIENVLEGDELRTDLKAVEAKIQ-ELGPEHILCLHSTTACFAPRVPDRL
Fly             KAITATGLVPVVIPCLIKGESLNTNVDLFREKIK-SLGVDSILCLYTTTSCFAPRNSDDI
Worm            KSIYHAGFEPIIVEPIRDRDSLITDVETVNRIIE-QRG-EEILCVMTTTSCFAPRSPDNV
Archaebacterium KATSFIGMRMRLVETVLDGDIVKVEVSDIEDAIRKEINENNNPVVLSTLTFFPPRKSDDI

Man             EELAVICANYDIPHIVNNAYGVQSSKCMHLIQQGARVGRIDAFVQSLDKNFMVPVGGAII
Mouse           EELAVICANYDIPHVVNNAYGLQSSKCMHLIQQGARVGRIDAFVQSLDKNFMVPVGGAII
Fly             AEVSKLSKQWQIPHLVNNAYGLQAKEIVNQLECANRVGRIDYFVQSSDKNLLVPVGSAIV
Worm            EAISAICAAHDVPHLVNNAYGLQSEETIRKIAAAHECGRVDAVVQSLDKNFQVPVGGAVI
Archaebacterium KEIAKICQDYDIPHIINGAYAIQNFYYIEKLKKALKY-RIDAVVSSSDKNLFTPIGGGII

Man             AGFNDSFIQEISKMYPGRASASPSLDVLITLLSLGSGNGYKKLLKERKEMFSYLSNQIKKL
Mouse           AGFNEPFIQDISKMYPGRASASPSLDVLITLLSLGCSGYRKLLKERKEMFVYLSTQLKKL
Fly             ASFNESVLHDVASTYAGRASGSQSLDVLMTLLSLGRNGFRLLFDQRGENFNYLRENLRKF
Worm            AAFKQNHIQSIAQSYPGRASSVPSRDLVLTLLYQGQSAFLEPFGKQKQMFLKMRRKLISF
Archaebacterium YTKDESFLKEISLTYPGRASANPIVNILISLLAIGTKDYLNLMKEQKECKKLLNELLEDL

Man             SEAYNERLLHTPHNPISLAMTLKTLDEHRDKAVTQLGSMLFTRQVSGARVVPLGSMQT-V
Mouse           AEAHNERLLQTPHNPISLAMTLKTIDGHHDKAVTQLGSMLFTRQVSGARAVPLGNVQT-V
Fly             AEPRGEIVIDSRFNSISLAITLATLAGDQMKSITKLGSMLHMRGVSGARVIVPGQNKT-I
Worm            AENIGECVYEVPENEISSAMTLSTIPPAKQ---TLFGSILFAKGITGARVVTSSQSKTTI
Archaebacterium AKKKGEKVLNV-ENPISSCITTK-------KDPLDVAGKLYNLRVTGPRGVRRND-----

Man             SGYTFRGFMSHTNNYPCAYLNAASAIGMKMQDVDLFIKRLDRCLKAVRKERSKES---DD
Mouse           SGHTFRGFMSHADNYPCAYLNAAAAIGMKMQDVDLFIKRLDKCLNIVRKEQTRASVVSGA
Fly             DGHEFLGK----------------------------------------------------
Worm            EGCEFINFGSHTTEQHGGYLNIACSVGMTDHELEELFTRLTSSYAKFVRELAKED---ER
Archaebacterium -----KFGTCYLKEYPYDYIVVNSAIGVKKEDIYKVIEKLDEVL----------------

Man             NYDKTEDVDIEEMALKLDNVLLDTYQDASS
Mouse           DRNKAEDADIEEMALKLDDVLGDVGQGPAL
Fly             ------------------------------
Worm            INSSGRRIPINE-SFDMEND----------
Archaebacterium ------------------------------
```

Figure 1 Amino acid sequence alignment of SLA/LP molecules from various species

same dominant epitope near the carboxyterminus of the SLA/LP molecule[21], which is conserved in all known variants (see above). Thus, autoimmunity to SLA/LP has features of a highly selected immune response and seems to be driven by the same mechanism in all patients. The T cell response to SLA/LP is so far not characterized. Whether SLA/LP autoimmunity is driven by SLA/LP itself or another protein with similar structure is not clear.

PATHOGENETIC RELEVANCE OF SLA/LP AUTOIMMUNITY

Because SLA/LP autoantibodies are strictly disease-specific it is tempting to speculate that they might be involved in the pathogenesis of AIH. However, the SLA/LP molecule is a cytoplasmic protein and it is not clear how SLA/LP autoantibodies may recognize an intracellular protein; a possible translocation of SLA/LP molecules to the cell surface has not been examined. Alternatively, liver cell damage may be mediated by SLA/LP-specific T lymphocytes; although specific T cells have not yet been described, a pathogenic role for specific T cells is likely, given the highly selected phenotype of SLA/LP autoantibodies. Be that as it may, it is also possible that autoimmunity to SLA/LP could be only an epiphenomenon of liver cell damage, and not involved in pathogenesis. However, our preliminary findings suggest that, at least in mice, hepatic inflammation and liver cell damage can be induced by autoimmunization to SLA/LP.

CONCLUSIONS

Antibodies to SLA/LP are highly specific diagnostic markers for AIH. SLA/LP autoimmunity is highly selected and uniform among AIH patients. The function of the SLA/LP molecule and the relevance of its variants is not known. The possible role of SLA/LP autoimmunity in the pathogenesis of AIH remains to be clarified.

References

1. Alvarez F, Berg PA, Bianchi FB et al. International Autoimmune Hepatitis Group Report: review of criteria for diagnosis of autoimmune hepatitis. J Hepatol. 1999;31:929–38.
2. Krawitt EL. Autoimmune hepatitis. N Engl J Med. 1996;334:897–903.
3. Lohse AW, Gerken G, Mohr H et al. Distinction between autoimmune liver diseases and viral hepatitis: clinical and serological characteristics in 859 patients. Z Gastroenterol. 1995;33:527–33.
4. Lunel F, Abuaf N, Frangeul L et al. Liver/kidney microsome antibody type 1 and hepatitis C virus infection. Hepatology. 1992;16:630–6.
5. Lohse AW, Obermayer-Straub P, Gerken G et al. Development of cytochrome P450 2D6-specific LKM-autoantibodies following liver transplantation for Wilson's disease – possible association with a steroid-resistant transplant rejection episode. J Hepatol. 1999;31:149–55.
6. Manns M, Gerken G, Kyriatsoulis A, Staritz M, Meyer zum Buschenfelde KH. Characterization of a new subgroup of autoimmune chronic active hepatitis by autoantibodies against a soluble liver antigen. Lancet. 1987;1:292–4.
7. Stechemesser E, Klein R, Berg PA. Characterization and clinical relevance of liver–pancreas antibodies in autoimmune hepatitis. Hepatology. 1993;18:1–9.
8. Kanzler S, Weidemann C, Gerken G et al. Clinical significance of antibodies to soluble liver antigen in autoimmune hepatitis. J Hepatol. 1999;31:635–40.
9. Wies I, Brunner S, Henninger J, Herkel J, Meyer zum Buschenfelde KH, Lohse AW. Identification of target antigen for SLA/LP autoantibodies in autoimmune hepatitis. Lancet. 2000;355:1510–15.
10. Baeres M, Herkel J, Czaja AJ et al. Establishment of standardised SLA/LP immunoassays: specificity for autoimmune hepatitis, worldwide occurrence, and clinical characteristics. Gut. 2002;51:259–64.
11. Vitozzi S, Djilali-Saiah I, Lapierre P, Alvarez F. Anti-soluble liver antigen/liver-pancreas (SLA/LP) antibodies in pediatric patients with autoimmune hepatitis. Autoimmunity. 2002; 35:485–92.
12. Kanzler S, Bozkurt S, Herkel J, Galle PR, Dienes HP, Lohse AW. [Presence of SLA/LP autoantibodies in patients with primary biliary cirrhosis as a marker for secondary autoimmune hepatitis (overlap syndrome)]. Dtsch Med Wochenschr. 2001;126:450–6.
13. Boberg KM. Prevalence and epidemiology of autoimmune hepatitis. Clin Liver Dis. 2002;6: 347–59.

14. Miyakawa H, Kawashima Y, Kitazawa E et al. Low frequency of anti-SLA/LP autoantibody in Japanese adult patients with autoimmune liver diseases: analysis with recombinant antigen assay. J Autoimmun. 2003;21:77–82.
15. Strassburg CP, Manns MP. Autoantibodies and autoantigens in autoimmune hepatitis. Semin Liver Dis. 2002;22:339–52.
16. Costa M, Rodriguez-Sanchez JL, Czaja AJ, Gelpi C. Isolation and characterization of cDNA encoding the antigenic protein of the human tRNP(Ser)Sec complex recognized by autoantibodies from patients withtype-1 autoimmune hepatitis. Clin Exp Immunol. 2000; 121:364–74.
17. Volkmann M, Martin L, Baurle A et al. Soluble liver antigen: isolation of a 35-kd recombinant protein (SLA-p35) specifically recognizing sera from patients with autoimmune hepatitis. Hepatology. 2001;33:591–6.
18. Gelpi C, Sontheimer EJ, Rodriguez-Sanchez JL. Autoantibodies against a serine tRNA–protein complex implicated in cotranslational selenocysteine insertion. Proc Natl Acad Sci USA. 1992;89:9739–43.
19. Kernebeck T, Lohse AW, Grötzinger J. A bioinformatical approach suggests the function of the autoimmune hepatitis target antigen soluble liver antigen/liver pancreas. Hepatology. 2001;34:230–3.
20. Atkins JF, Gesteland RF. The twenty-first amino acid. Nature. 2000;407:463–5.
21. Herkel J, Heidrich B, Nieraad N, Wies I, Rother M, Lohse AW. Fine specificity of autoantibodies to soluble liver antigen and liver/pancreas. Hepatology. 2002;35:403–8.

9
Role of microsomal antigens in autoimmune hepatitis

C. P. STRASSBURG

INTRODUCTION

The reactivity of B cells against intrinsic molecular structures of the hepatocyte, but also against other cellular compartments of the body, which leads to
the formation of autoantibodies, is a prominent feature in autoimmune liver
diseases. However, it is important to distinguish between serological autoimmunity on the one hand and genuine autoimmune disease on the other hand
(Table 1). While the former represent a feature frequently found in liver
diseases, but also in other infectious diseases, genuine autoimmune disease
lacks an identifiable, definite offending cause. The interest in both stems from
the possibility that mechanisms leading to serological autoimmunity and even
to transient effects of autoimmune disease, as in the example of drug-induced
immune-mediated liver disease, may serve as an aid to understanding the break
of tolerance encountered in autoimmune liver diseases.

Table 1 Serological autoimmunity and autoimmune liver diseases

Autoimmune liver diseases
    Autoimmune hepatitis
    Primary biliary cirrhosis
    Primary sclerosing cholangitis
    Overlap syndromes
Viral hepatitis C
Viral hepatitis D
Drug-induced hepatitis
Genetic liver disease

AUTOIMMUNE LIVER DISEASES

The diagnosis of the autoimmune liver diseases primary biliary cirrhosis (PBC) and autoimmune hepatitis (AIH) is reached by exclusion of other causes of chronic hepatic damage which include viral infection, metabolic abnormalities, genetic liver diseases and toxic exposure, as well as haemodynamic alterations[1]. One of the most striking features of AIH and PBC is the elevation of serum immunoglobulin levels and the detection of circulating autoantibodies. A third presumably (auto)immune-mediated condition of the liver is primary sclerosing cholangitis which is not characterized by female predominance, a well-defined autoantibody/autoantigen profile or a response to immunosuppression. The majority of research has focused on the characterization of autoantibodies associated with PBC and AIH. From a clinical point of view the two diseases differ in their classical biochemical presentation, AIH displaying a typical 'hepatitic' enzyme profile with elevated alanine aminotransferase (ALT) > aspartate aminotransferase (AST) in addition to an elevation of immunoglobulin G levels, and PBC displaying a typical cholestatic profile with an elevation of alkaline phosphatase (AP) and gamma-glutamyltransferase (γGT), as well as bilirubin. However, autoantibody profiles in both diseases are both distinguishing and characterized by a degree of overlap: PBC displays antimitochondrial autoantibodies (AMA) with are neither organ- nor cell type-specific, in addition to antinuclear autoantibodies (ANA) which are neither cell- nor disease-specific. AIH displays ANA, smooth muscle autoantibody (SMA) and liver–kidney microsomal (LKM) autoantibodies, as well as auto-antibodies against soluble liver antigen (SLA/LP). These five types of auto-antibodies are the major serological findings in AIH and PBC, contributing to serological diagnostics and providing a research window to explore autoanti-gens as clues to the immunopathogenesis of autoimmune liver diseases. Of note, ANA occur in both conditions and require subclassification. Similarily, LKM autoantibodies also occur in drug-induced, immune-mediated hepatitis as well as in virus infection with the hepatitis C and D viruses (Figure 1). These considerations point to an important distinction when analysing autoantigens in immune-mediated liver diseases. Many of the same autoantibodies, namely LKM-1, occur both as serological markers of virus or drug exposure-associated autoimmunity as well as in genuine autoimmune disease of the liver. Only in PBC and AIH are autoimmune liver diseases present, while all other manifestations of humoral autoimmunity are associated with a defined aetiol-ogy of chronic or transient liver disease. When all the data on the simultaneous occurrence of markers of viral infections (cytomegalovirus (CMV), measles, chronic hepatitis C (HCV), Epstein–Barr virus (EBV), and chronic hepatitis A (HAV) etc.) and AIH are taken into consideration there is little evidence to suggest that viruses are the cause of AIH. They may serve as triggers of AIH; however, detailed analysis of humoral autoreactivities with microsomal auto-antigens exhibit differences between AIH and virus-associated serological autoimmune phenomena.

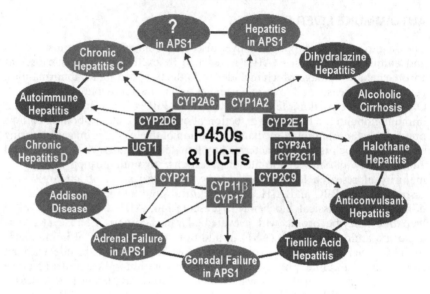

Figure 1 Heterogeneity and specificity of microsomal targets of autoantibodies in immune mediated liver diseases (viral, genetic, ideopathic)

ANALYSIS OF AUTOANTIBODIES IN AIH AND VIRUS INFECTION

The serological hallmark of AIH are autoantibodies detected by serological *in-vitro* assays[2]. For screening purposes, ANA, SMA, LKM and AMA continue to be determined by indirect immunofluorescence on rodent liver, stomach and kidney tissue. When these autoantibodies are undetectable, and the suspicion of autoimmune hepatitis remains, autoantibodies against cytosolic antigens, in particular the SLA/LP antigen, may be helpful.

Hepatitis C is associated with an array of extrahepatic manifestations, including mixed cryoglobulinaemia, membranoproliferative glomerulonephritis, polyarthritis, porphyria cutanea tarda, Sjörgen's syndrome and autoimmune thyroid disease[3-6]. Not surprisingly, numerous autoantibodies are found to be associated with chronic hepatitis C. Similar to AIH antinuclear, SMA, LKM and antithyroid antibodies are found with a high prevalence.

The examination of LKM autoantibodies in HCV patients revealed that, although anti-CYP2D6 titres are similar to titres in AIH type 2, differences exist regarding the epitopes recognized by LKM autoantibodies[7-10]. In patients with AIH type 2 the epitope of amino acids (aa) 257–269 is recognized with a significantly higher prevalence than in chronic hepatitis C[9]. In addition, the immune reaction seems to be more heterogeneous than in AIH, as indicated by recognized protein targets of 59 kDa and 70 kDa[11].

When the CYP2D6 molecule is analysed in detail for epitope reactivity of circulating autoantibodies in patients with AIH and HCV, four regions of interest emerge[12]. All sera from AIH as well as HCV patients which are positive for LKM-1 autoantibodies upon standard screening tests recognize the full length protein. The major region of reactivity resides between aa 257 and 269 (Figures 2 and 3). With respect to this epitope a differential pattern is observed, characterized by the binding of 64% of sera from AIH type 2 patients and only 24% of HCV patients. A second major epitope was found between aa 321 and 351 with a less convincing differential recognition pattern (AIH 20% and HCV 12%). Two minor epitopes are located between aa 373 and 389 and aa 419 and 429, recognized by under 10% of sera. Taken together none of these epitopes was accountable for the entire reactivity measurable with the full-length protein, most likely owing to the fact that these were *Escherichia coli* or similarly expressed linear fragments assayed with denaturing techniques such as Western blot. In a recent analysis by our group[12] we found that elimination of sequence from the amino terminus of the molecule up to aa 351 virtually eliminated all binding activity which was not reconstituted by any of the individual fragments located upstream of aa 351. This indicates that conformation-dependent epitopes are most likely responsible for the remainder of immunoreactivity with the CYP2D6 molecule. This was also substantiated by the finding that a larger fragment spanning aa 321–376 reacted with 76% of all AIH type 2 and HCV sera, representing the highest number of all epitope definitions found to date. The three-dimensional rendering of the CYP2D6 molecule was able to show that this novel epitope was located on the surface of the molecule distant from the previous major linear epitope between aa 257 and 269[12]. A sequence search and comparison analysis further showed that this area of the CYP molecule overlaps with the previously defined epitopes recognized by autoantibodies directed against CYP1A2, CYP2C9, CYP3A1 and CYP21B found in different drug-mediated autoimmune reactions, as well as in adrenal autoimmunity (CYP21B) (Figure 2). This interesting finding points to the usefulness of epitope analyses to elucidate common immunological mechanisms involved in the generation of autoimmunity. The model of drug-induced hepatitis is therefore linked to the patterns found in genuine AIH at the B-cell level.

From a practical point of view the differential recognition of autoepitopes in HCV and AIH type 1 has clinical implications. LKM autoantibodies in chronic hepatitis C seem to indicate an increased risk of exacerbation of the disease[9,13,14]. Dalekos et al. studied antibody titres and performed epitope mapping of LKM-1-positive sera from patients with chronic hepatitis C. Interestingly, a patient with a high LKM-1 titre and autoantibodies directed against an epitope of aa 257–269, which is preferentially recognized by patients with AIH type 2, showed exacerbation of the disease under interferon treatment. In contrast to other patients with HCV infection this patient further recognized a rarely detected epitope on the C-terminal third of the protein.

Recently another autoantibody was detected in patients infected with HCV. About 2% of HCV-positive sera in general and 7.5% of LKM-1-positive HCV sera recognize CYP2A6. This autoantibody appears to occur more frequently in HCV-infected patients with LKM-1 autoantibodies. Interestingly anti-

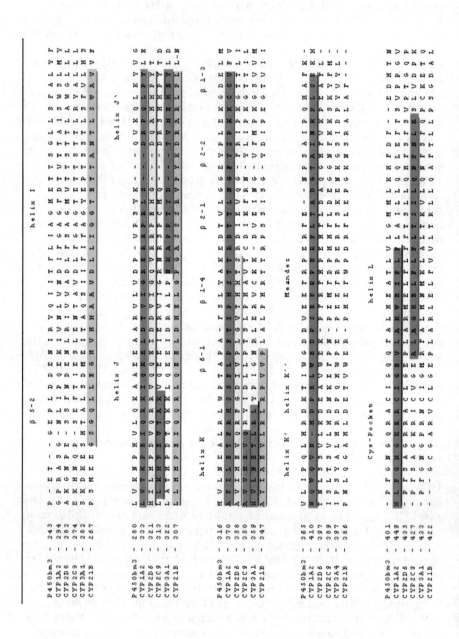

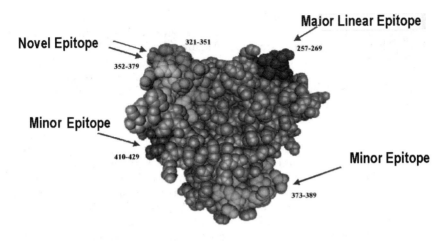

Figure 3 Threedimensional representation of the major CYP2D6 epitopes recognized by LKM-1 autoantibodies.

CYP2A6 autoantibodies are not detected in patients with AIH type 2, who exhibit high titres of LKM-1 autoantibodies. The clinical relevance of this finding remains to be determined[15]. Anti-CYP2A6 autoantibodies have also been detected in patients with the autoimmune polyglandular syndrome type 1 (APS-1).

LESSONS FROM DRUG-INDUCED IMMUNE-MEDIATED LIVER DISEASE

The characterization of autoantibodies directed against CYP and uridine diphosphate glucuronosyltransferases (UGT) have led to the definition of specificities which span a number of immune-mediated diseases (Figure 1). These include AIH, drug-induced hepatitis[16–18], autoimmune polyglandular syndrome type 1 (APS-1)[19], and chronic hepatitis C (HCV) and D (HDV)[20].

LKM autoantibodies against CYP1A2 and CYP2A6 are found in patients with APS-1 and hepatic involvement[21–23]. Anti-CYP2A6 autoantibodies also occur in HCV infection[14,15]. LM autoantibodies, which are characterized by an immunofluorescence pattern selectively staining the hepatocellular cytoplasm but not kidney, have been found to be directed against CYP1A2[24–26]. These autoantibodies are also found in APS-1 syndrome with hepatic involvement and occur in dihydralazine-induced hepatitis[27,28]. A second type of LKM autoantibodies, LKM-2, are directed against CYP2C9 and are induced in

Figure 2 (opposite)   Convergence of microsomal autoreactivity on overlapping epitope regions on different CYP molecules

ticrynafen-associated hepatitis[16,29]. A third group of LKM autoantibodies, LKM-3, were identified in 6–10% of patients with HDV by Crivelli et al. in 1983[30]. These autoantibodies are directed against family 1 UGT (UGT1A)[31,32], which are also a superfamily of drug-metabolizing proteins located in the endoplasmic reticulum[33,34]. LKM-3 autoantibodies have been identified in HDV infection but also in AIH type 2 patients[31,32]. They can also occur in LKM-1-negative and ANA-negative AIH.

A small percentage of patients treated with therapeutic drugs can develop severe hepatitis, which is characterized by lymphocytic liver infiltrations and autoantibodies directed against hepatic proteins. It is believed that drug-metabolizing enzymes, mainly CYP, create reactive metabolites, which in turn modify either the metabolizing CYP enzyme itself and/or other hepatic proteins[35–37]. In susceptible patients these modified proteins induce an immune response resulting in severe 'drug-induced hepatitis'[38]. Modified proteins preferentially include CYP, which in themselves are often targets for autoantibodies. As typical examples tienilic acid-induced hepatititis[39], dihydralazine hepatitis[27], halothane hepatitis[40] and anticonvulsant-induced hepatitis[41] have been characterized. It is debated whether alcoholic liver disease is caused in part by an autoimmune reaction against hepatic proteins, directed against both acetaldehyde- and hydroxyethyl-modified hepatic proteins[42,43]. It is suggested that metabolism of ethanol by CYP2E1 generates hydroxyethyl-radicals, which can represent targets of autoimmunity[43]. When all of the characterized autoepitopes in drug-induced immune-mediated liver disease are plotted and aligned with their corresponding CYP targets an interesting picture emerges, which indicates that humoral autoimmunity recognizes overlapping targets on the CYP molecule irrespective of the CYP isoform. This observation raises important issues. The epitopes between HCV-associated LKM-1 autoantibodies and AIH type 2-associated LKM-1 autoantibodies show amino acid specificity while they recognize the same portion of the CYP molecule targeted in dihydralazine, ticrynafen, alcoholic, anticonvulsant hepatitis and even in autoantibodies detectable in the APS-1 syndrome and in adrenal insufficiency (Figure 2). Although this may be a serendipitous finding it points attention towards a concrete molecular and submolecular structure which is involved in breaking tolerance in actual disease manifestations. Convergence on this molecule provides a promising research angle for elucidating the aetiology of autoimmune disease. This is particularly relevant since most models of AIH have not provided credible evidence of the actual target antigen. In this respect currently ongoing studies on a model of AIH based on a CYP2D6 mouse will be of great interest.

CONCLUSIONS

A considerable number of studies have contributed to the realization that the so-called hepatic autoantigens are neither organ-, cell- nor disease-specific with the exception of AMA, which show an unmatched specificity for PBC. They are mandatory for the diagnosis of autoimmune liver diseases and contribute to therapeutic efficacy and safety. Specialized analyses contribute to the definition and recognition of autoimmune liver diseases and to the diagnosis of virus-associated serological autoimmunity. The distinction of serological autoimmunity and true autoimmune disease is the most important task confronting the hepatologist in defining the patient's disease and in initiating treatment. As epitope mapping experiments clearly show, autoantibodies remain to represent one of the few pathophysiological windows permitting insights into the processes underlying autoimmune liver diseases.

References

1. Manns MP, Strassburg CP. Autoimmune hepatitis: clinical challenges. Gastroenterology. 2001;120:1502–17.
2. Strassburg CP, Obermayer-Straub P, Manns MP. Autoimmunity in liver diseases. Clin Rev Allergy Immunol. 2000;18:127–39.
3. Agnello V, Chung RT, Kaplan L. A role of hepatitis C virus infection in type II cryoglobulinemia. N Engl J Med. 1992;19:1490.
4. Cacoub P, Lunel-Fabiani F, Huong Du LT. Polyarteritis nodosa and hepatitis C infection. Ann Intern Med. 1992;116:605–6.
5. Haddad J, Deny P, Munz-Gotheil C, Pasero G, Bombardieri S, Highfield P. Lymphocytic sialadenitis of Sjögren's syndrome associated with chronic hepatitis c virus liver disease. Lancet. 1992;339:321–3.
6. Johnson RJ, Gretch DR, Yamabe C et al. Membranoproliferative glomerulonephritis associated with hepatitis C virus infection. N Engl J Med. 1993;18:465–70.
7. Yamamoto AM, Cresteil D, Homberg JC, Alvarez F. Characterization of anti-liver–kidney microsome antobody (anti-LKM1) from hepatitis C virus-positive and -negative sera. Gastroenterology. 1993;104:1762–7.
8. Ma Y, Peakman M, Lobo-Yeo A et al. Differences in immune recognition of cytochrome P4502D6 by liver kidney microsomal (LKM) antibody in autoimmune hepatitis and chronic hepatitis C virus infection. Clin Exp Immunol. 1994;97:94–9.
9. Muratori L, Lenzi M, Ma Y et al. Heterogeneity of liver/kidney microsomal antibody type 1 in autoimmune hepatitis and hepatitis C virus related liver disease. Gut. 1995;37:406–12.
10. Yamamoto AM, Johanet C, Duclos-Vallee JC et al. A new approach to cytochrome CYP2D6 antibody detection in autoimmune hepatitis type-2 (AIH-2) and chronic hepatitis C virus (HCV) infection: a sensitive and quantitative radioligand assay. Clin Exp Immunol. 1997;108:396–400.
11. Durazzo M, Philipp T, Van Pelt FN et al. Heterogeneity of liver–kidney microsomal autoantibodies in chronic hepatitis C and D virus infection. Gastroenterology. 1995;108: 455–62.
12. Sugimura T, Obermayer-Straub P, Kayser A et al. A major CYP2D6 autoepitope in autoimmune hepatitis type 2 and chronic hepatitis C is a three-dimensional structure homologous to other cytochrome P450 autoantigens. Autoimmunity. 2002;35:501–13.
13. Todros L, Touscoz G, D'Urso N et al. Hepatitis C virus-related chronic liver disease with autoantibodies to liver–kidney microsomes (LKM). Clinical characterization from idiopathic LKM-positive disorders. J Hepatol. 1991;13:128–31.
14. Dalekos GN, Wedemeyer H, Obermayer-Straub P et al. Epitope mapping of cytochrome P4502D6 autoantigen in patients with chronic hepatitis C during alpha-interferon treatment. J Hepatol. 1999;30:366–75.
15. Dalekos GN, Obermayer Straub P, Maeda T, Tsianos EV, Manns MP. Antibodies against cytochrome P4502A6 (CYP2A6) in patients with chronic viral hepatitis are mainly linked to hepatitis C virus infection. Digestion. 1998;59:S36.
16. Zimmerman HJ, Lewis JH, Ishak KG, Maddrey W. Ticrynafen-associated hepatic injury: analysis of 340 cases. Hepatology. 1984;4:315–23.
17. Beaune PH, Pessayre D, Dansette P, Mansuy D, Manns MP. Autoantibodies against cytochromes P450: role in human diseases. Adv Pharmacol. 1994;30:199–245.
18. Homberg JC, Abuaf N, Helmy-Khalil S et al. Drug induced hepatitis associated with anticytoplasmic organelle autoantibodies. Hepatology. 1985;5:722–5.

19. Obermayer-Straub P, Manns MP, editors. Autoimmunity in the Autoimmune Polyglandular Syndrome Type 1 After the Discovery of the Gene. Dordrecht: Karger, 2000.
20. Strassburg CP, Obermayer-Straub P, Manns MP. Autoimmunity in hepatitis C and D virus infection. J Viral Hepat. 1996;3:49–59.
21. Clemente MG, Obermayer-Straub P, Meloni A et al. Cytochrome P450 1A2 as the hepatocellular autoantigen in autoimmune polyendocrine syndrome type 1. J Hepatol. 1995;23:126.
22. Clemente MG, Meloni A, Obermayer-Straub P, Frau F, Manns MP, DeVirgiliis S. Two Cytochromes P450 are major hepatocellular autoantigens in autoimmune polyglandular syndrome type 1. Gastroenterology. 1998;114:324–8.
23. Gebre-Medhin G, Husebye ES, Gustafsson J et al. Cytochrome P450IA2 and aromatic L-amino acid decarboxylase are hepatic autoantigens in autoimmune polyendocrine syndrome type I. FEBS Lett. 1997;412:439–45.
24. Bourdi M, Larrey D, Nataf J et al. Anti-liver endoplasmic reticulum antibodies are directed against human cytochrome P-450IA2. A specific marker of dihydralazine hepatitis. J Clin Invest. 1990;85:1967–73.
25. Manns MP, Griffin KJ, Quattrochi L et al. Identification of cytochrome P450 IA2 as a human autoantigen. Arch Biochem Biophys. 1990;280:229–32.
26. Sacher M, Blümel P, Thaler H, Manns M. Chronic active hepatitis associated with vitiligo, nail dystrophy, alopecia and a new variant of LKM antibodies. J Hepatol. 1990;10:364–9.
27. Bourdi M, Gautier JC, Mircheva J et al. Anti-liver microsomes autoantibodies and dihydralazine induced hepatitis: specificity of autoantibodies and inductive capacity of the drug. Mol Pharmacol. 1992;42:280–5.
28. Clemente MG, Obermayer-Straub P, Meloni A et al. Cytochrome P450 1A2 is a hepatic autoantigen in autoimmune polyglandular syndrome type 1. J Clin Endocrinol Metab. 1997;82:1353–61.
29. Van Pelt F, Straub P, Manns M. Molecular basis of drug-induced immunological liver injury. Semin Liver Dis. 1995;15:283–300.
30. Crivelli O, Lavarini C, Chiaberge E et al. Microsomal autoantibodies in chronic infection with HBsAg associated delta (delta) agent. Clin Exp Immunol. 1983;54:232–8.
31. Philipp T, Durazzo M, Trautwein C et al. Recognition of uridine diphosphate glucuronosyl transferases by LKM-3 antibodies in chronic hepatitis D. Lancet. 1994;344:578–81.
32. Strassburg CP, Obermayer-Straub P, Alex B et al. Autoantibodies against glucuronosyltransferases differ between viral hepatitis and autoimmune hepatitis. Gastroenterology. 1996;111:1576–86.
33. Strassburg CP, Kneip S, Topp J et al. Polymorphic gene regulation and interindividual variation of UDP-glucuronosyltransferase activity in human small intestine. J Biol Chem. 2000;275:36164–71.
34. Tukey RH, Strassburg CP. Genetic multiplicity of the human UDP-glucuronosyltransferases and regulation in the gastrointestinal tract. Mol Pharmacol. 2001;59:405–14.
35. Kenna JG, Neuberger JM. Immunopathology and treatment of halothane hepatitis. Clin Immunother. 1995;3:108–24.
36. Kenna JG. Immunoallergic drug-induced hepatitis: lessons from halothane. J Hepatol. 1997;26:5–12.
37. Pohl LR, Pumford NR, Martin JL. Mechanisms, chemical structures and drug metabolism. Eur J Haematol. 1996;57:98–104.
38. Pessayre D. Toxic and immune mechanisms leading to acute and subacute drug induced liver injury. In: Miguet JP, Dhumeaux D, editors. Progress in Hepatology. Paris: John Libbey Eurotext, 1993:23–39.
39. Beaune P, Dansette PM, Mansuy D et al. Human anti-endoplasmic reticulum autoantibodies appearing in a drug-induced hepatitis are directed against a human liver cytochrome P-450 that hydroxylates the drug. Proc Natl Acad Sci USA. 1987;84:551–5.
40. Eliasson E, Kenna G. Cytochrome P450 2E1 is a cell surface autoantigen in halothane hepatitis. Mol Pharmacol. 1996;50:573–82.
41. Leeder JS, Riley RJ, Cook VA, Spielberg SP. Human anti-cytochrome P450 antibodies in aromatic anticonvulsant-induced hypersensitivity reactions. J Pharmacol Exp Ther. 1992; 263:360–7.

42. Albano E, Clot P, Morimoto M, Tomasi A, Ingelman-Sundberg M, French SW. Role of cytochrome P450 2E1-dependend formation if hydroxyethyl free radical in the development of liver damage in rats intragastrically fed with ethanol. Hepatology. 1996;23:155–63.

43. Clot P, Albano E, Eliasson E et al. Cytochrome P450 2E1 hydroxyethyl radical adducts as the major antigen in autoantibody formation among alcoholics. Gastroenterology. 1996; 111:206–16.

Section IV
Pathogenesis II

Chair: I.R. MACKAY and R. POUPON

10
Induction and destruction phases of primary biliary cirrhosis

T. K. MAO and M. E. GERSHWIN

INTRODUCTION

Primary biliary cirrhosis (PBC) is regarded as a model organ-specific auto-immune disease with pathology concentrated in the liver. Middle-aged women are primarily affected, and they tend to endure a chronic and progressive state of inflammation that leads to the destruction of their intrahepatic bile ducts. The serological hallmark of PBC is antimitochondrial antibodies (AMA) that recognize E2 components of 2-oxoacid dehydrogenase enzymes, with pyruvate dehydrogenase complex-E2 (PDC-E2) being the immunodominant autoantigen. The long-standing paradox in PBC is that all of these mitochondrial proteins are present in nucleated cells but the autoimmune attack is highly specific for the biliary epithelium.

In the past several decades significant advances have been made with regard to the characterization of the destruction phase of PBC, including histopathological analysis, as well as identification of the autoepitopes in PBC. Epidemiological studies have shown that the prevalence of PBC ranges from 5 to 402 per million, with the frequency of a family member developing disease significantly higher (100–1000-fold) than the general population. More recent analysis of disease concordance rates in monozygotic twins has reinforced the existence of a genetic component on susceptibility to PBC. One intriguing hypothesis regarding the induction of PBC is that environmental exposure of xenobiotic chemicals may alter self-proteins enough to break tolerance. Certainly, PBC is a complex autoimmune disease with genetic and environmental factors representing two critical aspects in uncovering the onset of disease. In this chapter present data regarding the induction and destruction phases of PBC will be discussed. In addition, we will also share our view on the potential impact oxidative stress on the immunogenicity of PDC-E2.

INDUCTION PHASE OF PBC

Genetic component

Since the early 1970s numerous case reports identifying family members with PBC have suggested that disease prevalence is increased within first-degree relatives[1,2]. Such isolated studies, which have hinted at a genetic component in the development of PBC, have subsequently been confirmed in comprehensive, geo-epidemiological studies of familial PBC. Cumulatively, the data from these studies displayed a familial prevalence of PBC in the range of 1–6%, suggesting that family members of PBC patients are much more likely to develop the disease than the general population, in which the proportion is reported to be in the neighbourhood of 5–402 per million[1,3].

The classical epidemiological strategy to delineate any genetic influences in the development of multifactorial diseases is through the study of concordance rates in twins. In most human autoimmune diseases the concordance rates in monozygotic twins (i.e. genetically identical) are generally less than 50%[4]. Most recently our laboratory has been able to identify 16 sets of twins, eight of which were confirmed to be monozygotes. Upon subsequent re-evaluation of patient charts the concordance rate in the monozygotic twins was 0.63 (five sets) in which both individuals were diagnosed with PBC, while none of the dizygotic sets was found concordant for PBC[5]. Hence, the concordance rate for PBC in identical twins is among the highest reported in autoimmune diseases. The dramatic discrepancy in the concordance rate of dizygotic twins compared with that of monozygotic twins supports the presence of multiple genes contributing to the genetic predisposition of PBC.

Infectious aetiology?

The fact that PBC patients have unusually high titres of autoantibodies directed to the mitochondria, even before the appearance of pathology, remains an anomaly that continues to perplex PBC investigators. Since it is believed that mitochondria in mammalian cells originated from aerobic bacteria, the favoured theory of tolerance breakdown involves some form of a mimic elicited by an infectious bacterial agent. Support of an infectious aetiology was demonstrated when several studies reported a high prevalence of bacteriuria with very frequent recurrence rates in females with PBC compared to other forms of chronic liver diseases[6–8]. AMA crossreact with a number of bacterial mitochondria, including *Escherichia coli*[9,10]. Due to the highly conserved nature of pyruvate dehydrogenase complex (PDC), the sequence of *E. coli* PDC-E2 contains similar lipoyl domains that are critical in the recognition of human PDC[9]. Many studies that have suggested an infectious aetiology merely involve correlating the presence of infectious agents with PBC via either serological examinations or amplification of microbial genes within liver tissue. Although a number of bacteria were found in livers and bile from patients with PBC[11,12], to date no consensus bacterium has been specifically associated with PBC. It may very well be that PBC patients display an aberrant immune response to non-specific bacterial infections that initiate or participate in the development of PBC. Such inappropriate responses of the innate immune system may explain the tendency of persistent urinary tract infections in females with PBC.

Chronic viral infection of the biliary epithelium has also been considered as the culprit for the induction of PBC. It has been argued that the antimitochondrial immune response is appropriately directed towards biliary epithelial cells that are aberrantly expressing a viral mimic with significant sequence homology to PDC-E2, while ignoring healthy tissues[13]. In particular, a recent study detected retroviral antibodies in PBC patients[14]. However, this reactivity failed to be disease-specific as sera from other biliary diseases also presented some reactivity against viral proteins. In a following investigation researchers from this laboratory also demonstrated the presence of a betaretrovirus in the liver and lymph nodes of some patients with PBC[15]. Moreover, the culture of normal biliary epithelial cells (BEC) in the presence of a homogenate of infected lymph nodes was shown to induce the expression of a PDC-E2-like antigen on the cell membrane[15]. However, it is important to note that our laboratory recently failed to detect any immunohistochemical or molecular evidence of mouse mammary tumour virus in lymphocytes or peripheral blood lymphocytes[16]. Our inability to recapitulate the data regarding the retroviral aetiology of PBC suggests that results of the previous study may be due to contamination.

Xenobiotic trigger?

One potential hypothesis regarding the aetiology of PBC is that xenobiotic exposure may conjugate to self-proteins and induce a change in the molecular structure of the native protein sufficient to induce an immune response. Such immune responses may then result in the recognition not only of the modified protein, but also the native form that can lead to the breakdown of tolerance. Hence, in the case of PBC, the ubiquitous nature of mitochondrial proteins may perpetuate the autoimmune response initiated by the xenobiotic-induced adduct leading to chronic inflammation. What makes this more compelling is that many xenobiotics are metabolized in the liver, and therefore the potential exists for liver-specific alteration of proteins. Table 1 shows the purported xenobiotic compounds known, to date, to have been associated with various autoimmune conditions.

Gershwin and co-workers have systematically modified a peptide encompassing the inner lysine-lipoyl domain of PDC-E2 with 18 different xenobiotic structures conjugated to the lysine residue in place of lipoic acid[17]. Sera from PBC patients were shown to react against three of these xenobiotic-modified autoepitopes significantly better than to the native domain[17]. Subsequently, immunization of rabbits with one of these compounds, 6-bromohexanoate, induced antimitochondrial antibodies (AMA) with reactivity against PDC-E2[18]. Although this model recognized the breakdown of tolerance at the serological level, histopathological features characteristic of PBC observed in humans were not reproduced. Such failure to produce PBC-like lesions within the liver lends support to the common notion that PBC is a complex disease that requires a genetic predisposition for pathogenesis. Moreover, the evalua-

Table 1 Role of xenobiotics in autoimmune conditions

Compound	Associated autoimmune condition	References
Mercury	Immune complex formation, glomerulonephritis	76, 77
Iodine	Autoimmune thyroiditis	78
Vinyl chloride	Scleroderma-like disease	79
Contaminated L-tryptophan	Eosinophilia myalgia syndrome	80, 81
Toxic oil	Scleroderma-like disease	76, 82
Silica	Rheumatoid arthritis, systemic lupis erythematosus (SLE), scleroderma	83
Halothane	Autoimmune hepatitis	84, 85
Convanine	SLE-like syndrome	86
6-Bromohexanoate	PBC	18

tion of rabbit T cell responses to 6-bromohexanoate was not pursued in this study. Nevertheless, existing data on the serological recognition of xenobiotics warrant further studies examining the capacity of xenobiotic-modified PDC-E2 to induce T cell activation in humans.

DESTRUCTION PHASE OF PBC

The serological hallmark of patients with PBC is that they exhibit high titres of AMA, with the immunodominant response directed at PDC-E2. Although this autoantigen is present throughout the body, histological examinations have characterized PBC as a chronic inflammatory response with lymphocytic infiltrates in the portal tracts of the liver that precipitate the destruction of the biliary epithelium. Naturally, autoimmune mechanisms have been implicated in the pathogenesis of this disease, yet the intimate roles of AMA, T cells, as well as BEC in the immunopathology of PBC remain speculative. While PDC-E2 is the major autoantigen in PBC, other members of the 2-oxoacid dehydrogenase complexes (2-OADC) are also targets of AMA, including the E2 subunit of the branched chain 2-oxoacid dehydrogenase complex (BCOADC-E2), the E2 subunit of the 2-oxoglutarate dehydrogenase complex (OGDC-E2), and the E3BP subunit of PDC[19–25] (Table 2).

Humoral immune responses

It is well established that PBC is characterized by the presence of autoantibodies that are reactive with both mitochondrial and nuclear antigens. The most common autoantibodies present in patients are directed against the lipoyl domains of PDC, BCOADC, and OGDC. Mitochondrial antigens are most frequently recognized with PDC-E2 present in over 90% of patient sera, followed by BCOADC-E2 (53–55%) and OGDC-E2 (39–88%)[19–25]. Nuclear autoantigens have also been thoroughly studied; however, antinuclear auto-antibodies (ANA) are unquestionably less common reactants. ANA are present in approximately 60% of patient sera with no significant difference in pre-

Table 2 Amino acid sequence of flanking lipoyl-lysine residues in the four mitochondrial proteins containing the lipoic acid binding domain

Autoantigen	Amino acid sequences													
PDC-E2 (inner lipoyl domain)	G	D	L	L	A	E	I	E	T	D	K*	A	T	I
PDC-E2 (outer lipoyl domain)	G	D	L	I	A	E	V	E	T	D	K*	A	T	V
E3BP	G	D	A	L	C	E	I	E	T	D	K*	A	V	V
OGDC-E2	D	E	V	V	K	E	I	E	T	D	K*	T	S	V
BCOADC-E2	F	D	S	O	C	E	V	Q	S	D	K*	A	S	V

*Specific lysine residue where lipoic acid is attached via an amide bond.

valence between AMA-positive and AMA-negative patients[26,27]. Nevertheless, ANA are clinically relevant in aiding diagnosis of AMA-negative cirrhosis exhibiting PBC-like pathology.

Interestingly, AMA are able to inhibit the enzymatic activity of PDC-E2, and this prompted investigators to ponder how these autoantibodies could react with an intracellular antigen in a manner influential to cell viability. This led to the development of the IgA hypothesis that attempts to explain the role of AMA in the disease process and concomitantly address the issue of tissue specificity. Polyimmunoglobulin receptor is known to be restricted to the basal side of epithelial surfaces, including biliary epithelial cells. Polymeric antibodies of IgA and IgM isotypes normally bind to the receptor and are internalized, transported through the cell and delivered to the mucosal surfaces of the body. It is therefore possible that polymeric AMA can have access to intracellular mitochondrial antigens during transcytosis and disrupt the metabolism of BEC. As a result, normal cellular turnover of the epithelium is disturbed and the immune system is subsequently exposed to an increased amount of self-mitochondrial antigens that can perpetuate the autoimmune response. Support of the IgA hypothesis was provided with the demonstration that IgA from patients co-localized with PDC-E2 inside an epithelial cell line[28]. In addition, it is well accepted that patients with PBC display a significant increase in immunoglobulin levels, particularly IgM[29–31]. Concentrations of monomeric IgA, polymeric IgA, and those of secretory IgA and IgM are also increased relative to normal and cholestatic liver disease controls[32,33]. These data support the idea that the trafficking of polymeric AMA through BEC facilitates the specific damage of the biliary epithelium observed in PBC.

Aberrant expression of PDC-E2 on the surface of BEC has also been proposed to facilitate its recognition by AMA, leading to biliary destruction via antibody-dependent cell cytotoxity. In support of this hypothesis, immunohistochemical analysis revealed that high-intensity staining of an antigen was recognized by anti-PDC-E2 antibodies specifically in BEC of patients[34,35]. It was subsequently shown, through the analysis of AMA binding to membrane fraction of purified BEC, that the antigen may be the E3BP, rather than PDC-E2[36]. Another possible route for the recognition of this antigen is the release of PDC onto the cell surface blebs during apoptosis. Recently, one study demonstrated that immunoreactive PDC was detected on the outer surface of the plasma membrane of cells undergoing apoptosis[37].

Extensive studies have mapped the autoepitopes recognized by B cells through the use of recombinant proteins, truncated constructs, and a combination of peptides. These studies have demonstrated that the immune reactivity of AMA is directed against a conformational epitope that includes a lipoyl domain[38-42]. These domains contain signature motifs of amino acids ETDKA, ETDK(T), and (QS)DKA wherein the lipoic acid is covalently attached to the ε group of lysine (K) via an amide bond. While all four mitochondrial autoantigens contain at least one lipoic acid domain, PDC-E2 has two (inner and outer) lipoyl domains. Of note, AMA react with the outer lipoyl domain (OLD) (amino acid sequence 1–90) of PDC-E2; however, they do so at a significantly lower (100-fold) titre. This implies that the predominant epitope lies within the inner lipoyl domain (ILD), specifically at residues 128–221 (93 amino acids)[22,38] (Table 3). Interestingly, the AMA response to lipoylated recombinant PDC-E2 is of substantially higher titre and affinity than to the unlipoylated antigen[43], suggesting that lysine-lipoate may be a part of the immunodominant epitope.

Table 3 B cell and T cell epitopes for PDC-E2

Epitope	Amino acid position	References
B-cell	1–90 (OLD), 128–221 (ILD)	22, 38
CD4 T-cell	163–176 (ILD)	52–54
CD8 T-cell	159–167 (ILD)	56

Cellular immune responses

The presence of lymphocytic infiltration in the portal tracts, coupled with the aberrant expression of MHC class II antigen on the surface of the biliary epithelium[44], have suggested that an intense autoimmune response is directed against BEC. It has been hypothesized that the destruction of biliary cells in PBC is mediated by autoreactive liver-infiltrating T cells through either cytotoxicity or cytokine production. Initial studies by Meuer et al. demonstrated that there was a marked enrichment of CD8$^+$ cytotoxic T cells in lymphocytes isolated from the liver of patients with PBC compared to peripheral blood[45]. Moreover, the T cells that have infiltrated around bile ducts are predominantly CD4$^+$ TCR $\alpha\beta^+$ with the CD4$^+$/CD8$^+$ ratio of 2–2.5[46]. However, some studies have indicated that CD8$^+$ T lymphocytes are particularly prominent in the liver[47].

Perhaps more important than phenotyping the infiltrating T lymphocytes is the identification of autoreactive T cell lines that proliferate in the presence of putative mitochondrial autoantigens. An investigation by Van de Water et al. demonstrated, for the first time, that T cell lines that reacted with PDC-E2 and

BCOADC-E2 were present in the liver of patients with PBC, but not in those of chronic active hepatitis controls[48]. Peripheral blood mononuclear cells (PBMC) from over 70% of patients with PBC showed a HLA class II-restricted proliferative response to PDC but not in controls[49]. More recently, Akbar and colleagues have discovered that the frequency is higher if dendritic cells are loaded with PDC prior to culture with PBMC; 100% of AMA-positive patients responded, as well as 12% of negative patients[50]. More detailed studies have revealed that the autoreactive T cell clones specific for PDC-E2 are heterogeneous with respect to TCR Vβ usage and directed to both the inner and outer lipoyl domains[51]. Subsequent epitope mapping studies have identified a dominant HLA DR4*0101 restricted T cell epitope spanning residues 163–176 of PDC-E2 and encompassing the ILD, the same region recognized by AMA[52–54] (Table 3). Furthermore, the CD4[+] T cell lines also reacted with the outer lipoyl domain of PDC-E2, as well as the lipoyl domain of OGDC-E2[55]. Frequency analysis indicates that such autoreactive T cells are sequestered in the liver compartment as opposed to the peripheral blood[55].

Lysis of BEC is considered a major effector mechanism by which the biliary epithelium is destroyed; therefore it is particularly important to characterize the autoreactive CD8[+] T cell response. Recently, the Gershwin laboratory has identified a HLA-A2 restricted epitope of PDC-E2 that was recognized by CD8[+] T cells derived from peripheral blood and liver of patients with PBC[56–58]. Similarly to the aforementioned CD4[+] T cells, the dominant autoepitope spans the lipoic acid-binding residue of the ILD of PDC-E2 (residues 159–167) and PDC-E2 specific CD8[+] T cells are enriched in the liver[56] (Table 3). More importantly, autoreactive CD8[+] T cells isolated from the liver of PBC patients display cytotoxic activity against autoepitope (amino acids 159–167) pulsed autologous cells, suggesting that they have at least the potential to be responsible for BEC cytotoxicity in vivo. In addition, these HLA class I-restricted T cells could be stimulated through cross-presentation of soluble PDC-E2, the efficiency of which is increased 10-fold when using immune complexes composed of PDC-E2 and affinity-purified antibodies from PBC patients.

It is important to note that the peptides used to elicit T cell responses in these studies are unlipoylated at the lysine residue. Hence, it would be interesting to examine the effects of post-translationally modified peptides on the stimulation of T cells. Since lipoylation of PDC-E2 enhanced its recognition by AMA, it is plausible that lipoic acid modification of the T cell epitopes can increase the binding affinity of peptide to MHC molecules, which can augment T cell responses. Future studies regarding the three-dimensional analysis of the interaction of lysine-derivatized PDC-E2 peptide with MHC molecules should be performed.

CD4[+] T helper (Th) cells are known to exert their effector mechanisms by releasing a dichotomy of cytokines that influences the adaptive immune immune response and hence are characterized as either Th1 or Th2 cells. The concept the Th1/Th2 paradigm suggests that Th1 cells secrete IL-2 or IFN-γ conducive for cellular immunity, while Th2 cells release IL-4, IL-5, IL-6, and IL-13 to promote humoral immunity[59]. Organ-specific autoimmune diseases, such as experimental autoimmune encephalomyelitis (EAE; an animal model for multiple sclerosis) and type 1 diabetes, have been generally regarded as a cellular-mediated disease with Th1 cells providing an appropriate cytokine milieu conducive for tissue damage mediated by autopathogenic T cells[60,61]. With regard to PBC, many studies have attempted to phenotypically characterize CD4$^+$ T cells by profiling their cytokine production. Berg et al. investigated the cytokine profiles in supernatants from non-stimulated PBMC of PBC patients and healthy controls, and discovered a shift in the Th1/Th2 ratio towards Th1 in patients (1:1) as compared to healthy controls (1:3)[62]. Van de Water et al. also found that CD4$^+$ T cell clones specific for PDC-E2 derived from PBC livers secreted either IL-2 or IFN-γ, but not Th2 cytokines, such as IL-4[51]. Another study using *in-situ* hybridization demonstrated that IFN-γ mRNA expression was more frequently detected than IL-4 expression in cirrhotic liver sections and the levels of IFN-γ correlated with the degree of bile duct destruction[63]. The predominance of a Th1-type cytokine profile was later supported in a study whereby, using RT-PCR, Nagano et al. showed high expression of IFN-γ and low levels of IL-10 (Th2 cytokine) from whole liver RNA[64].

Clustering of T cell and B cell autoepitopes

Standard immunology suggests that there exists an intimate relationship between B cells, CD4 Th cells, and CD8 cytotoxic lymphocytes (CTL) to provide an effective immune response. In particular, CD4 Th cells can offer critical support for both humoral and cellular-mediated responses. The clonal differentiation of these different subsets of lymphocytes must recognize the same antigen to provide an orchestrated immune response against the antigen. However, according to conventional mechanisms of antigen presentation, it is hard to imagine that all target antigens are present both endogenously and exogenously to generate clones of B cell, CD4 Th cells, and CD8 CTL.

With this in mind it was interesting to discover that PBC patients respond to a particular region of PDC-E2. This immunodominant region contains the inner lipoyl domain with the B cell, CD4 T cell, and CD8 T cell autoepitopes for PDC-E2 partially overlapping. The clustering of immunodominant T cell and B cell epitopes of PDC-E2 is not unusual, as a similar theme is observed with autoantigens in other autoimmune diseases, including multiple sclerosis[65,66] and type 1 diabetes mellitus[67,68]. According to conventional antigen presentation pathways, CD4 T cells recognize epitopes derived from exogenous proteins, whereas CD8 T cells are stimulated by endogenously derived antigens. Therefore, the antigen sources for the induction of CD4 and CD8 T cells are theoretically generated through different pathways, and may not even be derived from the same polypeptide molecule to explain the clustering of T cell and B cell epitopes. However, recent work from the Gershwin laboratory suggests that exogenous PDC-E2, possibly derived from apoptotic cellular turnover, may be presented to both CD4 and CD8 T cells. Kita et al. have shown that the induction of CD8 T cells is augmented by PDC-E2–autoantibody immune complexes (IC) cross-presented by dendritic cells[56]. Such autoantibodies that are bound to the immunodominant B cell epitope will also encompass the CD4 and CD8 T cell epitopes, and protect them from

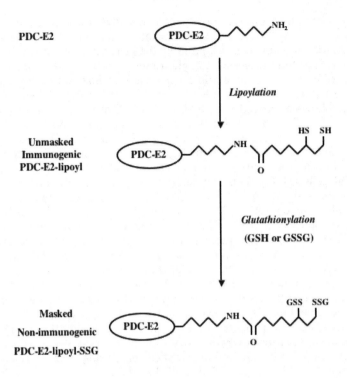

Figure 1 The role of the glutathionylation system on the immunogenicity of PDC-E2

degradation by proteases once the IC are efficiently internalized by dendritic cells. In this situation the immunodominant CD4 T cell epitope my be presented by conventional MHC class II pathways to CD4 T cells, whereas the CD8 epitope may be presented by MHC class I through cross-presentation[69].

The role of oxidative stress in the immunogenicity of PDC-E2

PBC is characterized by chronic cholestasis in which oxidative stress is non-specifically associated with chronic inflammation that leads to the generation of reactive oxygen species. Several studies examining the intracellular status of oxidative stress have shown that 4-hydroxynonenal, a by-product of lipid peroxidation, was immunohistochemically detected in the bile ducts of PBC patients[70–72]. Other markers for lipid peroxidation (i.e. 8-isoprostane and malondialdehyde) were also observed to be significantly elevated in plasma as compared to age- and sex-matched controls[73].

Despite these observations the role of oxidative stress specific to the pathogenesis of PBC remains unclear. However, one line of research has shown that AMA appeared to preferentially recognize PDC-E2 with reduced sulphydryl groups[74], suggesting redox levels may alter the immunogenicity of mitochondrial autoantigens and promote the generation of autoantibodies. In particular, one study suggested that glutathione levels may affect the immunogenicity of PDC-E2[75]. The authors demonstrated that, following apoptosis, HeLa, Jurkat T, and Caco-2 cells presented a form of PDC-E2 that becomes undetectable when probed with AMA. This loss of recognition of apoptotic cell-derived PDC-E2 by AMA is not due to disappearance or degradation, but rather a reversible structural change in the protein from which the epitope can be recovered. However, most notably, autoantibody recognition of PDC-E2 persists in apoptotic rat cholangiocytes, as well as human salivary gland epithelial cells (HSG). This immunoreactivity can be eliminated with the addition of oxidized glutathione (GSSG) to SDS-treated cholangiocyte cell lysates which render PDC-E2 non-antigenic when probed with AMA. Conversely, apoptotic HeLa cells pretreated with BSO (buthionine sulphoximine; a glutathione synthetase inhibitor), prior to apoptotic stimulus were unable to prevent the recognition by patient sera, thereby 'resurrecting' the antigen with the depletion of glutathione. The authors therefore concluded that the persistent recognition of PDC-E2 derived from cholangiocytes and HSG is due to a failure in these cells to covalently link PDC-E2 to glutathione during the course of apoptosis. These data indicate that the epithelial cells most frequently affected in patients with PBC are unique, in that they do not block or mask the immunoreactivity of AMA.

The use of glutathionylation as a critical processing event in determining recognition by AMA suggests that, while lipoylation of the lysines in PDC-E2 is required for enzymatic activity and predominates under normal conditions, the trade-off is exposing the immune system to a more immunogenic antigen than the unlipoylated or octanoylated PDC-E2 (Figure 1). Therefore, it appears that proteins containing lysine-lipoate must be highly regulated in order to prevent the inappropriate release of these proteins.

CONCLUSION

In the past several decades PBC research has primarily focused on the identification of autoantigens and the mapping of their autoepitopes. Epidemiological studies have hinted at a genetic component in the predisposition of disease, yet identification of pertinent genetic polymorphisms associated with PBC has proven to be elusive. Genetic diversity of the human population, as well as the exposure to various environmental factors (i.e. xenobiotics) have also complicated our understanding of PBC. Finally, oxidative stress can represent a specific feature observed in PBC by which depletion of glutathione stores may suppress the only purported regulatory system that prevents the accumulation of potentially autoreactive PDC-E2. Hence, the failure of this or other regulatory system(s) may overwhelm the immune system with immunogenic PDC-E2 that can initiate the breakdown of tolerance in a genetically susceptible individual.

References

1. Tsuji K, Watanabe Y, Van De Water J et al. Familial primary biliary cirrhosis in Hiroshima. J Autoimmun. 1999;13:171–8.
2. Tanaka A, Borchers AT, Ishibashi H, Ansari AA, Keen CL, Gershwin ME. Genetic and familial considerations of primary biliary cirrhosis. Am J Gastroenterol. 2001;96:8–15.
3. Kim WR, Lindor KD, Locke GR 3rd et al. Epidemiology and natural history of primary biliary cirrhosis in a US community. Gastroenterology. 2000;119:1631–6.
4. Gregersen PK. Discordance for autoimmunity in monozygotic twins. Are 'identical' twins really identical? Arthritis Rheum. 1993;36:1185–92.
5. Selmi C, Mayo MJ, Bach N et al. Primary biliary cirrhosis in monozygotic and dizygotic twins: genetics, epigenetics, and environment. Gastroenterology. 2004;127:485–92.
6. Burroughs AK, Rosenstein IJ, Epstein O, Hamilton-Miller JM, Brumfitt W, Sherlock S. Bacteriuria and primary biliary cirrhosis. Gut. 1984;25:133–7.
7. Butler P, Valle F, Hamilton-Miller JM, Brumfitt W, Baum H, Burroughs AK. M2 mitochondrial antibodies and urinary rough mutant bacteria in patients with primary biliary cirrhosis and in patients with recurrent bacteriuria. J Hepatol. 1993;17:408–14.
8. Butler P, Hamilton-Miller J, Baum H, Burroughs AK. Detection of M2 antibodies in patients with recurrent urinary tract infection using an ELISA and purified PBC specific antigens. Evidence for a molecular mimicry mechanism in the pathogenesis of primary biliary cirrhosis? Biochem Mol Biol Int. 1995;35:473–85.
9. Fussey SP, Ali ST, Guest JR, James OF, Bassendine MF, Yeaman SJ. Reactivity of primary biliary cirrhosis sera with *Escherichia coli* dihydrolipoamide acetyltransferase (E2p): characterization of the main immunogenic region. Proc Natl Acad Sci USA. 1990;87: 3987–91.
10. Baum H. Mitochondrial antigens, molecular mimicry and autoimmune disease. Biochim Biophys Acta. 1995;1271:111–21.
11. Tanaka A, Prindiville TP, Gish R et al. Are infectious agents involved in primary biliary cirrhosis? A PCR approach. J Hepatol. 1999;31:664–71.
12. Hiramatsu K, Harada K, Tsuneyama K et al. Amplification and sequence analysis of partial bacterial 16S ribosomal RNA gene in gallbladder bile from patients with primary biliary cirrhosis. J Hepatol. 2000;33:9–18.
13. Sutton I, Neuberger J. Primary biliary cirrhosis: seeking the silent partner of autoimmunity. Gut. 2002;50:743–6.
14. Mason AL, Xu L, Guo L et al. Detection of retroviral antibodies in primary biliary cirrhosis and other idiopathic biliary disorders. Lancet. 1998;351:1620–4.
15. Xu L, Shen Z, Guo L et al. Does a betaretrovirus infection trigger primary biliary cirrhosis? Proc Natl Acad Sci USA. 2003;100:8454–9.
16. Selmi C, Ross SR, Ansari AA et al. Lack of immunological or molecular evidence for a role of mouse mammary tumor retrovirus in primary biliary cirrhosis. Gastroenterology. 2004; 127:493–501.
17. Long SA, Quan C, Van de Water J et al. Immunoreactivity of organic mimeotopes of the E2 component of pyruvate dehydrogenase: connecting xenobiotics with primary biliary cirrhosis. J Immunol. 2001;167:2956–63.
18. Leung PS, Quan C, Park O et al. Immunization with a xenobiotic 6-bromohexanoate bovine serum albumin conjugate induces antimitochondrial antibodies. J Immunol. 2003; 170:5326–32.
19. Gershwin ME, Mackay IR, Sturgess A, Coppel RL. Identification and specificity of a cDNA encoding the 70 kd mitochondrial antigen recognized in primary biliary cirrhosis. J Immunol. 1987;138:3525–31.
20. Coppel RL, McNeilage LJ, Surh CD et al. Primary structure of the human M2 mitochondrial autoantigen of primary biliary cirrhosis: dihydrolipoamide acetyltransferase. Proc Natl Acad Sci USA. 1988;85:7317–21.
21. Fussey SP, Guest JR, James OF, Bassendine MF, Yeaman SJ. Identification and analysis of the major M2 autoantigens in primary biliary cirrhosis. Proc Natl Acad Sci USA. 1988;85: 8654–8.
22. Van de Water J, Gershwin ME, Leung P, Ansari A, Coppel RL. The autoepitope of the 74-kD mitochondrial autoantigen of primary biliary cirrhosis corresponds to the functional site of dihydrolipoamide acetyltransferase. J Exp Med. 1988;167:1791–9.

23. Surh CD, Roche TE, Danner DJ et al. Antimitochondrial autoantibodies in primary biliary cirrhosis recognize cross-reactive epitope(s) on protein X and dihydrolipoamide acetyl-transferase of pyruvate dehydrogenase complex. Hepatology. 1989;10:127–33.

24. Fregeau DR, Roche TE, Davis PA, Coppel R, Gershwin ME. Primary biliary cirrhosis. Inhibition of pyruvate dehydrogenase complex activity by autoantibodies specific for E1 alpha, a non-lipoic acid containing mitochondrial enzyme. J Immunol. 1990;144:1671–6.

25. Dubel L, Tanaka A, Leung PS et al. Autoepitope mapping and reactivity of autoantibodies to the dihydrolipoamide dehydrogenase-binding protein (E3BP) and the glycine cleavage proteins in primary biliary cirrhosis. Hepatology. 1999;29:1013–18.

26. Wesierska-Gadek J, Hohenuer H, Hitchman E, Penner E. Autoantibodies against nucleoporin p62 constitute a novel marker of primary biliary cirrhosis. Gastroenterology. 1996; 110:840–7.

27. Nakanuma Y, Harada K, Kaji K et al. Clinicopathological study of primary biliary cirrhosis negative for antimitochondrial antibodies. Liver. 1997;17:281–7.

28. Malmborg AC, Shultz DB, Luton F et al. Penetration and co-localization in MDCK cell mitochondria of IgA derived from patients with primary biliary cirrhosis. J Autoimmun. 1998;11:573–80.

29. Taal BG, Schalm SW, de Bruyn AM, de Rooy FW, Klein F. Serum IgM in primary biliary cirrhosis. Clin Chim Acta. 1980;108:457–63.

30. MacSween RN, Horne CH, Moffat AJ, Hughes HM. Serum protein levels in primary biliary cirrhosis. J Clin Pathol. 1972;25:789–92.

31. Newkirk MM, Klein MH, Katz A, Fisher MM, Underdown BJ. Estimation of polymeric IgA in human serum: an assay based on binding of radiolabeled human secretory component with applications in the study of IgA nephropathy, IgA monoclonal gammopathy, and liver disease. J Immunol. 1983;130:1176–81.

32. Kvale D, Schrumpf E, Brandtzaeg P, Solberg HE, Fausa O, Elgjo K. Circulating secretory immunoglobulins of the A and M isotypes in chronic liver disease. J Hepatol. 1987;4:229–35.

33. Soppi E, Granfors K, Leino R. Serum secretory IgA, IgA1 and IgA2 subclasses in inflammatory bowel and chronic liver diseases. J Clin Lab Immunol. 1987;23:15–17.

34. Joplin R, Lindsay JG, Hubscher SG et al. Distribution of dihydrolipoamide acetyltransferase (E2) in the liver and portal lymph nodes of patients with primary biliary cirrhosis: an immunohistochemical study. Hepatology. 1991;14:442–7.

35. Joplin R, Lindsay JG, Johnson GD, Strain A, Neuberger J. Membrane dihydrolipoamide acetyltransferase (E2) on human biliary epithelial cells in primary biliary cirrhosis. Lancet. 1992;339:93–4.

36. Joplin RE, Wallace LL, Lindsay JG, Palmer JM, Yeaman SJ, Neuberger JM. The human biliary epithelial cell plasma membrane antigen in primary biliary cirrhosis: pyruvate dehydrogenase X? Gastroenterology. 1997;113:1727–33.

37. Macdonald P, Palmer J, Kirby JA, Jones DE. Apoptosis as a mechanism for cell surface expression of the autoantigen pyruvate dehydrogenase complex. Clin Exp Immunol. 2004; 136:559–67.

38. Surh CD, Coppel R, Gershwin ME. Structural requirement for autoreactivity on human pyruvate dehydrogenase-E2, the major autoantigen of primary biliary cirrhosis. Implication for a conformational autoepitope. J Immunol. 1990;144:3367–74.

39. Cha S, Leung PS, Coppel RL, Van de Water J, Ansari AA, Gershwin ME. Heterogeneity of combinatorial human autoantibodies against PDC-E2 and biliary epithelial cells in patients with primary biliary cirrhosis. Hepatology. 1994;20:574–83.

40. Leung PS, Chuang DT, Wynn RM et al. Autoantibodies to BCOADC-E2 in patients with primary biliary cirrhosis recognize a conformational epitope. Hepatology. 1995;22:505–13.

41. Moteki S, Leung PS, Coppel RL et al. Use of a designer triple expression hybrid clone for three different lipoyl domains for the detection of antimitochondrial autoantibodies. Hepatology. 1996;24:97–103.

42. Moteki S, Leung PS, Dickson ER et al. Epitope mapping and reactivity of autoantibodies to the E2 component of 2-oxoglutarate dehydrogenase complex in primary biliary cirrhosis using recombinant 2-oxoglutarate dehydrogenase complex. Hepatology. 1996;23:436–44.

43. Quinn J, Diamond AG, Palmer JM, Bassendine MF, James OF, Yeaman SJ. Lipoylated and unlipoylated domains of human PDC-E2 as autoantigens in primary biliary cirrhosis: significance of lipoate attachment. Hepatology. 1993;18:1384–91.

44. Ballardini G, Mirakian R, Bianchi FB, Pisi E, Doniach D, Bottazzo GF. Aberrant expression of HLA-DR antigens on bile duct epithelium in primary biliary cirrhosis: relevance to pathogenesis. Lancet. 1984;2:1009–13.
45. Meuer SC, Moebius U, Manns MM et al. Clonal analysis of human T lymphocytes infiltrating the liver in chronic active hepatitis B and primary biliary cirrhosis. Eur J Immunol. 1988;18:1447–52.
46. Ishibashi H, Nakamura M, Shimoda S, Gershwin ME. T cell immunity and primary biliary cirrhosis. Autoimmun Rev. 2003;2:19–24.
47. Hoffmann RM, Pape GR, Spengler U et al. Clonal analysis of liver-derived T cells of patients with primary biliary cirrhosis. Clin Exp Immunol. 1989;76:210–15.
48. Van de Water J, Ansari AA, Surh CD et al. Evidence for the targeting by 2-oxo-dehydrogenase enzymes in the T cell response of primary biliary cirrhosis. J Immunol. 1991;146:89–94.
49. Lohr H, Fleischer B, Gerken G, Yeaman SJ, Meyer zum Buschenfelde KH, Manns M. Autoreactive liver-infiltrating T cells in primary biliary cirrhosis recognize inner mitochondrial epitopes and the pyruvate dehydrogenase complex. J Hepatol. 1993;18:322–7.
50. Akbar SM, Yamamoto K, Miyakawa H et al. Peripheral blood T-cell responses to pyruvate dehydrogenase complex in primary biliary cirrhosis: role of antigen-presenting dendritic cells. Eur J Clin Invest. 2001;31:639–46.
51. Van de Water J, Ansari A, Prindiville T et al. Heterogeneity of autoreactive T cell clones specific for the E2 component of the pyruvate dehydrogenase complex in primary biliary cirrhosis. J Exp Med. 1995;181:723–33.
52. Shimoda S, Nakamura M, Ishibashi H, Hayashida K, Niho Y. HLA DRB4 0101-restricted immunodominant T cell autoepitope of pyruvate dehydrogenase complex in primary biliary cirrhosis: evidence of molecular mimicry in human autoimmune diseases. J Exp Med. 1995;181:1835–45.
53. Shigematsu H, Shimoda S, Nakamura M et al. Fine specificity of T cells reactive to human PDC-E2 163-176 peptide, the immunodominant autoantigen in primary biliary cirrhosis: implications for molecular mimicry and cross-recognition among mitochondrial autoantigens. Hepatology. 2000;32:901–9.
54. Shimoda S, Nakamura M, Shigematsu H et al. Mimicry peptides of human PDC-E2 163-176 peptide, the immunodominant T-cell epitope of primary biliary cirrhosis. Hepatology. 2000;31:1212–16.
55. Shimoda S, Van de Water J, Ansari A et al. Identification and precursor frequency analysis of a common T cell epitope motif in mitochondrial autoantigens in primary biliary cirrhosis. J Clin Invest. 1998;102:1831–40.
56. Kita H, Lian ZX, Van de Water J et al. Identification of HLA-A2-restricted CD8(+) cytotoxic T cell responses in primary biliary cirrhosis: T cell activation is augmented by immune complexes cross-presented by dendritic cells. J Exp Med. 2002;195:113–23.
57. Kita H, Matsumura S, He XS et al. Quantitative and functional analysis of PDC-E2-specific autoreactive cytotoxic T lymphocytes in primary biliary cirrhosis. J Clin Invest. 2002;109:1231–40.
58. Matsumura S, Kita H, He XS et al. Comprehensive mapping of HLA-A0201-restricted CD8 T-cell epitopes on PDC-E2 in primary biliary cirrhosis. Hepatology. 2002;36:1125–34.
59. Mosmann TR, Sad S. The expanding universe of T-cell subsets: Th1, Th2 and more. Immunol Today. 1996;17:138–46.
60. Liblau RS, Singer SM, McDevitt HO. Th1 and Th2 CD4+ T cells in the pathogenesis of organ-specific autoimmune diseases. Immunol Today. 1995;16:34–8.
61. Lafaille JJ. The role of helper T cell subsets in autoimmune diseases. Cytokine Growth Factor Rev. 1998;9:139–51.
62. Berg PA, Klein R, Rocken M. Cytokines in primary biliary cirrhosis. Semin Liver Dis. 1997;17:115–23.
63. Harada K, Van de Water J, Leung PS et al. In situ nucleic acid hybridization of cytokines in primary biliary cirrhosis: predominance of the Th1 subset. Hepatology. 1997;25:791–6.
64. Nagano T, Yamamoto K, Matsumoto S et al. Cytokine profile in the liver of primary biliary cirrhosis. J Clin Immunol. 1999;19:422–7.
65. Pelfrey CM, Trotter JL, Tranquill LR, McFarland HF. Identification of a novel T cell epitope of human proteolipid protein (residues 40–60) recognized by proliferative and cytolytic CD4+ T cells from multiple sclerosis patients. J Neuroimmunol. 1993;46:33–42.

66. Honma K, Parker KC, Becker KG, McFarland HF, Coligan JE, Biddison WE. Identification of an epitope derived from human proteolipid protein that can induce autoreactive CD8[+] cytotoxic T lymphocytes restricted by HLA-A3: evidence for cross-reactivity with an environmental microorganism. J Neuroimmunol. 1997;73:7–14.

67. Panina-Bordignon P, Lang R, van Endert PM et al. Cytotoxic T cells specific for glutamic acid decarboxylase in autoimmune diabetes. J Exp Med. 1995;181:1923–7.

68. Wicker LS, Chen SL, Nepom GT et al. Naturally processed T cell epitopes from human glutamic acid decarboxylase identified using mice transgenic for the type 1 diabetes-associated human MHC class II allele, DRB1*0401. J Clin Invest. 1996;98:2597–603.

69. Ackerman AL, Cresswell P. Cellular mechanisms governing cross-presentation of exogenous antigens. Nat Immunol. 2004;5:678–84.

70. Paradis V, Kollinger M, Fabre M, Holstege A, Poynard T, Bedossa P. *In situ* detection of lipid peroxidation by-products in chronic liver diseases. Hepatology. 1997;26:135–42.

71. Kawamura K, Kobayashi Y, Kageyama F et al. Enhanced hepatic lipid peroxidation in patients with primary biliary cirrhosis. Am J Gastroenterol. 2000;95:3596–601.

72. Tsuneyama K, Harada K, Kono N et al. Damaged interlobular bile ducts in primary biliary cirrhosis show reduced expression of glutathione-*S*-transferase-pi and aberrant expression of 4-hydroxynonenal. J Hepatol. 2002;37:176–83.

73. Aboutwerat A, Pemberton PW, Smith A et al. Oxidant stress is a significant feature of primary biliary cirrhosis. Biochim Biophys Acta. 2003;1637:142–50.

74. Mendel-Hartvig I, Nelson BD, Loof L, Totterman TH. Primary biliary cirrhosis: further biochemical and immunological characterization of mitochondrial antigens. Clin Exp Immunol. 1985;62:371–9.

75. Odin JA, Huebert RC, Casciola-Rosen L, LaRusso NF, Rosen A. Bcl-2-dependent oxidation of pyruvate dehydrogenase-E2, a primary biliary cirrhosis autoantigen, during apoptosis. J Clin Invest. 2001;108:223–32.

76. Yoshida S, Gershwin ME. Autoimmunity and selected environmental factors of disease induction. Semin Arthritis Rheum. 1993;22:399–419.

77. Bagenstose LM, Salgame P, Monestier M. Murine mercury-induced autoimmunity: a model of chemically related autoimmunity in humans. Immunol Res. 1999;20:67–78.

78. Rose NR, Burek CL. Autoantibodies to thyroglobulin in health and disease. Appl Biochem Biotechnol. 2000;83:245–51; discussion 251–4, 297–313.

79. D'Cruz D. Autoimmune diseases associated with drugs, chemicals and environmental factors. Toxicol Lett. 2000;112–13:421–32.

80. Hess EV. Are there environmental forms of systemic autoimmune diseases? Environ Health Perspect. 1999;107(Suppl. 5):709–11.

81. Simat TJ, Kleeberg KK, Muller B, Sierts A. Synthesis, formation, and occurrence of contaminants in biotechnologically manufactured L-tryptophan. Adv Exp Med Biol. 1999; 467:469–80.

82. Rao T, Richardson B. Environmentally induced autoimmune diseases: potential mechanisms. Environ Health Perspect. 1999;107(Suppl. 5):737–42.

83. Steenland K, Goldsmith DF. Silica exposure and autoimmune diseases. Am J Ind Med. 1995;28:603–8.

84. Pumford NR, Halmes NC, Hinson JA. Covalent binding of xenobiotics to specific proteins in the liver. Drug Metab Rev. 1997;29:39–57.

85. Obermayer-Straub P, Strassburg CP, Manns MP. Autoimmune hepatitis. J Hepatol. 2000; 32:181–97.

86. Babu BR, Frey C, Griffith OW. L-arginine binding to nitric-oxide synthase. The role of H-bonds to the nonreactive guanidinium nitrogens. J Biol Chem. 1999;274:25218–26.

11
Disease models in primary biliary cirrhosis

D. E. JONES

INTRODUCTION

The immunological and pathological processes which characterize primary biliary cirrhosis (PBC) have been extensively studied in human subjects. These studies, which both predated and followed the identification of the mitochondrial antigens to which the characteristic autoreactive immune responses seen in PBC are directed, have given us considerable insight into the pathological processes that are occurring in patients who have PBC. The exclusive study of human patients with established disease does, however, have limitations when it comes to the study of key pathogenetic and therapeutic events. The principal limitations are with regard to the study of early pathogenetic events and with regard to the application of novel approaches to therapy.

Early disease events

PBC, in contrast to some of the other autoimmune diseases such as insulin-dependent diabetes mellitus (IDDM) and autoimmune hepatitis, has an indolent course. Patients can express antimitochondrial antibodies (AMA), the classical serological markers of disease[1] for a long period of time before the development of either typical biochemical features of the disease or classical symptoms[2,3]. This means that the earliest events in disease pathogenesis, including the triggering processes responsible for the initial breakdown of immune self-tolerance to pyruvate dehydrogenase complex (PDC), take place long before patients are diagnosed and can be studied.

Novel approaches to therapy

Currently available treatments such as ursodeoxycholic acid (UDCA) show some efficacy in terms of slowing of disease progression. There are, however, no currently available treatments which are able to arrest, let alone reverse, the progression to cirrhosis. The advent of newer therapeutic modalities and approaches gives rise to new agents of potential relevance to the treatment of

PBC. In many cases, however, intervening in immune regulatory processes holds at least theoretical possibilities of disease exacerbation. In the current ethical environment it is often difficult to develop and test such novel approaches to therapy without at least some prior application in relevant animal models confirming safety[4].

OVERVIEW OF APPROACHES TO MODELLING DISEASES IN EXPERIMENTAL ANIMALS

In broad terms animal models of any disease can be subdivided in terms of scope of the model and in terms of the mechanism of disease development. In terms of disease scope there is the potential for any disease model to fully replicate the human disease process in the context of an experimental animal. An example in the study of autoimmunity would be the non-obese diabetic (NOD) mouse which spontaneously develops breakdown of immune tolerance to the pancreatic islet cells, islet inflammation and islet cell destruction followed by the clinical manifestations of insulin-dependent diabetes. Alternatively, a disease model may replicate one or more aspects of human disease, but fall short of fully replicating the whole disease process. Again, an example in the field of autoimmunity would be experimental allergic encephalomyelitis (EAE), an acute inflammatory demyelinating condition resulting from sensitization of susceptible animal species and strains with myelin-derived antigens (most typically myelin basic protein (MBP)) or peptide epitopes there-derived. EAE has some features (such as acute inflammatory deyelination) which are redolent of multiple sclerosis. Other features, most notably the progressive fibrotic change associated with worsening disability, and the relapsing/remitting disease kinetic, are less well reproduced. The usefulness of such 'partial' models largely depends on the aspect of the human disease process which they are used to study. Clearly, for example, EAE is a useful model in which to study aspects of acute inflammatory demyelination and mechanisms of breakdown of immune self-tolerance to MBP and related antigen. EAE is a much less useful model in which to study progressive fibrosis arising secondary to such acute inflammatory change. A parallel situation arises in the context of animal models of PBC.

With regard to the mechanism of disease development there is a contrast between spontaneously arising disease models (the NOD mouse again being the classic example) in which the process arises either spontaneously or as a consequence of interaction with naturally occurring trigger factors, and induced models in which some form of host modification or exposure to specific artificial triggering factors is required. At the simplest level this manipulation would consist of using inbred animal strains which have typically been specifically bred to exaggerate characteristics such as susceptibility to the induction of autoimmune disease which render the strains experimentally useful (and it is worth noting that the 'spontaneous' NOD model is highly dependent on possession of a specific and abnormal genotype). At a more complex level specific recombinant technology can be applied to generate novel genotypes in which genes encoding proteins which normally regulate processes responsible for the development of autoimmunity, such as regulatory cytokines, are deleted ('knockouts') or over-expressed 'transgenics'.

The perfect disease model would be one in which the same triggering factors as those which are important in human disease, acting in animal species and strains expressing the same genetic susceptibility factors, induce the same pathological processes which induce the same disease. No such animal model exists, however, for any human disease. The usefulness of any model depends on the similarity between the pathological processes occurring in the human and animal model setting (and real problems can be encountered where apparently similar histological appearances can suggest related processes with real model relevance, when in fact highly divergent processes are responsible for the induction of these appearances in different species). Usefulness can also depend on the reasons for modelling the relevant disease (no matter how close a model is to human disease it will be of little use in the study of, for example, therapies if the manipulation required to induce the model independently interferes with the function of potential therapies (a problem encountered with the neonatally thymectomized mouse model used in PBC as outlined below)). Given the absence of the 'perfect' animal model of disease all models must, by definition, have flaws and/or limitations. These flaws and limitations can lead the unwary investigator down blind alleys. The wise investigator uses them for the unique opportunities they provide as an adjunct to human studies whilst always remaining aware of their limitations.

OVERVIEW OF APPROACHES TO MODELLING PBC

No single model has been described, either spontaneously arising or induced, in which the pathological features of PBC (both immunological and with regard to target cell biology) are fully replicated. A number of models have been described in which aspects of the pathological process are seemingly reproduced (including the breakdown of immune self-tolerance to the key autoantigen pyruvate dehydrogenase complex, PDC) and the development of the characteristic histological lesion of non-suppurative destructive cholangitis (NSDC). Intriguingly, there are two major failings in the experimental approaches adopted to model PBC to date. The first is in linking the characteristic breakdown of immune tolerance to self-PDC with the subsequent development of NSDC. This suggests that there may be an additional factor or change present in the target biliary epithelial cells (BEC) of PBC patients required for the development of target cell damage where tolerance breakdown to PDC occurs which has as yet not been replicated in any of the described animal models. The second is in translating NSDC into the characteristic biliary cirrhosis seen in advanced PBC. Again, this suggests the requirement for an additional factor (possibly genetic susceptibility to the development of liver fibrosis and its sequelae) absent in the modelled animals.

SPONTANEOUS PBC MODELS

Three spontaneous animal models of PBC have been proposed in the literature.

Faenza rabbits

A strain of domesticated rabbits from the Faenza region of Italy was identified in which adult animals (but not young of the same strain or a 'control' strain from elsewhere in Italy) developed NSDC with copper accumulation, together with biochemical features of cholestasis and AMA development[5]. The colony of animals developing NSDC lesions was subsequently reported as dying out.

Senescent C57BL/6NCrj mice

Female C57BL/6NCrj mice of over 18 months of age have been reported to develop NSDC in the context of low titre IgM (but not IgG) AMA[6]. Adoptive transfer of splenocytes from affected females into naive young female recipients was reported to result in recipient development of bile duct lesions.

Murine MLO infection

Normal mice inoculated into the eyelid with human mycoplasma-like organisms (MLO) develop a systemic disease, one feature of which is NSDC[7]. MLO-parasitized leucocytes are described in portal tract infiltrates which develop (albeit in a minority of infected animals). Serological responses and the breakdown of tolerance to key autoantigens were not studied.

As far as the author is aware none of the above observations have been replicated, and the pathogenetic mechanisms underpinning the development of NSDC in each of these animals have not been studied. Again, as far as the author is aware none of these models is the subject of any ongoing study.

STIMULATED PBC MODELS

GVHD-based models

Histological lesions showing a resemblance to the classical NSDC lesions of PBC have been described in a number of animal graft-versus-host disease (GVHD) models. In all such models the development of a lesion is dependent on the presence of mature α/β T cells in the graft which recognize and react to major and minor histocompatibility antigens expressed by the host. The host must, *de facto,* be unable to reject the graft. The subsets of reactive donor T cells, and the kinetics of disease development (acute versus chronic) depend to a significant degree on the strain combinations used (which determines the degree and nature of the histoincompatibility). One of the most widely used strain combinations for the study of NSDC is B10.D2→BALB/c (irradiated)[8–10]. The donor and recipient strains are matched at the MHC locus

(H-2^d) but differ at a number of minor histocompatibility loci (including Mls3). These animals develop a highly reproducible lesion of portal tract inflammation evolving into a marked inflammation accompanied by NSDC at day 14 and beyond. In addition to the morphological similarities to the BEC changes seen in PBC, BEC in the NSDC of GVHD show molecular features indicative of some similarities in underlying pathological processes between PBC and NSDC of GVHD. These include a mixed CD4$^+$ and CD8$^+$ infiltrate, NEC up-regulation of ICAM-1, BEC up-regulation of MHC class II and localized release of IFN-γ[11,12], all features typical of PBC[13–15]. The development of NSDC lesions in the GVHD model can be prevented by the use of agents directed at blocking LFA-1- and ICAM-1-mediated adhesion[12,16] (observations which exemplify the potential usefulness of animal models of disease for the early testing of potential therapeutic agents). The development of NSDC is obviously not restricted to minor histocompatibility antigen-mismatched animals, lesions also being seen in pairings with major histocompatibility mismatch[17,18].

Neonatal thymectomy models

It has long been recognized that animals thymectomized in the neonatal period are particularly susceptible to the development of autoimmune disease. The conventional understanding of the pathogenesis of this process is that such animals suffer from defects of peripheral immune tolerance mediated by a population of thymic emigrants emerging in the postnatal period. The most plausible candidates for the cells lacking in such animals are the family of CD4$^+$CD25$^+$ FoxP3 expressing T cells known to have a key peripheral immune regulatory function[19]. The depletion of CD4$^+$ CD25$^+$ 'Treg' cells can induce the development of autoimmune disease[20]. Moreover, other manipulations which result in decreased numbers of CD4$^+$ CD25$^+$ cells are similarly associated with susceptibility to the induction of autoimmune disease[21]. In the context of animal models of PBC, neonatally thymectomized A/J mice sensitized with porcine IBEC in adjuvant developed portal tract infiltrates in the context of AMA development[22]. Actual bile duct damage was, however, limited. The scale of the changes noted was significantly greater than those seen in control animals receiving either adjuvant alone or cells other than IBEC, or in animals receiving the same sensitization regime in the absence of neonatal thymectomy. Although intriguing in terms of the combination of histological and serological lesions seen, and the implications that the observations hold for the importance of CD4$^+$ CD25$^+$ regulatory T cells in the maintenance of normal immune tolerance to self-PDC, one of the limitations of this model is that it is inappropriate for the study of any therapeutic approaches which might function through the induction or augmentation of T cell regulatory circuits (including oral tolerance to antigen[23]).

Sensitisation-based models

The most productively used approach, and the one which most closely matches approaches adopted in other autoimmune disease models, is the sensitization of susceptible species and strains of animals with relevant antigens. The two principal approaches used have been sensitization of rabbits with PDC mimicking xenobiotics, and sensitization of the autoimmune susceptible SJL/J mouse with foreign, self and chemically modified self-PDC.

1. *Rabbit xenobiotic model*: Given the importance of lipoic acid in both the functioning of the pyruvate dehydrogenase complex and in the dominant B cell epitopes within both PDC-E2 and PDC-E3BP[24–27], one recent strand of research has focused on the lipoic acid cofactor covalently attached to both the PDC-E2 and PDC-E3BP chains at the centre of breakdown of immune tolerance to PDC. This rabbit modelling work has followed on from human studies performed by the same group, which suggested that the affinity of AMA for lipoylated human PDC-E2 is in fact lower than for synthetic molecules mimicking substituted forms of lipoic acid (xenobiotics)[28]. Of particular interest were halogenated xenobiotics including 6-bromohexano-ate. This strand of work was extended to look at the response of experi-mental rabbits to sensitization with 6-bromohexanoate conjugated to bovine serum albumin[29,30]. Rabbits sensitized with this xenobiotic-mounted anti-body response reactive with native PBC which demonstrated a full range of properties (including the ability to inhibit PDC function *in vivo*) typical of PBC-associated AMA. Perhaps surprisingly, given the work performed by this group highlighting the importance of the autoreactive T cell response to PDC in the pathogenesis of PBC, the question of breakdown of T cell tolerance to self-PDC has not been addressed in this model; nor has that of the development of histological changes of NSDC.

2. *SJL/J mouse model*: The autoimmune susceptible SJL/J strain of mouse has proved a highly useful setting in which to study key questions regarding the mechanisms of breakdown of immune tolerance to PDC, including, in particular, the interrelationship between breakdown of B cell and T cell tolerance; an issue not addressed in any other model or, indeed, in human disease. SJL/J mice are normally fully tolerant of PDC, mounting neither antibody nor T cell responses when sensitized with mouse PDC (mPDC)[24,31,32]. When sensitized with mammalian PDC of non-mouse origin (bovine (b)PDC) they mount a high-titre antibody response fully cross-reactive with mPDC[31,33]. This antibody response is mounted rapidly, and is sustained in nature. The striking extent of the immunogenicity of bPDC (which has a very high degree of homology with mPDC) is indicated by the fact that sensitization with bPDC induces AMA even in the absence of adjuvant. These simple studies indicate that, perhaps surprisingly, the induction of AMA is easy to achieve. Over the short term following sensitization, however, breakdown of B cell tolerance to self-PDC is not accompanied by breakdown of T cell tolerance[31]. T cell tolerance break-down is not seen until over 20 weeks following sensitization, and appears to arise as a result of the process of epitope spreading. T cell tolerance breakdown appears, therefore, to be in marked contrast to B cell tolerance breakdown, a highly restricted event. Critically, the natural history of T cell tolerance breakdown can be modified by manipulation of the sensitization

regime. We believe that the nature of the manipulations able to accelerate T cell tolerance breakdown provides important insight into the pathogenesis of PBC.

Of particular note is that co-sensitization of animals with both bPDC and mPDC results in rapid breakdown of both B cell and T cell tolerance to self-PDC[32]. In these experiments the simultaneous presence of an active antobody/B cell response reactive with self-PDC and the self-antigen itself appears to be critical for promoting the breakdown of T cell tolerance. The explanation for the importance of presence of the reactive B cell response appears to be that activated B cells reactive with self-PDC actually play a critical role in presenting self-antigen to potentially autoreactive T cells[34,35]. It is intriguing to note, in the light of the apparent ability of lipoic acid mimicking xenobiotics to break B cell tolerance to self-PDC[29], that exposure to mPDC modified by attachment of biotin, a molecule which shows significant structural homology with lipoic acid, is able to induce antibody responses reactive with native self-PDC and to replace the need for the foreign, bPDC component in the co-sensitization model[36]. This observation would suggest that it is the induction of the B cell response reactive with self-PDC which is critical for setting a permissive environment for T cell tolerance breakdown to occur, rather than the precise route by which this B cell response comes about, which is important. This raises the interesting possibility that different mechanisms giving rise to a B cell response reactive with self-PDC (e.g. exposure to chemically modified PDC, PDC mimicking xenobiotics and foreign PDC in the context of infection with a PDC-containing microorganism) might act as the key pathogenetic factor in different PBC patients[34]. This concept warrants further exploration.

Another factor which appears to play an important role in driving the development of tolerance breakdown to self-PDC in susceptible animals is the inflammatory microenvironment. Although complete Freund's adjuvant (CFA) is effective at promoting T cell tolerance breakdown in appropriately sensitized animals, oligodeoxynucleotides containing a CpG motif (CpG ODN) are even more efficacious[37]. CpG ODN are ligands for TLR9, an activation pathway of particular relevance for augmenting the ability of B cells to act as professional APC[38,39]. This is of particular relevance given the postulated role for B cells to act as professional APC for the breakdown of T cell tolerance to self-PDC. The activation of status of APC, and in particular the degree of responsivity to TLR9, appears to unperpin the increased susceptibility of the SJL/J mouse to the induction of autoimmune disease[40].

SJL/J mice induced to break T cell tolerance to self-PDC show some histological features redolent of PBC with, in particular, the development of NSDC[32,33,36,37]. Strikingly, the development of bile duct destruction (as opposed to bile duct damage), ductular reaction and biliary fibrosis is not seen. The kinetics of development of NSDC appear to mirror those of development of breakdown of T cell tolerance to PDC (as opposed to the far more accelerated kinetic of development of B cell tolerance breakdown), suggesting a potential contribution by autoreactive T cells to the process (a view supported by preliminary T cell adoptive transfer studies). It is clear, however, that at least part of the process of NSDC results from the adjuvant regimes required to promote T cell tolerance breakdown[33,41] (a non-specific targeting effect previously noted in other studies[42]. Because of the complexity of the aetiological process underpinning NSDC development it is the author's belief that the greatest value of this model is in providing a setting in which to explore mechanisms of tolerance breakdown to PDC and the therapeutic reimposition of tolerance[4] rather than the downstream events of target cell damage.

WHAT ANIMAL MODELS TELL US ABOUT THE PATHOGENESIS OF PBC

Although it is clear that there is no definitive animal model of PBC, the models that have been described provide us with important and unique insights into individual aspects of disease pathogenesis. What conclusions can be drawn from such animal work?

1. NSDC represents a final common pathway for immune-mediated injury rather than a specific entity: The absence of exposure to allogeneic cells in the vast majority of PBC patients means that, quite patently, a process analogous to GVHD cannot be directly responsible for NSDC in PBC. What the GVHD studies do suggest, however, is that the portal tract represents a powerful potential target for immune-mediated injury. Moreover, the shared features between GVHD and PBC NSDC, including up-regulation of MHC and adhesion molecules, suggest that inflammatory cell targeting and ingress into the portal tract, localized cytokine release, up-regulation of MHC and adhesion molecules on the BEC surface and, ultimately, BEC damage represent a pattern response to locally targeted inflammatory/immune responses. In simple terms NSDC of PBC and GVHD look similar because, although the trigger for disease is different, the resulting pathway of damage and the response of the target tissue to that damage is the same. This concept of NSDC representing a pattern of response rather than a discrete disease process can potentially be expanded into the human setting. It may be that the similarities in aspects of the histological appearances between PBC, PSC, ductopenic rejection, sarcoid and human GVHD represent an equivalent patterned response. It is interesting to note in this respect that human patients with GVHD have been reported to express serum AMA[43].

2. Peripheral T cell tolerance mechanisms play a role in the normal prevention of breakdown of immune self-tolerance to PDC: Neonatally thymectomized mice which lack $CD4^+$ $CD25^+$ regulatory T cells are particularly susceptible to the induction of AMA. APC stimulation via the TLR pathway (known to suppress $CD4^+$ $CD25^+$ regulatory pathways[44] augments the breakdown of immune tolerance to self-PDC in murine models. Moreover, the scale of breakdown of self-tolerance to PDC in sensitization models is greatest in SJL/J mice, a strain with a notably high baseline state of APV activation which can be mimicked in less susceptible strains by TLR ligand exposure[40]. These observations both clearly point to a key role for $CD4^+$ $CD25^+$ Treg T

cells in the peripheral regulation of immune self-tolerance to PDC. This view, that peripheral regulatory mechanisms may be of real importance for the normal regulation of autoreactivity to self-PDC and the sequelae of its breakdown, and that the portal tracts appear to represent a particular focus for T cell-driven inflammatory processes, is supported by the observation that mice of two mutant strains showing marked immune dysregulation, the MRL/lpr mouse and the aly/aly mouse, show among many other changes, the development of portal tract inflammatory infiltrates[45,46].

3. The induction of antibodies reactive with self-PDC is easy to achieve but does not, in isolation, appear to be significant. It is the much more restricted event of T cell tolerance breakdown which appears to hold the key to disease: Perhaps the biggest single surprise to come out of the body of animal modelling work carried out in PBC is the ease with which it is possible to induce autoantibodies reactive with self-PDC. This observation appears to be at clear odds with the characteristics of PDC (highly conserved throughout evolution, present in all nucleated cells) which would lead us to predict that tolerance breakdown would be hard to achieve. The breakdown of B cell tolerance to PDC can result from exposure to non-self PDC, to chemically modified self-PDC and to xenobiotics mimicking aspects of self-PDC. In each case the induced antibodies have all the properties characteristic of AMA. Of equal significance to the observation of the ease with which B cell tolerance breakdown can be induced is that of the relative difficulties encountered in inducing T cell tolerance breakdown. Taken together these observations would support the view emerging from human studies which suggests that the critical barrier to the development of PBC is the highly restricted event of breakdown of T cell tolerance to self-PDC rather than the looser state of B cell tolerance.

The fascinating question to arise from these observations is why PDC, despite all its characteristics which should favour tolerance, is so immunogenic. The answer would appear to be the unique properties associated with the lipoic acid co-factor. The unique biochemical problem of electron transfer has been 'solved' in the case of the 2-OADC by the unique system of lipoic acid transfer. The size, position, properties and, perhaps above all, susceptibility to modification and mimicking of lipoic acid appear to combine to make it the 'weak link' in physiological immune tolerance to protein complexes that have evolved to adopt this 'solution'.

CONCLUSIONS

PBC does not arise spontaneously in any animal species or strain; nor can the disease be induced in its entirety in animals. In this sense the 'search' for an animal model of the disease could be regarded as having failed. Studies of *aspects* of the pathogenetic processes thought to be important in PBC pathogenesis, most notably the breakdown of immune tolerance to self-PDC, have, however, generated important insights which have changed our thinking regarding disease pathogenesis. These models have allowed us to test hypotheses generated by studies of human disease using experimental protocols which would simply not be possible in humans. The ongoing use of animal models of *aspects* of disease pathogenesis and, increasingly, treatment is therefore fully justified as an adjunct to ongoing studies of human disease.

References

1. Jones DEJ. Primary biliary cirrhosis. Autoimmunity. 2004;37:325–8.
2. Mitchison HC, Bassendine MF, Hendrick A et al. Postitive antimitochondrial antibody but normal alkaline phosphatase: is this primaruy biliary cirrhosis? Hepatology. 1986;6:1279–84.
3. Metcalf JV, Mitchinson HC, Palmer JM, Jones DEJ, Bassendine MF, James OFW. Natural history of early primary biliary cirrhosis. Lancet. 1996;348:1399–402.
4. Jones DEJ, Palmer JM, Robe AJ, Kirby JA. Oral tolerisation to pyruvate dehydrogenase complex as a potential therapy for primary biliary cirrhosis. Autoimmunity. 2002;35:537–44.
5. Tison V, Callea F, Morisi C. Spontaneous primary biliary cirrhosis in rabbits. Liver. 1982;2:152–61.
6. Kanda K, Onji M, Ohta Y. Spontaneous occurrence of autoimmune cholangitis in senescent mice. J Gastroenterol Hepatol. 1993;8:7–14.
7. Johnson L, Wirostko E, Wirostko W. Primary biliary cirrhosis in the mouse: induction by human mycoplasma-like organisms. Int J Exp Pathol. 1990;71:701–12.
8. Howell CD, Yoder T, Claman HN, Vierling JM. Hepatic homing of mononuclear inflammatory cells isolated during murine chronic graft-vs-host disease. J Immunol. 1989;143:476–83.
9. Howell CD, Li J, Ropper E, Kotzin BL. Biased liver T cell receptor Vbeta repertoire in a murine graft-versus-host disease model. J Immunol. 1995;155:2350–8.
10. Chen W, Howell CD. Oligoclonal expansion of T cell receptor V beta 2 and 3 cells in the livers of mice with graft-versus-host disease. Hepatology. 2002;35:23–9.
11. Howell CD, Yoder T, Vierling JM. Suppressor function of liver mononuclear cells isolated during murine chronic graft-vs-host disease II. Role of prostaglandins and interferon-gamma. Cell Immunol. 1992;140:54–66.
12. Howell CD, Li J, Chen W. Role of intercellular adhesion molecule-1 and lymphocyte function associated antigen-1 during nonsuppurative destructive cholangitis in a mouse graft-versus-host disease model. Hepatology. 1999;29:766–76.
13. Bjorkland A, Festin R, Mendel-Hartvig I, Nyberg A, Loff L, Totterman T. Blood and liver infiltrating lymphocytes in primary biliary cirrhosis: increase in activated T and natural killer cells and recruitment of primed memory T cells. Hepatology. 1991;13:1106–111.
14. Leon MP, Bassendine MF, Gibbs P, Thick M, Kirby JA. Immunogenicity of biliary epithelium: study of the adhesive interaction with lymphocytes. Gastroenterology. 1997;112:968–77.
15. Harada K, Van de Water J, Leung PS et al. *In situ* nucleic acid hybridisation of cytokines in primary biliary cirrhosis: predominance of the Th1 subset. Hepatology. 1997;25:791–6.
16. Kimura T, Suzuki K, Inada S et al. Monoclonal antibody against lymphocyte function-associated antigen 1 inhibits the formation of primary biliary cirrhosis-like lesions induced by murine graft-versus-host reaction. Hepatology. 1996;24:888–94.
17. Saitoh T, Fujiwara M, Nomoto M, Kamimura T, Ishihara K, Asakura H. Histologic studies on the hepatic lesions induced by graft-versus-host reaction in MHC class II dispirate hosts compared with primary biliary cirrhosis. Am J Pathol. 1989;135:301–7.
18. Saitoh T, Fujiwara M, Asakura H. L3T4+ T-cells induce hepatic lesions resembling primary biliary cirrhosis in mice with graft-versus-host reactions due to major histocompatibility complex class II disparity. Clin Immunol Immunopathol. 1991;59:449–61.
19. Sharp C, Thompson C, Samy ET, Noelle R, Tung KSK. CD40 ligand in pathogenesis of autoimmune ovarian disease of day 3-thymectomised mice: implication for CD40 ligand antibody therapy. J Immunol. 2003;170:1667–74.

20. McHugh RS, Shevach EM. Depletion of CD4$^+$CD25$^+$ regulatory T cells is necessary, but not sufficient for induction of organ-specific autoimmune disease. J Immunol. 2002;168: 5979–83.

21. Nishibori T, Tanabe Y, Su L, David M. Impaired development of CD4$^+$ CD25$^+$ regulatory T-cells in the absence of STAT1: increased susceptibility to autoimmune disease. J Exp Med. 2004;199:25–34.

22. Kobashi H, Yamamoto K, Yoshioka T, Tomita M, Tsuji T. Nonsuppurative cholangitis induced in neonatally thymectomized mice: a possible animal model for primary biliary cirrhosis. Hepatology. 1994;19:1424–30.

23. Zhang X, Izikson L, Liu L, Weiner HL. Activation of CD25(+)CD4(+) regulatory cells by oral antigen administration. J Immunol. 2001;167:4245–3.

24. Yeaman SJ, Kirby JA, Jones DEJ. Autoreactive responses to pyruvate dehydrogenase complex in the pathogenesis of primary biliary cirrhosis. Immunol Rev. 2000;174:238–49.

25. Quinn J, Diamond AG, Palmer JM, James OFW, Bassendine MF, Yeaman SJ. Lipoylated and unlipoylated domains of human PDC-E2 as autoantigens in primary biliary cirrhosis: significance of lipoate attachment. Hepatology. 1993;18:1384–91.

26. Palmer JM, Jones DEJ, Quinn J, McHugh A, Yeaman SJ. Characterisation of the autoantibody responses to recombinant E3 binding protein (protein X) of pyruvate dehydrogenase in primary biliary cirrhosis. Hepatology. 1999;30:21–6.

27. Bruggraber SF, Leung PS, Amano K et al. Autoreactivity to lipoate and a conjugated form of lipoate in primary biliary cirrhosis. Gastroenterology. 2003;125:1705–13.

28. Long SA, Quan C, Van de Water J et al. Immunoreactivity of organic mimeotopes of the E2 component of pyruvate dehydrogenase: connecting xenobiotics with primary biliary cirrhosis. J Immunol. 2001;167:2956–63.

29. Leung PS, Quan C, Park O et al. Immunization with a xenobiotic 6-bromohexanoate bovine serum albumin conjugate induces anti-mitochondrial antibodies. J Immunol. 2003; 170:5326–32.

30. Amano K, Leung PSC, Xu Q et al. Xenbobiotic-induced loss of tolerance in rabbits to the mitochondrial autoantigen of primary biliary cirrhosis is reversible. J Immunol. 2004;172: 6444–52.

31. Jones DEJ, Palmer JM, Yeaman SJ, Kirby JA, Bassendine MF. Breakdown of tolerance to pyruvate dehydrogenase complex in experimental autoimmune cholangitis a murine model of primary biliary cirrhosis. Hepatology. 1999;30:65–70.

32. Jones DEJ, Palmer JM, Bennett K et al. Investigation of a mechanism for accelerated breakdown of immune-tolerance to the primary biliary cirrhosis associated autoantigen, pyruvate dehydrogenase complex. Lab Invest. 2002;82:211–19.

33. Jones DEJ, Palmer JM, Kirby JA et al. Experimental autoimmune cholangitis: a mouse model of immune-mediated cholangiopathy. Liver. 2000;20:351–6.

34. Jones DEJ. Pathogenesis of primary biliary cirrhosis. J Hepatol. 2003;39:639–48.

35. Robe AJ, Palmer JM, Kirby JA, Jones DEJ. Breakdown of T-cell tolerance to the primary biliary cirrhosis (PBC) autoantigen pyruvate dehydrogenase complex is a B-cell driven process. Gut. 2004;53:A12.

36. Palmer JM, Robe AJ, Burt AD, Kirby JA, Jones DEJ. Covalent modification as a mechanism for the breakdown of immune tolerance to pyruvate dehydrogenase complex in the mouse. Hepatology. 2004;39:1583–92.

37. Jones DEJ, Palmer JM, Burt AD, Robe AJ, Kirby JA. Bacterial motif DNA as an adjuvant for the breakdown of immune self-tolerance to pyruvate dehydrogenase complex. Hepatology. 2002;36:679–86.

38. Constant SL. B lymphocytes as antigen-presenting cells for CD4$^+$ T-cell priming in vivo. J Immunol. 1999;162:5696–703.

39. Bourke E, Bosisio D, Golay J, Polentarutti N, Mantovani A. The toll-like receptor repertoire of human B lymphocytes: inducible and selective expression of TLR9 and TLR10 in normal and transformed cells. Blood. 2003;102:956–63.

40. Waldner H, Collins M, Kuchroo VK. Activation of antigen-presenting cells by microbial products breaks self-tolerance and induces autoimmune disease. J Clin Invest. 2004;113: 990–7.

41. Sasaki M, Allina J, Odin JA et al. Autoimmune cholangitis in the SJL/J mouse is antigen non-specific. Develop Immunol. 2002;9:103–11.

42. Karanikas V, Mackay IR, Rowley MJ, Veitch B, Loveland BE. Hepatic portal tract infiltrates in mice immunized with syngeneic lymphoid cells: connotations for models of autoimmune liver disease. J Gastroenterol Hepatol. 1995;10:491–7.
43. Siegert W, Stemerowicz R, Hopf U. Antimitochondrial antibodies in patients with chronic graft-versus-host disease. Bone Marrow Trans. 1992;10:221–7.
44. Pasare C, Medzhitov R. Toll pathway-dependent blockade of $CD4^+CD25^+$ T cell-mediated suppression by dendritic cells [comment]. Science. 2003;299:1033–6.
45. Tsuneyama K, Kono N, Hoso M et al. Aly/aly mice: a unique model of biliary disease. Hepatology. 1998;27:1499–507.
46. Tsuneyama K, Nose M, Nisihara M, Katayanagi K, Harada K, Nakanuma Y. Spontaneous occurrence of chronic non-suppurative destructive cholangitis and anti-mitochondrial antibodies in MRL/lpr mice: possible animal model for primary biliary cirrhosis. Pathol Int. 2001;51:418–24.

12
Infectious aetiology of primary biliary cirrhosis?

M. F. BASSENDINE

INTRODUCTION

Primary biliary cirrhosis (PBC) is generally considered to be an autoimmune disease as it is characterized by the presence of antimitochondrial antibodies (AMA) and autoreactive T cells[1]. Tissue-damaging autoreactivity is a rarity requiring a complex series of events to occur. The favoured hypothesis for the aetiology of PBC is that environmental factors trigger the disease is genetically susceptible individuals. This theory is supported by epidemiological evidence of geographic patterns in the prevalence of disease[2] with the existence of clusters[3,4], as well as discordance among monozygotic twins[5]. The possibility that the environmental factor(s) involved in the multi-step process of break-down in self-tolerance are infectious agents has been generated by a number of studies, but even the best evidence is only circumstantial.

MOLECULAR MIMICRY

AMA have long been recognized as the serological hallmark of PBC but are found prior to disease onset, suggesting they are induced by an initial environmental, possibly infectious, insult. Prior to the molecular characterization of the mitochondrial autoantigens it was recognized that AMA crossreact with proteins found in microorganisms including yeasts[6–9]. This suggested that 'molecular mimicry' may be involved in the development of autoreactivity, a mechanism whereby the immune system presents microbial peptides that share homology with self, leading to a promiscuous antibody and cell-mediated immune response. This paradigm has been investigated in more detail since the molecular characterization of the mitochondrial autoantigens.

The immunodominant epitope recognized by AMA in patients with PBC is located within the lipoyl domains of each of the 2-oxo acid dehydrogenase complexes; *viz.* pyruvate dehydrogenase complex (PDC)-E2, 2-oxoglutarate dehydrogenase complex (OGDC)-E2, branched-chain oxo acid dehydrogenase complex (BCOADC)-E2 and the E3 binding protein (E3BP) of PDC[10]. These complexes are involved in energy metabolism in the cell and are highly conserved in evolution. It is thus not surprising that AMA from PBC patients crossreact with homologous E2 in an ever-expanding list of microorganisms.

Escherichia coli (E. coli)

E. coli, a common cause of urinary tract infections (UTI), is one of the bacteria that some studies have implicated in the pathogenesis of PBC. It has been documented that *E. coli* UTI and septicaemia can result in transient AMA associated with elevated hepatobiliary enzymes[11]. It has long been proposed that the stimulus for AMA production in PBC is chronic UTI[12–14]. However a higher rate of UTI in PBC in controversial; some studies[15] have confirmed the observation with an odds ratio of 2.12, while others have refuted it[16,17]. Additional support for the notion that *E. coli* infection is involved in the induction of PBC-specific autoimmunity is provided by the recent demonstration of an association between recurrent UTI and antibodies to sp-100[18], an antinuclear antibody found in a minority of PBC patients. These workers have also suggested that *E. coli* proteins other than the homologous 2-oxo acid dehydrogenase complex E2, that are more likely to be recognized as non-self by the immune system, could be the triggers for PBC. By interrogating protein databases 10 non-PDC-E2 microbial sequences with a high degree of homology to the immunodominant epitope of PDC-E2 (amino acids 212–226) have been identified; six in *E. coli* – 47% of sera from PBC patients had antibodies against these *E. coli* antigens, again possibly implicating *E. coli* in the induction of autoimmunity[19]. However, the theory that *E. coli* UTI might trigger an antibody response that develops into the autoimmune response seen in PBC is not supported by other studies (Table 1). Thus AMA from PBC patients crossreact effectively with yeast PDC-E2 but less effectively with E2 from *E. coli*.[20], and are only weakly inhibitory of *E. coli* PDC compared to mammalian PDC in a functional enzyme assay[21] that has been shown to have >99% specificity for PBC[22]. Furthermore, immunogenetic analysis of a human monoclonal AMA with specificity for the inner lipoyl domain of PDC-E2 reveals that epitope shifting occurs during B cell affinity maturation; thus the final specificity of AMA may be different to germline reactivity, a finding which argues against bacterial molecular mimicry in PBC initiation[23].

In addition to these conflicting studies of molecular mimicry at the autoantibody (B lymphocyte) level, recent important studies have examined T cell crossreactivity. It has been shown that some T cell lines established from the PDC-E2 reactive T cells crossreact to *E. coli* or other bacterial PDC-E2 peptides[24]. Thus T cell clones selected by human PDC-E2 163–176 crossreact to peptides derived from the *E. coli* OGDC-E2 and the frequency of cross-reactivity is much higher than the frequency of crossreactivity to *E. coli* PDC E2[24,25]. The reverse has also been demonstrated – T cell clones specific for *E. coli* OGDC-E2 have been found to react promiscuously to a number of human mitochondrial autoantigens, including PDC-E2, E3BP, OGDC-E2 and BCOADC-E2[26]. This lends support to the argument that bacterial agents, either directly or through environmental modulation, can lead to T cell autoreactivity. Another interesting observation is that peptide 159–167 of

Table 1 Some of the infectious agents that have been implicated in the aetiology of primary biliary cirrhosis, showing that the evidence is controversial for most agents or confirmatory studies are currently lacking

Infectious agent	Evidence in favour (reference)	Evidence against (reference)
Escherichia coli	Stemerowicz et al., 1988[9] Butler et al., 1993[13] Butler et al., 1995[14] Tanimoto et al., 2003[26] Bogdanos et al., 2004[19]	Fussey et al., 1991[20] Teoh et al., 1994[21] Potter et al., 2001[23]
Mycobacteria	Klein et al., 1993[29] Vilagut et al., 1994[30]	O'Donohue et al., 1994[31] Jones et al., 1997[34] O'Donohue et al., 1998[32] Tanaka et al., 1999[33]
Helicobacter	Nilsson et al., 2000[37] Dohmen et al., 2002[35] Nilsson et al., 2003[36]	Tanaka et al., 1999[33] Durazzo et al., 2004[38]
Chlamydia	Abdulkarim et al., 2004[39]	Leung et al., 2003[40]
Propionibacterium acnes	Harada et al., 2001[41]	
Novosphingobium aromaticivorans	Selmi et al., 2003[42]	
Betaretrovirus	Xu et al., 2003[58] Xu et al., 2004[59]	Selmi et al., 2004[61]

PDC-E2 has been shown to induce specific MHC class-I-restricted CD8$^+$ CTL lines from 10/12 HLA-A2(+) PBC patients after *in-vitro* stimulation with antigen-pulsed dendritic cells (DC). However, if the DC are pulsed with soluble PDC-E2/human monoclonal antibody complex, autoantigen is presented at a higher relative efficiency, implicating a role for autoantibodies, and possibly crossreacting antibodies initiated by microorganisms, in the generation of autoreactive T cells[27]. This is in keeping with the concept that prolonged antigen/immune complex delivery via DC is a crucial factor for the conversion of transient autoimmunity to manifest autoimmune disease[28].

Mycobacteria

Mycobacteria, a cause of hepatic granuloma, constitute another microorganism that has been implicated by some studies in the pathogenesis of PBC. Mycobacteria were investigated partially because granulomatous portal lesions are characteristic of PBC. It has also been documented that patients with pulmonary tuberculosis (TB) have antibodies that crossreact with the mitochondrial M2 autoantigen; in this study none of the patients with *E. coli* infections was positive, suggesting a higher titre of crossreactive antibody in

TB[29]. Another study reported crossreactivity of anti-*Mycobacterium gordonae* antibodies with mammalian PDC-E2 and BCOADC-E2, suggesting that this particular mycobacterium may play a potential pathogenic role in PBC[30]. It was subsequently shown that these crossreacting antibodies to the human autoantigen are directed against one of the most immunogenic components of mycobacteria, the heatshock 65kDa protein[30]. However, the theory that mycobacterial proteins might trigger an immune response that develops into the autoimmune response seen in PBC is not supported by other studies (Table 1). Antibodies to the 65 kDa mycobacterial protein were found in patients with other chronic liver diseases[31]. Mycobacterial DNA was not detected in liver sections of patients with PBC[32,33] and CD4[+] T cell proliferative responses to mycobacterial proteins were similar in PBC compared to normal and chronic liver disease controls[34].

Other bacteria

Other bacteria have been implicated in inducing autoimmune responses in PBC. For example, a positive correlation has been found between the titre of anti-PDC autoantibodies and the titre of anti-*Helicobacter pylori* antibodies in PBC patients[35]. In addition an increased prevalence of serum antibodies to non-gastric *Helicobacter* species has been reported in patients with autoimmune liver disease including PBC[36], and gene sequences of *Helicobacter* species have been detected in human liver tissue in cholestatic liver diseases including PBC[37]. Again this suggestion has been refuted by other studies (Table 1) which have found no association between seroprevalence of *H. pylori* infection and PBC[38], and *Helicobacter* sequences in only 1/29 PBC liver tissues by PCR[33]. Another controversial suggestion is that there is a specific association of *Chlamydia pneumoniae* with PBC. One study has found *C. pneumoniae* antigen and RNA in liver tissue of patients with PBC[39], while another has confirmed that PBC sera react to *Chlamydia* lysates, but PCR amplification of the *Chlamydia*-specific 16S rRNA gene was negative in 25/25 PBC livers, and *Chlamydia* antigens were not detected in liver tissue by immunohistochemistry[40] (Table 1). Another study on PBC liver tissue has identified *Propionibacterium acnes* in epithelioid granuloma in PBC, raising the possibility that this bacterium is involved in the pathogenesis of the granulomas found in the vicinity of the damaged bile ducts in PBC[41].

Recently a newly defined bacterial strain, *Novosphingobium aromaticivorans*, has been implicated as a good candidate to induce molecular mimicry in PBC on the basis that PBC sera crossreact with two homologous *N. aromaticivorans* antigens at 100–1000-fold higher titre than against *E. coli* proteins, and evidence that this ubiquitous microorganism can be found in 25% of human faecal samples, both PBC patients and controls[42]. This unique bacterium has the capability to metabolize chemical compounds including oestrogens, so it could be hypothesized that prolonged exposure to *N. aromaticivorans* antigens, possibly modified by xenobiotic exposure, in the gut may be involved in the induction of PBC.

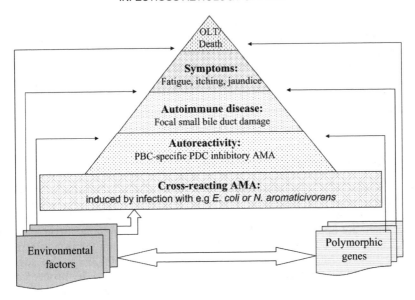

Figure 1 Putative natural history of primary biliary cirrhosis indicating that crossreactive antimitochondrial antibodies (AMA) may be initially induced by exposure to homologous E2 components of the 2-oxo acid dehydrogenase complexes of microorganisms, but that only a minority of these individuals will develop autoreactivity with PBC-specific PDC-inhibitory AMA. This may require a complex interaction of environmental factors with multiple host genes. Progression of the disease may be regarded as a pyramid in which some patients with autoreactivity develop tissue damage, some of whom progress with time to symptomatic disease, and subsequently require liver transplantation or die of liver failure. The ongoing role of infectious agents in the process is controversial

GUT-DERIVED BACTERIAL ANTIGENS

Evidence is accumulating for a possible role of gut-derived antigens in the induction of PBC. Recent data suggest that gastrointestinal permeability is increased in PBC[43]. The finding of a high prevalence of antibodies to calreticulin of the IgA class in PBC, as well as in yersiniosis and alcoholic liver disease, may reflect heightened reactivity of the gut-associated immune system[44]. This group has recently suggested that the emergence of antibodies to non-species-specific bacterial proteins in PBC may simply reflect up-regulation of natural antibodies in autoimmune liver diseases in general[45]. If exposure to environmental agents via the gut is involved in PBC, then the demonstration of crossreactivity of PBC sera with antigens found in lower plants, such as algae, fungi and ferns[46], could even imply that plants could be invoked as the environmental trigger!

Further evidence in support of a defect in intestinal immune defence in PBC is provided by the early report that deposits of lipid A, a common antigenic component of the cell wall in Gram-negative bacteria, were found in liver tissue of 11/21 patients with PBC[47]. These workers hypothesized that biological effects of lipid A, such as its adjuvant and mitogenic effects on lymphocytes, with stimulation of IgM, complement activation and induction of mediators such as interferon gamma, may contribute to the development of PBC. Subsequently it has been reported that there is a high prevalence of IgM anti-lipid A in PBC patients[48], that total IgM levels significantly correlate with anti-lipid A antibody levels, and that those patients with antibodies to lipid A have more frequent and more severe lesions of cholangitis[49]. This supports an aetiological role of gut-derived endotoxin antigen in the pathogenesis of PBC.

TISSUE-DAMAGING AUTOIMMUNITY

In PBC an early event is the development of AMA[50], which may arise, as discussed above, as a result of molecular mimicry. A recent epidemiological study of the prevalence of AMA in Japanese corporate workers has suggested that there are approximately 336 472 AMA-positive people in Japan aged 30–59 years, but <3000 symptomatic PBC patients, inferring the rate of symptomatic PBC among AMA-positive persons to be about 0.73%[51]. This suggests that an additional and/or prolonged trigger is required to proceed from crossreactive AMA, to PBC-specific PDC inhibitory autoreactivity[21,22], to manifest autoimmune tissue damage (Figure 1). Targeting of the biliary epithelium in PBC may be explained by abnormal expression of mitochondrial autoantigens on the surface of salivary and biliary epithelial cells (BEC)[52–54]. This is found early in the natural history of PBC and is localized predominantly around the biliary epithelium[55,56]. This aberrant surface expression of PDC has been reported to be induced in normal BEC by factor(s) contained in lymph nodes from patients with PBC[57], and data were subsequently presented to indicate that this transmissible factor was a human betaretrovirus[58], which shares close homology with murine mammary tumour virus (MMTV)[59]. These workers have reported in PBC that the retroviral proteins co-localize to cells, demonstrating aberrant autoantigen expression, with a higher viral burden in lymphoid tissue. This retrovirus was thus considered an attractive candidate for providing another 'hit' in the multistage process of tolerance breakdown, raising the question of whether antiretroviral agents might be of some benefit in PBC[60]. However, a role of this betaretrovirus with homology to MMTV has now been seriously questioned by recent data which have been unable to recapitulate the earlier study, and have raised concerns of viral contamination[61].

In addition, other mechanisms have been proposed to account for aberrant expression of autoantigen on BEC and tissue damage, including transcytosis of secretory IgA[62–64], with IgA AMA causing apoptotic BEC death via caspase activation[65].

In conclusion, it is still questionable whether the environmental factor(s) that may trigger PBC in genetically susceptible individuals are microorganisms or xenobiotics. An ingested xenobiotic, modifying gut flora in an individual with altered intestinal permeability and a defect in immunity, would provide a unifying multi-hit hypothesis. The interaction of the intestinal/liver immune system in discriminating between harmful and beneficial gut antigens[66] warrants further study in PBC.

References

1. Bassendine MF. The molecular basis of primary biliary cirrhosis. In: M Trauner and PLM Jansen, editors. Molecular Pathogenesis of Cholestasis. Georgetown, New York: Landes Bioscience and Kluwer Academic Plenum Publishers; 2004:221–46.
2. Metcalf JV, James OFW. The geoepidemiology of primary biliary cirrhosis. Semin Liver Dis. 1997;17:13–22.
3. Prince MI, Chetwynd A, Diggle P, Jarner M, Metcalf JV, James OF. The geographical distribution of primary biliary cirrhosis is a well-defined cohort. Hepatology. 2001;34:1083–8.
4. Abu-Mouch S, Selmi C, Benson GD et al. Geographic clusters of primary biliary cirrhosis. Clin Dev Immunol. 2003;10:127–31.
5. Selmi C, Mayo MJ, Bach N et al. Primary biliary cirrhosis in monozygotic and dizygotic twins: genetics, epigenetics and environment. Gastroenterology. 2004;127:485–92.
6. Sayers T, Baum H. Possible cross-reaction of human anti-mitochondrial antibody with membrane vesicles of *Paracoccus dinitrificans*. Biochem Soc Trans. 1976;4:138–9.
7. Lindenborn-Fotinos J, Baum H, Berg PA. Mitochondrial antigens in primary biliary cirrhosis: further characterisation of the M2 antigen by immunoblotting, revealing species and non-species specific determinants. Hepatology. 1985;5:763–9.
8. Ghadiminejad I, Baum H. Evidence for the cell-surface localization of antigens cross-reacting with the 'mitochondrial antibodies' of primary biliary cirrhosis. Hepatology. 1987;7:743–9.
9. Stemerowicz R, Hopf U, Moller B et al. Are antimitochondrial antibodies in primary biliary cirrhosis induced by R(rough)-mutants of enterobacteriaceae? Lancet. 1988;ii:1166–9.
10. Bassendine MF, Jones DEJ, Yeaman SJ. Biochemistry and autoimmune response to the 2-oxoacid dehydrogenase complexes in primary biliary cirrhosis. Semin Liver Dis. 1997;17:49–60.
11. Ohno N, Ota Y, Hatakeyama S et al. A patient with *E-coli*-induced pyelonephritis and sepsis who transiently exhibited symtoms associated with primary biliary cirrhosis. Intern Med. 2003;42:1144–8.
12. Rosenstein IJ, Hazelhurst GR, Burroughs AK, Epstein O, Sherlock S, Brumfitt W. Recurrent bacteriuria and primary biliary cirrhosis: ABO blood group, PI blood group, and secretor status. J Clin Pathol. 1984;37:1055–8.
13. Butler P, Valle F, Hamilton-Miller JM, Brumfitt W, Baum H, Burroughs AK. M2 mitochondrial antibodies and urinary rough mutant bacteria in patients with primary biliary cirrhosis and patients with recurrent bacteriuria. J Hepatol. 1993;17:408–14.
14. Butler P, Hamilton-Miller J, Baum H, Burroughs AK. Detection of M2 antibodies in patients with recurrent urinary tract infection using an ELISA and purified PBC specific antigens. Evidence for a molecular mimicry mechanism in the pathogenesis of primary biliary cirrhosis. Biochem Mol Biol Int. 1995;35:473–85.
15. Parikh-Patel A, Gold EB, Worman H, Krivy KE, Gershwin ME. Risk factors for primary biliary cirrhosis in a cohort of patients in the United States. Hepatology. 2001;33:16–21.
16. Floreani A, Bassendine MF, Mitchison HC, Freeman R, James OFW. No specific association between primary biliary cirrhosis and bacteriuria. J Hepatol. 1989;8:201–7.
17. O'Donohue J, Workman MR, Rolando N, Yates M, Philpott-Howard J, Williams R. Urinary tract infections in primary biliary cirrhosis and other chronic liver diseases. Eur J Clin Microbiol Infect Dis. 1997;16:743–6.
18. Bogdonos DP, Baum H, Butler P et al. Association between the primary biliary cirrhosis specific anti-sp100 antibodies and recurrent urinary tract infection. Dig Liver Dis. 2003;35:801–5.
19. Bogdanos DP, Baum H, Grasso A et al. Microbial mimics are major targets of cross-reactivity with human pyruvate dehydrogenase in primary biliary cirrhosis. J Hepatol. 2004;40:31–9.
20. Fussey SP, Lindsay JG, Fuller C et al. Autoantibodies in primary biliary cirrhosis: analysis of reactivity against eukaryotic and prokaryotic 2-oxo acid dehydrogenase complexes. Hepatology. 1991;13:467–74.

21. Teoh KL, Rowley MJ, Zafirakis H et al. Enzyme inhibitory autoantibodies to pyruvate dehydrogenase complex in primary biliary cirrhosis: applications of a semiautomated assay. Hepatology. 1994;20:1220–4.

22. Jois J, Omagari K, Rowley MJ, Anderson J, Mackay IR. Enzyme inhibitory antibody to pyruvate dehydrogenase: diagnostic utility in primary biliary cirrhosis. Ann Clin Biochem. 2000;37:67–73.

23. Potter KN, Thomson RK, Hamblin A, Richards SD, Lindsay SG, Stevenson FK. Immunogenetic analysis reveals that epitope shifting occurs during B-cell affinity maturation in primary biliary cirrhosis. J Mol Biol. 2001;306:37–46.

24. Shimoda S, Nakamura M, Shigematsu H et al. Mimicry peptides of human PDC-E2 163-176 peptide, the immundominant T-cell epitope of primary biliary cirrhosis. Hepatology. 2000;31:1212–16.

25. Shigematsu H, Shimoda S, Nakamura M et al. Fine specificity of T cells reactive to human PDC-E2 163-176 peptide, the immunodominant autoantigen in primary biliary cirrhosis: implications for molecular mimicry and cross-recognition among mitochondrial autoantigens. Hepatology. 2000;32:901–9.

26. Tanimoto H, Shimoda S, Makanura M et al. Promiscuous T cells selected by *Escherichia coli*: OGDC-E2 in primary biliary cirrhosis. J Autoimmun. 2003;20:255–63.

27. Kita H, Lian ZX, Van de Water J et al. Identification of HLA-A2-restricted CD8(+) cytotoxic T cell responses in primary biliary cirrhosis: T cell activation is augmented by immune complexes cross-presented by dendritic cells. J Exp Med. 2002;195:113–23.

28. Ludwig B, Junt T, Hengartner H, Zinkernagel RM. Dendritic cells in autoimmune diseases. Curr Opin Immunol. 2001;13:657–662.

29. Klein R, Wiebel M, Engelhart S, Berg PA. Sera from patients with tuberculosis recognise the M2a-epitope (E2-subunit of pyruvate dehydrogenase) specific for primary biliary cirrhosis. Clin Exp Immunol. 1993;92:308–16.

30. Vilagut L, Vila J, Vinas O et al. Cross-reactivity of anti-*Mycobacterium gordonae* antibodies with the major mitochondrial autoantigens in primary bilary cirrhosis. J Hepatol. 1994;21: 673–7.

31. O'Donohue J, McFarlane B, Bomford A, Yates M, Williams R. Antibodies to atypical mycobacteria in primary biliary cirrhosis. J Hepatol. 1994;21:887–9.

32. O'Donohue J, Fidler H, Garcia-Barcelo M, Nouri-Aria K, Williams R, McFadden J. Mycobacterial DNA not detected in liver sections from patients with primary biliary cirrhosis. J Hepatol. 1998;28:433–8.

33. Tanaka A, Prindiville TP, Gish R et al. Are infectious agents involved in primary biliary cirrhosis? A PCR approach. J Hepatol. 1999;31:664–71.

34. Jones DEJ, Palmer JM, Leon MP, Yeaman SJ, Bassendine MF, Diamond AG. T-cell responses to tuberculin-purified protein derivative in primary biliary cirrhosis: evidence for defective T-cell function. Gut. 1997;40:277–83.

35. Dohmen K, Shigematsu H, Miyamoto Y, Yamasaki F, Irie K, Ishibashi H. Atrophic corpus gastritis and *Helicobacter pylori* infection in primary biliary cirrhosis. Dig Dis Sci. 2002;47: 162–9.

36. Nilsson I, Kornilovs KI, Lindgren S, Ljungh A, Wadstrom T. Increased prevalence of seropositivity for non-gastric *Helicobacter* species in patients with autoimmune liver disease. J Med Microbiol. 2003;52:949–53.

37. Nilsson HO, Taneera J, Castedal M, Glatz E, Olsson R, Wadstrom T. Identification of *Helicobacter pylori* and other *Helicobacter* species by PCR, hybridization, and partial sequencing in human liver samples from patients with primary sclerosing cholangitis or primary biliary cirrhosis. J Clin Microbiol. 2000;38:1072–6.

38. Durazzo M, Rosina F, Premoli A et al. Lack of association between seroprevalence of *Helicobacter pylori* infection and primary biliary cirrhosis. World J Gastoenterol. 2004;10: 3179–81.

39. Abdulkarim AS, Petrovic LM, Kim WR, Angulo P, Lloyd RV, Lindor KD. Primary biliary cirrhosis:an infectious disease caused by *Chlamydia pneumoniae*? J Hepatol. 2004;40:380–4.

40. Leung PSC, Park O, Matsumura S, Ansari AA, Coppel RL, Gershwin ME. Is there a relation between *Chlamydia* infection and primary biliary cirrhosis? Clin Dev Immunol. 2003;10:227–33.

41. Harada K, Tsuneyama K, Sudo Y, Masuda S, Nakanuma Y. Molecular identification of bacterial 16S ribosomal RNA gene in liver tissue of primary biliary cirrhosis: is *Propionibacterium acnes* involved in granuloma formation? Hepatology. 2001;33:530–6.

42. Selmi C, Balkwill DL, Invernizzi P et al. Patients with primary biliary cirrhosis react against a ubiquitous xenobiotic-metabolizing bacterium. Hepatology. 2003;38:1250–7.

43. Di Leo V, Venturi C, Baragiotta A, Martines D, Floreani A. Gastroduodenal and intestinal permeability in primary biliary cirrhosis. Eur J Gastroenterol Hepatol. 2003;15:967–73.

44. Kreisel W, Siegel A, Bahler A et al. High prevalence of antibodies to calreticulin of the IgA class in primary biliary cirrhosis: a possible role of gut-derived bacterial antignes in its aetiology? Scand J Gastroenterol. 1999;34:623–8.

45. Roesler KW, Schmider W, Kist M et al. Identification of [beta]-subunit of bacterial RNA-polymerase – a non-species specific bacterial protein – as target of antibodies in primary bilairy cirrhosis. Dig Dis Sci. 2003;48:561–9.

46. Lang P, Klein R, Becker EW, Berg PA. Distribution of the PBC-specific- (M2) and the naturally occurring mitochondrial antigen – (NOMAg) systems in plants. Clin Exp Immunol. 1992;90:509–16.

47. Hopf U, Stemerowicz R, Rodloff A et al. Relation between *Escherichia coli* R(rough)-forms in gut, lipid A in liver, and primary biliary cirrhosis. Lancet. 1989;ii:1419–21.

48. Ide T, Sata M, Nakano H, Susuki H, Tanikawa K. Increased serum IgM class anti-lipid A antibody and therapeutic effect of ursodeoxycholic acid in primary biliary cirrhosis. Hepato-Gastroenterology. 1997;44:1569–73.

49. Ballot E, Bandin O, Chazuilleres O, Johanet C. Immune response to lipopolysaccharide in primary biliary cirrhosis and autoimmune diseases. J Autoimmun. 2004;22:153–8.

50. Metcalf JV, Mitchison HC, Palmer JM, Jones DEJ, Bassendine MF, James OFW. Natural history of early primary biliary cirrhosis. Lancet. 1996;348:1399–402.

51. Shibata M, Onozuka Y, Morizane T et al. Prevalence of antimitochondrial antibody in Japanese corporate workers in Kanagawa prefecture. J Gastroenterol. 2004;39:255–9.

52. Joplin R, Gordon Lindsay J, Johnson GD, Strain A, Neuberger J. Membrane dihydroli-poamide acetyltransferase (E2) on human biliary epithelial cells in primary biliary cirrhosis. Lancet. 1992;339:93–4.

53. Van de Water J, Turchany J, Leung PS et al. Molecular mimicry in primary biliary cirrhosis. Evidence for biliary epithelial expression of a molecule cross-reactive with pyruvate dehydrogenase complex-E2. J Clin Invest. 1993;91:2653–64.

54. Joplin R, Wallace LL, Lindsay JG, Yeaman SJ, Palmer JM, Neuberger JM. Pyruvate dehydrogenase-X (PDH-X) is associated with the biliary epithelial cell (BEC) plasma membrane in primary biliary cirrhosis (PBC). Hepatology. 1996;24:165A.

55. Nakanuma Y, Tsuneyama K, Kono N, Hoso M, Van de Water J, Gershwin ME. Biliary epithelial expression of pyruvate dehydrogenase complex in primary biliary cirrhosis:an immunohistochemical and immunoelectron microscopic study. Hum Pathol. 1995;26:92–8.

56. Joplin R, Wallace LL, Johnson GD et al. Subcellular localisation of pyruvate dehydrogen-ase dihydrolipoamide acetyltransferase in human intrahepatic biliary epithelial cells. J Pathol. 1995;176:381–90.

57. Sadamoto T, Joplin R, Keogh A, Mason A, Carman W, Neuberger J. Expression of pyruvate dehydrogenase complex PDC-E2 on biliary epithelial cells induced by lymph nodes from primary biliary cirrhosis. Lancet. 1998;352:1595–6.

58. Xu L, Shen Z, Guo L et al. Does a betaretrovirus infection trigger primary biliary cirrhosis? Proc Natl Acad Sci USA. 2003;100:8454–9.

59. Xu L, Sakalian M, Shen Z, Loss G, Neuberger J, Mason A. Cloning the human betaretrovirus proviral genome from patients with primary biliary cirrhosis. Hepatology. 2004;39:151–6.

60. Poupon R, Poupon RE. Retrovirus infection as a trigger for primary biliary cirrhosis? Lancet. 2004;363:260.

61. Selmi C, Ross SR, Ansari AA et al. Lack of immunological or molecular evidence for a role of mouse mammary tumor retrovirus in primary biliary cirrhosis. Gastroenterology. 2004; 127:493–501.

62. Reynoso-Paz S, Leung PSC, Van de Water J et al. Evidence for a locally driven mucosal response and the presence of mitochondrial antigens in saliva in primary biliary cirrhosis. Hepatology. 2000;31:24–9.

63. Palmer JM, Doshi M, Kirby JA, Bassendine MF, Yeaman SJ, Jones DEJ. Secretory autoantibodies in primary biliary cirrhosis. Clin Exp Immunol. 2000;122:1–7.
64. Ikuno N, Mackay IR, Jois J, Omargari K, Rowley MJ. Antimitochondrial autoantibodies in saliva and sera from patients with primary biliary cirrhosis. J Gastroenterol Hepatol. 2001;16:1390–4.
65. Matsumura S, Van de Water J, Leung P et al. Caspase induction by IgA antimitochondrial antibody: IgA-mediated biliary injury in primary biliary cirrhosis. Hepatology. 2004;39: 1415–22.
66. Mowat AM. Anatomical basis of tolerance and immunity to intestinal antigens. Nature Rev Immunol. 2003;3:331–41.

13
Role of genetics in immunopathogenesis

P. T. DONALDSON

INTRODUCTION

For most common diseases, including heart disease, diabetes, hypertension, cancer, and both autoimmune and viral liver disease, genes are likely to determine an individual's 'disease risk'[1–3]. However, the diseases autoimmune hepatitis (AIH), primary biliary cirrhosis (PBC) and primary sclerosing cholangitis (PSC) are not single-gene Mendelian diseases, they are genetically 'complex'. They are for the most part multifactorial, polygenic disorders, where possession of a specific genetic variant (mutation or polymorphism) at a locus (an allele) simply increases or reduces the risk of disease (a trait) developing[2]. Moreover, possession of a disease-promoting allele (or disease-causing mutation, DCM) does not by itself lead to disease (incomplete penetrance); other disease-promoting alleles and/or encounter with specific environmental factors may be necessary for disease expression. In addition, not all of the DCM identified in complex disease are involved in disease initiation; many polymorphisms and mutations act to determine the clinical phenotype of a disease, for example: severity, progression and response to treatment.

Understanding the genetic basis of autoimmune liver disease will offer new approaches to both disease diagnosis and patient management, but the most important impact will be in understanding disease pathogenesis. Historically immunogenetics has been mostly concerned with the human major histocompatibility complex (MHC), especially the classical MHC genes that encode the human leucocyte antigens (HLA), which are essential elements in the formation of the 'immune synapse' for T cell immune responses (acquired immunity). However, an increasing number of investigators are now focusing on non-classical MHC genes and also on non-MHC immunogenes that affect both innate and acquired immunity to self and foreign antigens.

Studies in liver disease have detailed MHC-encoded susceptibility to a number of different autoimmune and also viral liver diseases[3]. These studies highlight important differences and similarities between diseases, and also implicate different pathways in disease genesis. In this chapter I will present some essential background, and then I will use three different examples from the current literature to demonstrate how immunogenetic data may focus our attention on immunopathogenesis in each of the three autoimmune liver diseases named above. In doing this I hope to promote a broader understanding of the subject, and indicate the way forward for new investigations in genetically complex autoimmune liver diseases. The subject of complex liver disease has been widely reviewed; therefore I will try not to repeat material that is available elsewhere[3-7].

ESSENTIAL BACKGROUND

Complex diseases are often seen as the 'poor relation' in genetics compared with single-gene (mostly) Mendelian diseases. As a consequence many 'card-carrying' geneticists have distanced themselves from this field, and immunogenetics became a forgotten discipline peopled by 'tissue (HLA) typers' and their ilk. It is important to understand this because we often forget that immunogenetics belongs to genetics, and consequently we should ask the same questions when we set out to identify immunogenes in 'complex disease' that we would apply to any study of any genetic disease.

There are two basic strategies we can use to identify DCM in complex diseases: linkage and association analysis. However, before we can begin to look for genes we need to establish the evidence for a heritable component.

MEASURING THE HERITABLE COMPONENT IN COMPLEX DISEASE

To determine the best and most pragmatic means of investigating a complex disease we need to have some estimate of the heritable component. Simple measures of heritability include: concordance in monozygotic and dizygotic twins, the degree of familial aggregation and calculations such as sibling relative risk (λs). In autoimmune liver disease no such data exist for type 1 AIH or PSC. PBC is the exception: the λs for PBC is 10.5 (see ref. 8), concordance in monozygotic twins may be as high as 75% (though this estimate is based on only eight twin pairs and may be biased), and familial prevalence may be as high as 6.4% (i.e. approximately 1 in 20 patients may have an affected family member). These figures are similar to those quoted for other auto-immune conditions.

STRATEGIES TO IDENTIFY GENES IN AUTOIMMUNE LIVER DISEASE

Autoimmune liver diseases are relatively rare and mostly adult-onset disorders; consequently they do not lend themselves to family-based genetic studies. This rules out linkage analysis, as a potential method for gene mapping. Association analysis is for the most part the only practical option. Furthermore, as there are relatively few children with these diseases, intra-familial association methods such as transmission disequilibrium testing (TDT) and sib-pair analysis are also out of the question. This leaves only case–control association analysis. To the 'card carrying' geneticist, case–control association studies provide 'a useful complimentary strategy once a region of the genome has been identified by linkage analysis'[2]. In other words this is not considered to be the gold standard method. Despite this view in complex diseases, case–control association analysis has been more successful than linkage-based studies. Recent discussion of fruits of whole genome scanning studies reveal that many linkage-based studies on large family collections have not produced results[9].

The problem with case–control association studies is that they are too often irreproducible[10]; consequently they have fallen into disrepute, but recently they are regaining favour. The pros and cons have been widely reviewed. The central issue is study design and especially statistical analysis and planning[10]. It is now recognized that well-designed case–control studies provide a powerful tool for gene mapping in complex disease. However, since we are dealing with mostly multifactorial diseases, where there may be several DCM, we need to design studies that will confidently detect low-risk DCM (i.e. with odds ratios of 1.5 or less) and it is thought unlikely that more than a few DCM will have large risk ratios (odds ratios of 2 or more)[11,12].

One region of the genome where the DCM appear to have large effects is the MHC. There are several possible reasons for this.

1. Most MHC associations relate to extended haplotypes – each of which may carry several disease-promoting alleles (i.e there may be multilocus involvement – with haplotypes acting as 'disease-promoting cassettes'). This, however, can be a disadvantage, because understanding the role of individual MHC genes within these haplotypes is often complicated by the strong linkage disequilibrium between some MHC-encoded gene loci. For example: linkage disequilibrium within the HLA 8.1 haplotype makes it difficult to map susceptibility between the HLA B locus and the *DQB1* locus.

2. Studies of HLA and some other MHC genes are not restricted to individual single-nucleotide polymorphisms (SNP), but consider the multiple variations at each locus. Thus, when we assign a particular HLA *DRB1* allele, for example, we are considering the sum of variation at that locus.

3. The majority of the MHC gene variation is functional and the genes within the MHC play important roles in both acquired and innate immunity (see below). In contrast many studies of non-classical MHC genes (for example *TNFA*) or of non-MHC immunogenes concentrate on individual SNP rather than on SNP haplotypes. This approach assumes that, where there are multiple SNP within a gene, only one of them (the one being studied) has any functional significance.

SELECTION OF CANDIDATES

Candidate genes for genetic studies are selected on the basis of their location if linkage data are available, or more frequently on prior knowledge of similar (related) diseases, knowledge of the disease pathogenesis, or knowledge of the gene and its function. All of the latter options involve a degree of speculation. The rationale for considering findings from other similar diseases or syndromes is based on the principle that not all DCM will be disease-specific. A considerable number will be non-specific DCM, that promote disease through activation of common or shared pathways. For example, fibrosis is a common end-stage pathway in liver disease. It therefore seems reasonable to speculate that there may be a common profibrotic genotype in patients with rapidly progressing fibrosis irrespective of the underlying liver disease[12]. Other examples include the *CTLA4* gene in autoimmunity[13] and the cytokine gene cluster on chromosome 5q in inflammatory bowel disease (IBD)[14].

In addition data from animal models, especially 'knock-outs' and 'knock-ins', may also be informative in selecting candidate genes for investigation. Microarray analysis may also provide useful clues, and in the absence of sufficient families for linkage-based whole genome scans, whole genome association analysis (WGAA) based on regularly spaced SNP is being considered as a method by which to identify potential disease genes and areas of the genome for further study[15].

The experts also advise that, when selecting mutations or polymorphisms to be studied, one should be mindful of both their function and biological relevance[10,16]. In the context of this chapter it is also important to remember that most biological systems have a degree of built-in redundancy to cope with relative differences in key components, and that biological pathways are complex. A good example of such a system, relevant to immunopathology, is the proinflammatory cytokine interleukin-1[3] and the importance of studying whole biological systems is highlighted in a recent 'blueprint' for genomic research by Francis Collins et al. at the NIH, Human Genomic Research Institute[17].

AN INTRODUCTION TO IMMUNOGENETICS

Immunogenetics concerns that branch of genetics that deals with the genes which regulate the immune response. Interest in immunogenetics of disease was born out of the interest in the 1960s in the role of HLA in transplant acceptance and rejection. However, the idea that the immune response is under genetic control predates the clinical need to study HLA, and its origins can be traced back to earlier in the twentieth century. Current and up-to-date nomenclature for HLA system can be found at www.anthonynolan.org.uk. The structure, function and organization of the human MHC is adequately covered elsewhere in both published reviews and on the Internet, and I will not duplicate that information here (see refs 3–7, 18 and http://www.anthonynolan.org).

Immunogenetics is not restricted to HLA or the MHC. The ABO blood groups, immunoglobulin and complement gene polymorphisms have long been included in the list of immunogenes, but in the last decade of the twentieth century, as we began to understand the extent of polymorphism in the human genome, we realized that almost any gene that encodes an immune active product can act as an 'immune response gene.'

The immunogenes that have been studied in autoimmune liver disease include several members of the immunoglobulin superfamily (HLA, immuno-globulin and T cell receptor genes); the MHC-encoded complement genes (*C2, C4A, C4B* and factor B or Bf); the MHC-encoded cytokine genes (*TNFA* and *TNFB*); non-MHC encoded cytokines genes (including *IL1A, IL1B, IL1RN, IFNG* (the currently identified genetic polymorphisms in the cytokine genes are summarized at www.pam.bris.ac.uk/services/GAI/cytokine4)), accessory molecules (including cytotoxic T-lymphocyte antigen-4, *CTLA4*), mannose-binding lectin (MBL); natural resistance-associated macrophage protein-1 (*NRAMP1*), caspase-8 (*CASP8*), vitamin D receptor (*VDR*), apolipoprotein-E (*APOE*); Fas (*TNFRSF6*), and matrix metalloproteinase-3 (*MMP3*). Of these the associations with MHC-encoded genes are the most consistent and convincing (Tables 1 and 2). Therefore, the three examples I have chosen all involve the MHC, but all are different.

HLA *DRB1* ALANINE/LYSINE-71 IN TYPE 1 AUTOIMMUNE HEPATITIS: IS THIS EVIDENCE FOR A MAJOR ROLE OF T CELL IMMUNITY IN THE GENESIS OF TYPE 1 AIH?

Type 1 AIH is a classical autoimmune disease in as much as the majority of patients are female, have strong genetic associations, particularly with the HLA 8.1 haplotype (formerly known as the A1-B8-DR3 haplotype) and respond to therapy with corticosteroids. In European patients and their North American cousins three different HLA haplotypes are found to be associated with this disease: two (the HLA 8.1 haplotype characterized by *DRB1*0301* and the *DRB1*0401-DQA1*03-DQB1*0301* haplotype) are found at an increased frequency and one (the *DRB1*1501-DQA1*0102-DQB1*0602* haplotype) is found at a reduced frequency, in patients versus matched controls (Table 1)[6]. Studies in other populations indicate similar genetic associations and help to identify the *DRB1* locus as the major susceptibility/resistance locus in this disease. The *DRB1* association in type 1 AIH points to the MHC-class II antigen-presenting pathway and events in T cell activation as the keys to understanding disease initiation. This fits well with the characteristics of type 1 AIH as a predominantly T cell-mediated disease. Furthermore, comparative analysis of the DNA sequence of the two major susceptibility alleles (*DRB1*0301* and *DRB1*0401*) versus the disease resistance allele (*DRB1*1501*) suggests that the key amino acid for disease susceptibility/resistance is that found at position-71 of the DRβ polypeptide. Thus, HLA alleles encoding lysine at position-71 are associated with an increased risk of disease (OR = 4.65) and those encoding alanine are associated with a reduced risk (OR = 0.52).

This model illustrates how an immunogenetic association can promote our understanding of the pathogenesis of a disease. In this case the data strongly support the hypothesis that relative differences in T cell responses and acquired immunity are the key to understanding the pathogenesis of type 1 AIH. However, the lysine-71 model does not tell us much about the autoantigen in type 1 AIH and is not universally accepted. An alternative model based on

Table 1 Summary of associations with HLA haplotypes in autoimmune liver disease

Disease	Population	Haplotype	Risk ratio
PBC	Japan	**DRB1*0803**-DQ3-DPB1*0501	N.A.
	NEC-Europe and NBC-USA	**DRB1*0801**-DQA1*0401-DQB1*0402	8.16
PSC	NEC-Europe	B8-**MICA*008**-TNFA*2-DRB3*0101-**DRB1*0301**-DQB1*0201	2.69
		DRB3*0101-DRB1*1301-DQA1*0103-**DQB1*0603**	3.8
		MICA*008-DRB5*0101-DRB1*1501-DQA1*0102-**DQB1*0602**	1.52
		DRB4*0103-DRB4*0401-DQA1*03-**DQB1*0302**	0.26
		DRB4*0103-DRB1*0701-DQA1*0201-**DQB1*0303**	0.15
		MICA*002	0.15
AIH	NEC-Europe and NBC-USA	A1-B8-MICA*008-TNFA*2-DRB3*0101-DRB1*0301-DQB1*0201	4.6–5.51
		DRB4*0103-**DRB1*0401**-DQA1*0301-DQB1*0301	3.3–3.7
		DRB5*0101-**DRB1*1501**-DQA1*0102-DQB1*0602	0.32–0.4
	Japan	Bw54-DRB4*-**DRB1*0405**-DQA1*0301-DQB1*0401	
	Argentina and Brazil	Adults: DRB4*0101-**DRB1*0405**	10.4
		Children: DRB3*0101-**DRB1***1301-DQA1*0103-DQB1*0603	16.3
		Children: DRB3*0101-**DRB1***0301-DQA1*0501-DQB1*0201	3.0

NEC, northern European Caucasoid.

All studies are based on adult cases unless otherwise marked.

Protective haplotypes have risk values <1.

Primary susceptibility/resistance alleles are in **bold** on each haplotype.

132

Table 2 Summary of key HLA haplotypes associated with acute (self-limiting versus protracted (chronic) infection in viral hepatitis

Virus	Population	Haplotype/allele
HAV	Argentina	Children: *DRB3*0101-**DRB1*1301**-DQA1*0103-DQB1*0603*
HBV	Gambia, acute infection	*DRB3*0301-**DRB1*1302**-DQA1*0102-DQB1*0501* *DRB3*0301-**DRB1*1302**-DQA1*0102-DQB1*0604* *DRB3*0301-DRB1*1301-DQA1*0103-DQB1*0603*
HCV	NEC-Europe, acute infection	*DRB3*0101-DRB1*1101-DQA1*0501-**DQB1*0301** DRB3*0101-DRB1*1104-DQA1*0501-**DQB1*0301** DRB4*0103-DRB1*0401-DQA1*03-**DQB1*0301** **DRB1*0101**-DQA1*0101-DQB1*0501*
HCV	NEC-USA Black-USA	***DRB1*0101**-DQA1*0101-DQB1*0501* ***DRB1*0101**-DQA1*0101-DQB1*0501*
HCV	NEC-Europe, chronic infection	*DRB5*0101-DRB1*1501-DQA1*0102-DQB1*0602* DRB3*0101-**DRB1*0301**-DQA1*0501-DQB1*0201* DRB4*0103-DRB1*0701-DQA1*0201-DQB1*0201*

NEC, northern European Caucasoid.

Summary represents key selected studies only. All studies based on adult cases unless marked otherwise.

valine/glycine dimorphism at position-86 of the DRβ polypeptide has been proposed for patients from Argentina and Brazil[19], but this later model does not fit the European/North American data[6]. There appears to be some support for DRβ-71 from Japan, although in Japan the key amino acid is arginine[20]. Interestingly arginine has similar properties to lysine, being a polar, highly charged amino acid.

It is of course possible that these different models of disease susceptibility are equally valid. It has been suggested that different genetic associations based on the *DRB1* locus may have arisen in different populations because of selection pressure exerted through exposure to endemic viruses. For example: in South America where hepatitis A virus is endemic, persistent HAV infection is associated with carriage of the HLA *DRB1*1301* allele[21], the same allele that characterises the majority of children who develop type 1 AIH in that population[19]. If there are different HLA associations in different populations, then we may eventually come to see HLA associations as the molecular footprints, left by the prevailing environmental triggers that precipitate type 1 AIH. This hypothesis is not entirely speculative. Viral triggers have been identified in individual cases of type 1 AIH.

However, H. Mencken said: 'for every complex human problem there is a neat and simple answer that is wrong'. He is quite correct in this supposition. In simplifying the story of MHC associations in type 1 AIH we overlook a number of facts. Our knowledge of the MHC in type 1 AIH is based on relatively few subjects and is incomplete, as many MHC genes have not yet been investigated. Early observations suggest that European and North

American patients with *DRB1*0301* or the extended haplotype (shorthand HLA 8.1) have quite different clinical characteristics to those with the *DRB1*0401* haplotype[6], even though both haplotypes carry *DRB1* alleles that encode lysine-71. This latter observation suggests that other genes within the MHC (or closely linked genes) must also be active and may modify the clinical phenotype in type 1 AIH.

Yet, despite my reservations, this example does prove the principle that a genetic association may point the way to understanding disease pathogenesis. This model is in keeping with type 1 AIH as a T cell-mediated disease. The possibility of overlap between susceptibility to viral hepatitis and susceptibility to type 1 AIH, in some populations at least, is most interesting. It remains to be determined whether HLA associations do tell us more about susceptibility to potential infectious triggers than they do about the self-antigens, and whether my premise concerning the immunopathology of type-1 AIH is correct.

MHC-CLASS I CHAIN-LIKE ALLELES AND SUSCEPTIBILITY AND RESISTANCE TO PSC: IS THIS EVIDENCE FOR A MAJOR ROLE FOR THE INNATE IMMUNE RESPONSE IN PSC?

The strongest HLA associations reported in autoimmune liver disease are those for PSC and type 1 AIH (Table 1). However, despite apparent similarities between these two diseases, the reported associations appear to map to different gene loci within the MHC. Thus in contrast to type 1 AIH (above) in PSC, MHC-encoded susceptibility appears to be involving either a combination of DR, DQ and *MICA* alleles or perhaps *MICA* alone[5,22].

Association studies have identified six differentially distributed haplotypes in PSC compared to healthy controls (Table 1). Three are associated with an increased risk of disease and three with a reduced risk of disease. One of the latter *MICA*002* is responsible for a very significant reduction in the risk of disease (OR = 0.15)[23]. This effect is very strong for a genetic association in a disease with very little obvious evidence of a heritable component. Two of the susceptibility haplotypes, the HLA8.1 haplotype and the *DRB1*1501-DQA1*0102-DQB1*0602* haplotype, each carry the *MICA*008* allele, but this allele is absent from the third susceptibility haplotype (*DRB1*1301-DQA1*0103-DQB1*0603*). Though it does not therefore explain all of the MHC-encoded susceptibility to PSC, there is a very high frequency of homozygosity for *MICA*008* in PSC, and this alone may explain the genetic associations with the first two haplotypes. It is important to note that there are no other common MHC alleles on these two haplotypes other than *Cw*0701*, which is not a major risk factor for PSC[5,22].

The *MICA*008* and *MICA*002* associations may have important implications in terms of immune pathogenesis in PSC. The potential involvement of the MIC genes indicates a more prominent role for the innate immune response[23]. MICα molecules are expressed exclusively on gastrointestinal and thymic epithelia and are also seen in non-diseased liver[24]. They may be induced by stress and heat shock and have been identified as a ligand for γδ T cells, natural killer (NK) (CD56[+]) cells and T cells expressing the NKG2D-DAP10 activatory receptor[25], all of which are abundant in the liver[26]. Increased numbers of γδ and NK cells have been documented in PSC livers[27,28]. If PSC were to arise as a result of an infection this would provide the catalyst for heat shock induction of MICα on biliary epithelium, leading to the activation of intrahepatic γδ and NK cells, subsequent cytokine secretion and cytolytic effector functions. Although the functional significance of the *MICA*008* allele has not been established, this allele carries a short tandem repeat sequence with a premature stop codon, and it has been suggested that this may lead to aberrant or unstable expression of the MICα protein[29]. In individuals homozygous for *MICA*008* this may lead to a loss of function and promote persistent immune activation through this pathway leading to autoimmunity, or it may fail to activate, with the consequence of persistent infection, and an increased likelihood of collateral damage and subsequent elevation in the risk that self (auto) antigens will be recognized.

This discussion presents a simplified picture. Three of the MHC haplotypes associated with PSC do not carry either *MICA*008* or *MICA*002*. The full story is clearly more complicated. However, this too fits with the characteristics of PSC. For example, PSC does not fulfil the classical criteria for an autoimmune disease (organ-specific autoantibodies, female preponderance and responsiveness to corticosteroid therapy); there is an almost universal overlap with IBD (mostly ulcerative colitis) and a very high incidence of malignancy, all of which are potential indications that this is a heterogeneous syndrome with several interrelated pathologies.

Yet despite these complications this example also indicates how an immunogenetic association can inform the debate on the pathogenesis of a disease. The association with the MHC-encoded *MICA* alleles may indicate a strong, but not exclusive, role for non-classical T or NK cells in the pathogenesis of PSC, and may also be important in understanding malignancy in this disease. Once again it remains to be determined whether the hypotheses above are correct. However, whether they are confirmed or refuted, at least there are, thanks to the genetic studies, hypotheses to be tested.

THE MHC AND PBC A MIXTURE OF DRB1 AND COMPLEMENT: IS THIS EVIDENCE FOR AN INFECTIOUS AETIOLOGY IN THIS DISEASE?

At first glance the MHC appears to have a lesser role in PBC than in type 1 AIH or PSC. There is no association with the HLA 8.1 haplotype, an association which was at one time considered to be a necessary hallmark of autoimmunity. Studies of HLA in PBC[4] do report a weak but significant association with HLA DR8 (*DRB1*0801* in Europeans and *DRB1*0803* in Japan). The risk of disease for Europeans with the *DRB1*0801* allele may be up to 8.16 times greater than for those without. However, in Europeans this association accounts for only 15–25% of PBC patients. This may indicate that the true association lies elsewhere along the chromosome, with *DRB1*0801* acting as a linkage marker. There are several likely candidates, only one of which has been studied. The MHC-encoded complement genes *C2*, *C4A*, *C4B* and *Bf* are often overlooked in immunogenetic studies because they are difficult to genotype. However, two early studies did include them[30,31]. The first, which was based on 33 PBC patients, found a four-fold increased risk of disease associated with the *C4B*2* allele[30]. The second, based on 25 patients, suggested that those with *C4A*Q0* and *C4B*2* were at increased risk of PBC[31]. Interestingly *C4B*2* and *DR8* may be found on the same haplotype. Though these studies are small they do indicate a potential role for the complement genes. Had these two studies been based on larger numbers, and the distribution of genotypes remained the same as reported above, the influence of *C4B*2* on disease risk would have been much more dramatic than that seen for *DRB1*0801*.

The MHC class III encoded complement genes are particularly interesting in both autoimmune and viral liver disease. The liver is a major site of complement production. The complement proteins C2, C4 and Bf are involved in early stages of both the classical and alternative complement cascades. In particular, C4 and C2 catalyse the production of C3 and C5 convertases, which convert C3 to C3b and C5 to C5b, respectively. This is the critical step that initiates the formation of the membrane attack complex whilst the C3a and C5a fragments promote chemotaxis (Bf has a similar function in the alternative pathway). The genes encoding the two isoforms of C4, *C4A* and *C4B*, show considerable polymorphism, with more than 35 alleles identified. In addition the number of *C4* genes on a haplotype can vary. A deleted *C4A* is the most common of the non-expressed *C4A* 'null' alleles (which are all labelled *C4A*Q0*). C4 deficiency is associated with impaired antibody responses to T cell-dependent antigens, and individuals with C2 and C4 deficiency are prone to systemic lupus-like diseases[32], especially those with homozygous deficiencies of C4 (and C1q) which account for 75% of cases of systemic lupus erythematosus (SLE), and homozygous deficiencies of C2 (33% of cases)[33]. However, the effect of partial (heterozygous) deficiencies is less clear. The two isoforms of C4 differ in their efficiency of covalent binding to antigens and antibody. *C4A* binds more efficiently to amino groups and antigen–antibody complexes, while *C4B* prefers hydroxyl groups found on red cells. Therefore, *C4A* is more effective in immune complex clearance, whereas *C4B* has a greater haemolytic activity. A relative deficiency, arising from possession of different *C4B* alleles for example, may lead to persistence of harmful pathogens and slower immune complex clearance, which may also prolong exposure to neo-antigen or cryptic self-antigens, increasing the risk of autoimmunity. Such a simple relationship could explain the MHC association in PBC, and may be particularly relevant in light of the numerous reports linking both viral and bacterial diseases with disease susceptibility[34–37].

However, this is entirely speculative, and recent studies have suggested that there may be other, particularly protective, MHC associations in PBC (unpublished data, October 2004), suggesting that there is more to uncover in respect to the MHC in PBC.

Nevertheless, this example also works. Overall these findings are not in keeping with the picture of PBC as a T cell-mediated autoimmune disease. Studies of the T cell response in PBC have also failed to link DR8 alleles with presentation of key immunogenic peptides[38]. Instead the responsive T-cells appear to be restricted by the HLA *DRB4*0101* allele, which encodes DRw53, a very common HLA antigen found in more than 50% of most populations, but which is not associated with susceptibility to the disease. This latter observation is not informative in terms of the pathogenesis of PBC, other than in a negative sense.

Overall the immunogenetic data presented here suggest that immunological investigations in PBC should not be restricted to the activities of T cells, and illustrates how genetic association studies can illuminate alternative avenues for investigation of disease pathogenesis.

RELATIONSHIP BETWEEN MHC AND VIRAL HEPATITIS: A CLUE TO SHARED PATHOLOGIES?

The relationship between host HLA and viral hepatitis has been widely explored. In most cases the risk of infection is not itself genetically determined, but host genes do play a role in determining outcome following exposure. Thus individuals who have the *DRB1*1301* allele are more prone to persistent hepatitis A virus (HAV) infection[21], those with *DRB1*1302* are more prone to persistent hepatitis B virus (HBV) infection[39] and those with *DQB1*0301* haplotypes are more likely to have self-limiting hepatitis C virus (HCV) infection[40,41].

It is particularly pertinent in respect to this chapter to note that the key HLA haplotypes in viral infection are also implicated in AIH and PSC (Tables 1 and 2), and although there is no absolute correlation between haplotypes that promote viral persistence and those which promote autoimmune disease, the overlap is intriguing. Thus *DRB1*0301* appears in some studies of HCV to promote viral persistence[42] and *DRB1*1301* may promote persistent HAV infection[3,21], whereas *DRB1*1501*, which protects from type 1 AIH in Europeans[6], promotes HCV persistence in most studies (albeit weakly), and *DRB1*0701*, which may protect from PSC[22], may promote HCV persistence. Taken together this evidence provides a basis for speculation regarding the overlap between autoimmunity and viral disease, though at present there are insufficient data to take this further. It is hoped that, as we accumulate more data on all of these diseases, these relationships may become clearer, and that they will inform the debate on immune pathogenesis in genetically complex autoimmune liver disease.

CONCLUSIONS

Overall we are still at the beginning of the process of unravelling the genetic basis of these 'complex' autoimmune liver diseases. Much more work is required before we can use this knowledge to understand the pathogenesis of autoimmune liver disease. However, I hope that the reader can see how studies of genes can inform the debate on disease pathogenesis, and how the hypotheses above may help to establish a framework for future studies of disease pathogenesis.

References

1. Hirschhorn JN, Lohmueller K, Bryne E, Hirschhorn K. A comprehensive review of genetic association studies. Genet Med. 2002;4:45–61.
2. Haines JL, Pericak-Vance MA. Overview of mapping common and genetically complex disease genes. In: Haines JL, Pericak-Vance MA, editors. Approaches to Gene Mapping in Complex Diseases. New York: Wiley, 1998:1–6.
3. Donaldson PT. Recent advances in clinical practice: genetics of liver disease: immunogenetics and disease pathogenesis. Gut. 2004;53:599–608.
4. Jones DEJ, Donaldson PT. Genetic factors in the pathogenesis of primary biliary cirrhosis. Clin Liver Dis. 2003;7:841–64.
5. Donaldson PT, Norris S. Immunogenetics in PSC. Balliere's Best Pract Res Clin Gastroenterol. 2001;15:611–27
6. Donaldson PT. Genetics in autoimmune hepatitis. Semin Liver Dis. 2002;22:353–63.
7. Donaldson PT. Genetics of autoimmune and viral liver diseases; understanding the issues. J Hepatol. 2004;41:327–32.
8. Jones DE, Watt FE, Metcalf JV et al. Familial primary biliary cirrhosis reassessed: a geographically-based population study. J Hepatol. 1999;30:402–7
9. Carlson CS, Eberle MA, Kruglyak L, Nickerson DA. Mapping complex disease loci in whole-genome association studies. Nature. 2004;429:446–52.
10. Colhoun H, McKeigue PM, Smith GD. Problems of reporting genetic associations with complex outcomes. Lancet. 2003;361:865–72.
11. Farrell M. Quantitative genetic variation: a post-modern view. Human Mol Genet. 2004; 13(review issue 1):R1–7.
12. Bataller R, North KE, Brenner DA. Genetic polymorphisms and the progression of liver fibrosis: a critical appraisal. Hepatology. 2003;37:493–503.
13. Ueda H, Howson JMM, Esposito L et al. Association of the T-cell regulatory gene CTLA4 with susceptibility to autoimmune disease. Nature. 2003;423:506–11.
14. Ahmed T, Satsangi J, McGovern D et al. The genetics of inflammatory bowel disease. Aliment Pharmacol Ther. 2001;15:731–48.
15. Ozaki K, Ohnishi Y, Iida A et al. Functional SNPs in the lymphotoxin-alpha gene that are associated with susceptibility to myocardial infarction. Nature Genet. 2002;32:650–4.
16. Anonymous. Freely associating. Nature Genet. 1999;22:1–2.
17. Collins FS, Green ED, Guttmacher AE and Guyer MS on behalf of the Human Genome Research Institute, USA. A vision for the future of genomics research: a blueprint for the genomic era. Nature. 2003;422:835–47.
18. Horton R, Wilming L, Rand V et al. Gene map of the extended human MHC. Nature Rev. 2004;5:889–98.
19. Pando M, Lariba J, Fernadez GC et al. Paediatric and adult forms of type 1 autoimmune hepatitis in Argentina: evidence for differential genetic predisposition. Hepatology. 1999; 30:1374–80.
20. Seki T, Ota M, Furuta S et al. HLA class II molecules and autoimmune hepatitis susceptibility in Japanese patients. Gastroenterology. 1992;103:1041–7.
21. Fainboim L, Velasco MCC, Marcos CY et al. Protracted, but not acute, hepatitis A virus infection is strongly associated with HLA-DRB1*1301, a marker for paediatric autoimmune hepatitis. Hepatology. 2001:33:1512–17.
22. Donaldson PT, Norris S. Evaluation of the role of MHC class II alleles, haplotypes and selected amino acid sequences in primary sclerosing cholangitis. Autoimmunity. 2002;35: 555–64.
23. Norris S, Kondeatis E, Collins R et al. Mapping MHC-encoded susceptibility and resistance in primary sclerosing cholangitis: the role of MICA polymorphism. Gastroenterology. 2001;120:1475–82.
24. Groh V, Bahram S, Bauer S et al. Cell stress regulated human major histocompatibility complex class I gene-regulated in gastrointestinal epithelium. Proc Natl Acad Sci USA. 1996;93:12445–50.
25. Bauer S, Groh V, Wu J et al. Activation of NK cells and T cells by NKG2D, a receptor for stress inducible MICA. Science. 1999;285:727–9.
26. Norris S, Doherty DG, McEntee G et al. Natural T cells in the human liver: cytotoxic lymphocytes with dual T cell and natural killer cell phenotype and function are

phenotypically heterogeneous and include Vα24JαQ and γδ T cell receptor bearing cells. Hum Immunol. 1999;60:20–31.

27. Martins EBG, Graham AK, Chapman RW et al. Elevation of γδ T lymphocytes in peripheral blood and livers of patients with primary sclerosing cholangitis and other autoimmune disease. Hepatology. 1996;23:988–99.
28. Hata K, van Thiel DH, Herberman RB et al. Phenotypic and functional characteristics of lymphocytes isolated from liver biopsy specimens from patients with active liver disease. Hepatology. 1992;15:816–23.
29. Fodil N, Pellet P, Laloux L et al. MICA haplotypic diversity. Immunogenetics. 1999;49: 557–60.
30. Briggs DC, Donaldson PT, Hayes P et al. A major histocompatibility complex class III allotype C4B2 associated with primary biliary cirrhosis. Tissue Ant. 1987;29:141–5.
31. Manns MP, Bremm A, Schneider PM et al. HLA DRw8 and complement C4 deficiency as risk factors in primary biliary cirrhosis. Gastroenterology. 1991;101:1367–73.
32. Fielder AH, Walport MJ, Batchelor JR. Family study of the major histocompatibility complex in patients with systemic lupus erythematosus: importance of null alleles of C4A and C4B in determining disease susceptibility. Br Med J. 1983;286:425–8.
33. Pickering MC, Perraudeau M, Walport MJ. HLA and systemic vasculitidies, systemic lupus erythematosus and Sjögren's syndromme. In: Warrens A, Lechler R, editors. HLA in Health and Disease. London: Academic Press, 2000:327–64.
34. Xu L, Shen Z, Gou L et al. Does a beta-retrovirus infection trigger primary biliary cirrhosis? Proc Natl Acad Sci USA. 2003;100:8454–9.
35. Vilagut L, Vila J, Vinas O et al. Cross reactivity of anti-*Mycobacterium gordonae* antibodies with the major mitochondrial auto antigens in PBC. J Hepatol. 1994;21:673–7.
36. Butler P, Hamilton-Miller J, Baum H et al. Detection of M2 antibodies in patients with recurrent urinary tract infection using ELISA and purified PBC specific antigens. Evidence for a molecular mimicry mechanism in the pathogenesis of primary biliary cirrhosis. Biochem Mol Biol Int. 1995;35:473–5.
37. Burroughs AK, Roesenstein IJ, Epstein O et al. Bacteriuria and primary biliary cirrhosis. Gut. 1984;25:133–7.
38. Shimoda S, Nakamura M, Ishibashi H et al. HLA DRB4*010-restricted immunodominant T cell autoepitope of pyruvate dehydrogenase complex in primary biliary cirrhosis: evidence of molecular mimicry in human autoimmune diseases. J Exp Med. 1995;181: 1835–45.
39. Thursz MR, Kwiatkowsky D, Allsopp CEM et al. Association between an MHC class II allele and clearance of hepatitis B virus in the Gambia. N Engl J Med. 1995;332:1065–9.
40. Alric L, Fort M, Izopet J-P et al. Study of host- and virus-related factors associated with spontaneous hepatitis C virus clearance. Tissue Ant. 2000;56:154–8.
41. Cramp M, Carucci P, Underhill J et al. Association between HLA class II genotype and spontaneous clearance of hepatitis C viremia. J Hepatol. 1998;29:207–13.
42. McKiernan SM, Hagan R, Curry M et al. The MHC is a major determinant of viral status, but not fibrotic stage, in individuals infected with hepatitis C. Gastroenterology. 2000;118: 1124–30.

Section V
Diagnosis

Chair: V.J. DESMET and M.P. MANNS

14
Liver histology in autoimmune hepatitis – diagnostic implications

H. P. DIENES

INTRODUCTION

Autoimmune hepatitis is a chronic progressive necroinflammatory liver disease mostly occurring in female individuals and leading to liver-related death within 6 months of diagnosis in as many as 40% of patients not treated. Cirrhosis develops in at least 40% of survivors, thus presenting an example of progressive liver disease. Pathogenesis is based on a complex interaction between triggering factors such as viral infections, autoantigens, genetic predispositions and immunoregulatory networks[1]. Aetiology is very complex but general agreement has been achieved that self-reactive CD4[+] T helper lymphocytes are the common critical factor in autoimmune hepatitis and responsible for autodestruction of the liver with accompanying immune mechanisms by CD8[+] lymphocytes and lymphocytes of the humoral immune response with B cells and plasma cells. So the disease was found to be associated with other autoimmune syndromes and it has been agreed that autoimmune hepatitis does not represent a homogeneous entitiy[2], but rather a syndrome characterized by immunogenetic associations of HLA-A1, B8DR3 or DR4[3], a spectrum of autoantibodies against several antigens of liver cells, elevation of liver enzymes as a token of liver cell destruction and general symptoms of liver failure.

Histopathology of the liver in autoimmune hepatitis reflects a general understanding of the underlying immune T cell-mediated mechanisms in destruction of liver cells, thus being a powerful tool in diagnosis and a refined instrument in the understanding of mechanisms of liver cell destruction[4,5].

DIAGNOSIS OF AUTOIMMUNE HEPATITIS (AIH)

The diagnosis of AIH is a clinical diagnosis based on history, family background, biochemical tests, serological and immunogenetic assays, as well as histopathology and response to therapy[2].

So far there is no pathognomonic diagnostic test, and a broad repertoire of autoantibodies has to be checked to establish diagnosis. The conventional spectrum of autoantibodies includes ANA, SMA and LKM1, that altogether are present in up to two-thirds of patients[6].

New autoantibodies continue to be characterized, since they may enhance diagnostic precision and may be useful as prognostic markers. These include autoantibodies to actin, ASGPR, SLA and LC1[7]. However, all these auto-antibodies are present only in a certain percentage of patients, and so far none of them has proved to be organ- and disease-specific at the same time. Applying the whole spectrum of autoantibody tests, however, still leaves a group of so-called seronegative AIH making up about 10% of patients. Especially in these patients histopathology is important for establishing the diagnosis.

In clinical practice the diagnosis of AIH may be straightforward in female patients of younger ages with the spectrum of autoantibodies such as ANA, SLA, SMA and elevation of liver enzymes such as transaminases, gamma-GT and AP[6]. However, since there are no specific markers for this disease, and about 10–20% of patients lack the presence of autoantibodies, establishing the diagnosis seems to be difficult. In order to fix the diagnosis with more reliability a scoring system has been suggested by the International Autoimmune Hepatitis Group that has been widely accepted[8]. Items on the scoring system include gender, autoantibodies, genetic background and histopathology, including features such as interface hepatitis, lymphoplasmocytic infiltrates and rosetting of liver cells. Included in the scoring system also is the response to steroid therapy.

Mainly based on different groups of autoantibodies AIH has been sub-divided in three types[2]:

Type 1: classic or lupoid AIH represents the most common form of AIH and is associated with ANA and SMA. Most patients show association with HLA-DR3 and DR4. Patients are younger and there is a higher frequency of treatment failure.

Type 2: AIH is characterized by anti-LKM1, and more common in Europe than in the United States.

Type 3: AIH is the least established form of the disease and is characterized by the presence of anti-SLA/LP in serum. Patients are similar to those of type 1 regarding laboratory features, and they respond well to corticosteroids.

HISTOPATHOLOGY OF AIH

Interface hepatitis synonymous with severe degrees of piecemeal necrosis is a constant but non-discriminating feature of AIH[4]. Portal tracts are densely infiltrated by lymphocytic or lymphoplasmocytic infiltration, with spilling over to parenchyma giving a periportal dominance of the inflammatory infiltrates. Interface hepatitis is not disease-specific; however, its absence should throw doubt on the diagnosis of AIH. The infiltrate consists of hepatic mesenchymal cells containing gamma-globulin-positive cells, lymphocytes, plasma cells to a variable degree and histiocytes that typically accompany these cells. They surround and engulf individual dying hepatocytes at the portal/parenchymal interface, extending to the lobule, and there giving the feature of spotty necrosis[5] (see Table 1).

Table 1 Histopathology of AIH

Main features
Portal/periportal inflammation, mainly lymphocytes
Interface hepatitis
Rosetting of hepatocytes
Lobular necrosis bound to lymphocytes/emperipolesis

Additional features
Hepatocellular polymorphism
Plasma cells
Reactive cholangitis
Centrilobular necrosis

The essential pattern of AIH consists in dense lymphocytic infiltrates originating in portal tracts as tertiary lymphoid centres extending into the lobule onto target cells. Panlobular (acinar) hepatitis is present less commonly, but it is part of the histological spectrum and may occur in acute-onset disease or in AIH that has relapsed after corticosteroid withdrawal. Depending on the grade of inflammatory activity bridging necrosis may develop, or confluent necrosis. In contrast, fibrous or granulomatous cholangitis and changes such as granulomas, siderosis, copper deposits and steatosis are incompatible with a definite diagnosis. These changes suggest other diagnoses or variant syndromes of AIH.

Plasma cells may be an intrinsic part of the process; however, their specificity in the inflammatory infiltrate does not preclude the diagnosis. The presence of plasma cells as constituants of the inflammatory infiltrate is variable and a percentage between 30% and 60% is given.

Hepatocytes in the lobule display a polymorphic picture with ballooning degeneration or pycnic cell necrosis.

Extensive interface hepatitis, panlobular hepatitis, bridging or massive necrosis and collapse are all features of increasing disease severity that may lead to fibrosis and loss of hepatocyte function. Liver cell regeneration in the form of thickening of the hepatic plates and rosette formation may be present as a token of simultaneous liver cell regeneration.

Bile ducts may be involved in the inflammatory process[10,11] and reactive cholangitis does not argue against the diagnosis of AIH. Lobular ducts may be reduced to a certain extent; however, overt bile duct destruction with disrupture of the basement membrane is not a genuine feature of AIH but may indicate a form of overlap syndrome. Ductular reaction is regarded as a feature of liver regeneration and may become quite conspicuous; however, it does not preclude the diagnosis of AIH.

The histological pattern at presentation may predict prognosis in untreated disease. Interface hepatitis progresses to cirrhosis in about 70% of patients within 5 years; however, individuals with this histological feature have a normal 5-year life expectancy. In contrast bridging hepatic necrosis is associated with an 82% frequency of cirrhosis and a 5-year mortality of 45% as has been reported in those patients not treated. Similar consequences occur in patients with massive necrosis at presentation and cirrhosis with active inflammation[9].

Other histological manifestations of AIH include perivenular necrosis and giant syncytial multinucleated hepatocytes that are regarded as non-specific reactions to injury that are associated with drug exposure, viral infection and autoimmune diseases.

Histopathology does not differentiate between the three groups of AIH based on clinical and serological features; with type 1 characterized by antibodies to ANA, type 2 AIH characterized by autoantibodies against LKM1, or type 3 AIH that is characterized by autoantibodies against SLA/LP. Histopathology cannot differentiate between the three groups; however, grading of the necroinflammatory activity is in general highest in type 1, so-called lupoid AIH[4].

VARIANTS OF AIH

Variants of the course and histopathology of AIH include three groups: the so-called seronegative type of autoimmune hepatitis, autoimmune hepatitis of recent onset and autoimmune hepatitis presenting with centrizonal injury.

There is no single autoantibody that is specific or pathognomonic for AIH and diagnosis is based on the spectrum of autoantibodies that may be present in the patient. Analysing a large cohort of 150 patients with AIH Lohse et al.[12] definded a group that had an established diagnosis of AIH based on history, HLA background, immunoglobulin elevation and response to therapy that, however, lacked autoantibodies in the serum. This group made up 9% of patients, and histopathology did not differ from the common type of AIH. In another study Ayata et al. found an incidence of 22% of autoimmune aetiology in cases of cryptogenic cirrhosis and all of these patients were negative for autoimmune antibodies[13] (see Table 2).

Table 2 'Seronegative' AIH

- Ten out of 154 patients with confirmed autoimmune hepatitis were without autoantibodies (8%)[12]

- Cryptogenic cirrhosis; 22% were of autoimmune aetiology; five of six cases were negative for autoantibodies[13]

- Histopathology of seronegative autoimmune hepatitis has typical features: interface hepatitis, rosetting of hepatocytes, conspicuous emperipolesis, plasma cells to a variable degree

Table 3 Acute/recent-onset AIH

L Burgart, AJ Czaja (Am J Surg Pathol, 1996)	G Nikias, AJ Czaja (J Hepatol. 1994)
Most patients with AIH who undergo biopsy early in its clinical course will have histological evidence of chronic liver disease	AIH with an acute presentation is indistinguishable by clinical and laboratory features from that with a chronic presentation, and is probably a pre-existing subclinical disease that is unmasked by disease progression or an abrupt exacerbation

Table 4 AIH: histopathology with centrilobular necrosis

Authors	No. of patients	Autoantibodies	Histopathology
Te et al (Gut, 1997)[18]	1	ANA 1:260	Unaffected
Pratt et al. (Gastroenterology, 1997)[17]	4	Low levels of ANA	Only centrilobular necrosis
Singh et al. (Am J Gastroenterol., 2002)[19]	1	None mentioned	Zone 3 necrosis
Misdraji et al. (Am J Surg Pathol., 2004)[20]	6	ANA 3×, SMA 1×	Zone 3 necrosis, mild portal infiltration

A separate category had been suggested by Czaja and co-workers[14], which they termed recent-onset AIH (see Table 3). He and his co-workers investigated a group of patients with acute illness who were diagnosed as having definite AIH. Liver biopsies of these patients, however, showed classical features of a chronic disease, considerable necroinflammatory leasons and additionally evidence of chronic liver disease, despite the lack of correlating clinical chronicity. So the authors concluded from their study that AIH runs a chronic course *per se* with bouts of acute disease that may start clinically as an acute disease. Histopathology in these cases does not differ from patients with chronic AIH with an acute flare-up[15,16].

In recent years several reports have described a new variant of AIH that is characterized by histological lesions of prominent centrilobular necrosis, often sparing portal tracts or displaying only mild portal extension and inflammatory infiltration[17–20] (see Table 4). Altogether 12 patients were reported in the literature, with small cohorts or case reports. Not all of the patients met the full criteria of AIH when applying strict criteria as recommended by the International Group. In these cases especially zone 3 necroses induced by drug intake and side-effects should be considered.

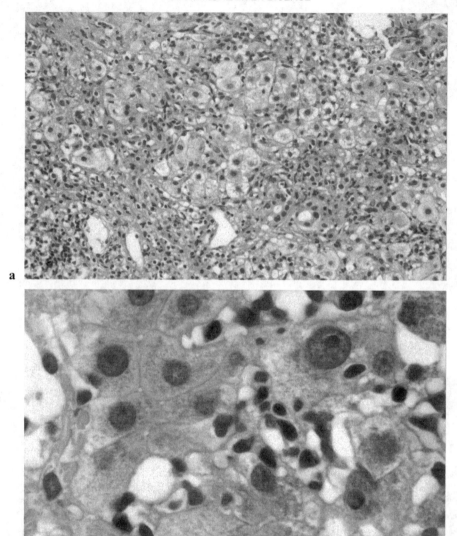

a

b

Figure 1 **A**: Autoimmune hepatitis with acute flare-up: there is severe interface hepatitis with abundant lymphocytic infiltrates in portal and periportal area and rosetting of hepatocytes (× 280; HE). **B**: In the lobule conspicuous emperipolesis with drop-out of hepatocytes is typical for autoimmune hepatitis (× 600; HE)

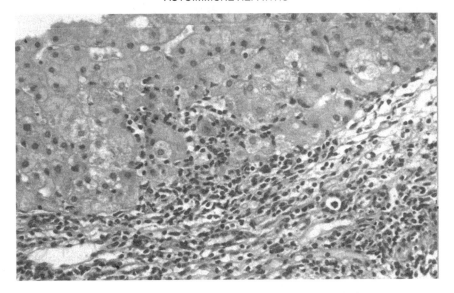

Figure 2 Chronic hepatitis in a patient taking alpha-methyldopa: there is severe portal inflammation with interface hepatitis including so-called piecemeal necroses (× 240; HE)

DIFFERENTIAL DIAGNOSIS OF AIH AND HISTOPATHOLOGY

There is a considerable number of drugs in clinical use that may lead to chronic hepatitis. Since in these cases viral markers are absent the disease may be confounded with AIH obscuring the real aetiology. There are also drugs that may induce not only liver injury but also autoantibodies such as ANA or smooth muscle antibodies, thus mimicking autoimmune liver disease[21] (see Table 5). To confirm the diagnosis of drug-induced hepatitis not only a detailed history of the patient is necessary, but histopathology may give some hints as to the causative factors when the biopsy displays increased numbers of eosinophils, small non-fibrosing granulomas, conspicuous zonal binding or a mixed pattern of inflammation including cholangitis.

The differential diagnosis of AIH also includes cases with overlapping of PBC or PSC[22,23].

The term PBC/AIH overlap syndrome is controversial, and a case of true overlap should be diagnosed only if a definite diagnosis of AIH and PBC can be established with certainty. Histopathology shows the most salient features of bile duct lesions with disrupture of basement membrane, epithelioid granulomas as well as considerable lobular hepatitis, rosetting of liver cells and conspicuous emperipolesis (see Table 6). Comparing three groups of patients (those with only AIH, those with only PBC and patients with clinical and histopathology of true overlap syndromes) we could show that patients with overlap show at the same time an inflammatory score equalling that of AIH and a score of bile duct lesions in the range of patients with PBC[11].

Table 5 Chronic drug-induced hepatitis with presence of autoantibodies (adapted from ref. 21)

Type I: few cases (< 5)

Benzarone	ASMA
Diclofenac	ANA
Ecstasy	ANA
Fenofibrate	ANA
Germander	ANA, ASMA
Papaverine	ANA, ASMA
Pemoline	ANA
Propylthiouracil	ANA

Type II: syndrome resembling AIH-2 or acute hepatitis

Dihydralazine	Anti-CYP1A2
Tienilic acid (ticrynafen)	Anti-CYP2C9
Halothane	Anti-carboxylesterase anti-protein disulphide isomerase
Iproniazid	AMA6

Table 6 Histological features of PBC/AIH OLS

PBC	AIH
Lymphoepithelial lesions of bile duct epithelia	Severe interface hepatitis
Epitheloid granulomas	Rosetting of liver cells
Disrupture of basement membrane	Conspicuous emperipolesis
Dense portal lymphocytic infiltrates	Continuous necroinflammatory activity

Patients with a diagnosis of AIH may also suffer from PSC, as has been reported by several authors[24,25]. This combination seems to be more frequent in children, as shown by Vergani et al. These children seem to run a special course and respond better to treatment, so that the term autoimmune cholangitis, as suggested by Vergani, seems to be justified. In adults most patients do not show a concomitant occurrence of both diseases, but rather show a sequential course, with PSC being the initial disease[26].

SUMMARY

In clinical practice liver biopsy is taken for the diagnosis of AIH to exclude other diseases, to determine the stage of the disease thus providing a significant component of the diagnosis. Histopathology displays typical but no pathognomonic features. Several variants of AIH include so-called seronegative AIH that displays typical histopathology as well as acute onset AIH that is characterized by an acute flare-up of chronic hepatitis. So-called AIH with centrizonal prominent necroses is a histopathological diagnosis and should be

confirmed by clinical serological parameters. The differential diagnosis of AIH includes drug-induced chronic hepatitis and with overlap syndrome of AIH and PBC and AIH with PSC.
The diagnosis of AIH is always a primary clinical diagnosis supported by biochemistry, serological tests and histopathology.

References

1. Czaja AJ, Manns MP, McFarlane IG, Hoofnagle JH. Autoimmune hepatitis: the investigational and clinical challenges. J Hepatol. 2000;31:1194–2000.
2. Czaja AJ, Freese DK. Diagnosis and treatment of autoimmune hepatitis. Hepatology. 2002; 36:479–97.
3. Czaja AJ, Donaldson PT. Genetic susceptibilities for immune expression and liver cell injury in autoimmune hepatitis. Immunol Rev. 2000;174:250–9.
4. Bach N, Thung SN, Schaffner F. The histological features of chronic hepatitis C and autoimmune chronic hepatitis: a comparative analysis. Hepatology. 1992;15:572–7.
5. Dienes HP, Popper H, Manns M, Baumann W, Thoenes W, Mayer zum Büschenfelde KH. Histologic features in autoimmune hepatitis. Z Gastroenterol. 1989;27:325–30.
6. Manns MP, Strassburg CP. Autoimmune hepatitis: clinical challenges. Gastroenterology. 2001;120:1502–17.
7. Kanzler S, Bozkurt S, Herkel J, Galle PR, Dienes HP, Lohse AW. Nachweis von SLA/LP-Autoantikörpern bei Patienten mit primär biliärer Zirrhose als Marker für eine sekundäre autoimmune Hepatitis (Overlapsyndrom). Dtsch Med Wochenschr. 2001;126.
8. Alvarez F, Berg PA, Bianchi FB et al. International Autoimmune Hepatitis Group report: review of criteria for diagnosis of autoimmune hepatitis. J Hepatol. 1999;31:929–38.
9. Czaja AJ, Carpenter HA. Autoimmune hepatitis. In: MacSween RNM, Burt AD, Portmann BC, Ishak KG, Scheuer MD, Anthony PP, editors. Pathology of the Liver. London: Churchill Livingstone, 2002:415–34.
10. Czaja AJ, Carpenter HA. Autoimmune hepatitis with incidental histologic features of bile duct injury. Hepatology. 2001;34:659–65.
11. Lohse AW, Meyer zum Büschenfelde KH, Franz B, Kanzler S, Gerken G, Dienes HP. Characterization of the overlap syndrome of primary biliary cirrhosis (PBC) and autoimmune hepatitis: evidence for its being a hepatic form of PBC in genetically susceptible individuals. Hepatology. 1999;29:1078–84.
12. Lohse AW, Gerken G, Mohr H et al. Relation between autoimmune liver diseases and viral hepatitis: clinical and serological characteristics in 859 patients. Z Gastroenterol. 1995;33: 527–33.
13. Ayata G, Gordon FD, Lewis WD et al. Cryptogenic cirrhosis: clinicopathologic findings at and after liver transplantation. Hum Pathol. 2002;33:1098–104.
14. Burgart LJ, Batts KP, Ludwig J, Nikias GA, Czaja AJ. Recent-onset autoimmune hepatitis – biopsy findings and clinical correlations. Am J Surg Pathol. 1995;19:699–708.
15. Maggiore G, Porta G, Bernard O et al. Autoimmune hepatitis with initial presentation as acute hepatic failure in young children. J Pediatr. 1990;116:280–2.
16. Nikias GA, Batts KP, Czaja AJ. The nature and prognostic implications of autoimmune hepatitis with an acute presentation. J Hepatol. 1994;21:866–71.
17. Pratt DS, Fawaz KA, Rabson A, Dellelis R, Kaplan MM. A novel histological lesion in glucocorticoid-responsive chronic hepatitis. Gastroenterology. 1997;113:664–8.
18. Te HS, Koukoulis G, Ganger DR. Autoimmune hepatitis: a histological variant associated with prominent centrilobular necrosis. Gut, 1997;41:269–71.
19. Singh R, Nair S, Farr G, Mason A, Perrillo R. Acute autoimmune hepatitis presenting with centrizonal liver disease: case report and review of the literature. Am J Gastroenterol. 2002; 97:2670–3.
20. Misdraji J, Thiim M, Graeme-Cook FM. Autoimmune hepatitis with centrilobular necrosis. Am J Surg Pathol. 2004;28:471–8.
21. Zimmerman HJ, Ishak KG. Hepatic injury due to drugs and toxins. In: MacSween RNM, Burt AD, Portmann BC, Ishak KG, Scheuer MD, Anthony PP, editors. Pathology of the Liver. London: Churchill Livingstone, 2002:621–710.

22. Terracciano LM, Patzina RA, Lehmann FS et al. A spectrum of histopathological findings in autoimmune liver disease. Am J Clin Pathol. 2000;114:705–11.
23. Heathcote EJ. Overlap of autoimmune hepatitis and primary biliary cirrhosis: an evaluation of a modified scoring system. Am J Gastroenterol. 2002;97:1090–91.
24. Abdo A, Bain VG, Kichian K, Lee SS. Evolution of autoimmune hepatitis to primary sclerosing cholangitis: a sequential syndrome. Hepatology. 2002;36:1393–9.
25. Gohlke F, Lohse AW, Dienes HP et al. Evidence for an overlap syndrome of autoimmune hepatitis and primary sclerosing cholangitis. J Hepatol. 1996;24:699–705.
26. Gregorio GV, Portmann B, Karani J et al. Autoimmune hepatitis/sclerosing cholangitis overlap syndrome in childhood: a 16-year prospective study. Hepatology. 2001;33:544–53.

15
Autoimmune hepatitis: clinical and laboratory diagnosis

A. J. CZAJA

INTRODUCTION

The diagnosis of autoimmune hepatitis has been codified by an international panel which first developed diagnostic criteria in 1992[1] and updated them in 1999[2] (Table 1). These criteria must now be applied to all patients suspected of having the disease[3]. The propensity for an acute, and rarely fulminant, presentation was recognized; the requirement for 6 months of disease activity to establish chronicity was waived; and lobular hepatitis was included among the histological manifestations. Cholestatic histological changes, including bile duct injury and ductopenia, were deemed incompatible with the diagnosis, but mild biliary changes within the background of classical interface hepatitis did not preclude the disease[4,5].

DEFINITE AND PROBABLE DIAGNOSES

Autoimmune hepatitis is characterized as a necroinflammation of the liver of unknown cause. The *definite* diagnosis requires the exclusion of other similar diseases; laboratory findings that indicate substantial immune reactivity; and histological findings of interface hepatitis[1,2] (Table 1). Autoimmune hepatitis is not a viral syndrome; nor is it associated with risk factors for viral infection (blood transfusions, intravenous drug use), alcohol-related injury, or drug toxicity. Serum γ-globulin levels must exceed 1.5-fold the upper limits of normal, and autoantibody titres must exceed 1:80. Low levels of autoantibodies are diagnostic only in children[1,2].

Wilson disease, drug-induced hepatitis (especially minocycline toxicity), chronic hepatitis C, and cholestatic syndromes with autoimmune features can resemble autoimmune hepatitis, and they must be excluded in all patients[1-3]. A cholestatic form of autoimmune hepatitis is unrecognized, and patients with pruritus, hyperpigmentation, xanthelasmas, or disproportionately elevated serum alkaline phosphatase levels have other diagnoses[6-9].

Table 1 International criteria for the diagnosis of autoimmune hepatitis

Definite AIH	Probable AIH
Normal α-1 AT phenotype	Partial α-1 AT deficiency
Normal caeruloplasmin level	Non-diagnostic caeruloplasmin/copper levels
Normal iron and ferritin levels	Non-diagnostic iron and/or ferritin changes
No active hepatitis A, B and/or C infection	No active hepatitis A, B and/or C infection
Daily alcohol <25 g/day. No recent hepatotoxic drugs	Daily alcohol <50 g/day. No recent hepatotoxic drugs
Predominant serum AST/ALT abnormality	Predominant serum AST/ALT abnormality
Globulin, γ-globulin or IgG level ≥1.5 times upper limit of normal	Hypergammaglobulinaemia of any degree
ANA, SMA, or anti-LKM1 ≥1:80 in adults and ≥1:20 in children; no AMA	ANA, SMA or anti-LKM1 ≥1:40 in adults; other autoantibodies
Interface hepatitis, moderate to severe	Interface hepatitis, moderate to severe
No biliary lesions, granulomas or prominent changes suggestive of another disease	No biliary lesions, granulomas or prominent changes suggestive of another disease

AIH, autoimmune hepatitis; α-1 AT, alpha-1 anti-trypsin; ANA, antinuclear antibodies; SMA, smooth muscle antibodies; anti-LKM1, antibodies to liver/kidney microsome type 1; AMA, antimitochondrial antibodies; IgG, serum immunoglobulin G level.

A *probable* diagnosis is justified when findings are compatible with autoimmune hepatitis but insufficient for a definite diagnosis[1,2] (Table 1). Patients who lack conventional autoantibodies but who are seropositive for investigational markers, such as antibodies to asialoglycoprotein receptor (ASGPR), soluble liver antigen/liver pancreas (SLA/LP), actin, or liver cytosol type 1 (LC1), are classified as having probable disease.

DIAGNOSTIC SCORING SYSTEM

A scoring system proposed by the International Autoimmune Hepatitis Group accommodates the diverse manifestations of autoimmune hepatitis and renders an aggregate score that reflects the net strength of the diagnosis before and after corticosteroid treatment[1,2] (Table 2). Each component of the syndrome is weighed; discrepant features are discounted; and biases associated with isolated inconsistencies are prevented.

The scoring system was developed as a research tool to ensure comparable study populations in clinical trials, and it is not a discriminative diagnostic index. The components of the scoring system are not unique to autoimmune hepatitis, and the scoring system was not developed by head-to-head comparisons with other liver diseases. Consequently, it should not be used to distinguish autoimmune hepatitis from other liver conditions[10].

The sensitivity of the scoring system for autoimmune hepatitis ranges from 97% to 100%, and its specificity for excluding chronic hepatitis C ranges from 66% to 92%[2]. The major weakness of the scoring system has been in

Table 2 International scoring system for diagnosis of autoimmune hepatitis

Gender	Female	+2	HLA	DR3 or DR4	+1	
AP:AST (or ALT) ratio	>3	−2	Immune disease	Thyroiditis, colitis, others	+2	
	<1.5	+2				
γ-Globulin or IgG level above normal	>2.0	+3	Other markers	Anti-SLA/LP, actin, LC1, pANCA	+2	
	1.5–2.0	+2				
	1.0–1.5	+1				
	<1.0	0				
ANA, SMA, or anti-LKM1 titres	>1:80	+3	Histological features	Interface hepatitis	+3	
	1:80	+2		Plasmacytic	+1	
	1:40	+1		Rosettes	+1	
	<1:40	0		None of above	−5	
				Biliary changes	−3	
				Other features	−3	
AMA	Positive	−4	Treatment response	Complete	+2	
				Relapse	+3	
Viral markers	Positive	−3				
	Negative	+3				
Drugs	Yes	−4	Pretreatment score			
	No	+1	Definite diagnosis	>15		
			Probable diagnosis	10–15		
Alcohol	<25 g/day	+2	Post-treatment score			
	>60 g/day	−2	Definite diagnosis	>17		
			Probable diagnosis	12–17		

AP:AST (or ALT) ratio, ratio of alkaline phosphatase level to aspartate or alanine aminotransferase level; anti-SLA/LP, antibodies to soluble liver antigen/liver pancreas; anti-LC1, antibodies to liver cytosol type 1; pANCA, perinuclear antineutrophil cytoplasmic antibodies; IgG, immunoglobulin G; ANA, antinuclear antibodies; SMA, smooth muscle antibodies; anti-LKM1, antibodies to liver/kidney microsome type 1; AMA, antimitochondrial antibodies; HLA, human leukocyte antigen.

distinguishing autoimmune hepatitis from cholestatic syndromes with auto-immune features[11,12]. The scoring system excludes autoimmune hepatitis in only 45–65% of patients with these disorders[2].

AGE OF ONSET AND GENDER PREDISPOSITION

Autoimmune hepatitis afflicts all ages and genders, and it has a global distribution. Earlier reports that autoimmune hepatitis had a bimodal age distribution between ages 10 years and 30 years and between 40 years and 50 years were probably affected by referral patterns to tertiary medical centres[13–15]. Current experiences suggest that autoimmune hepatitis occurs as commonly across all age ranges and that it may be underdiagnosed in the elderly[15–17] (Figure 1). It is a diagnosis that can and should be made in infants[18].

Seventy-eight per cent of patients are female, and the female:male ratio[19] is 3.5. Women are distinguished from men with the disease by higher frequencies

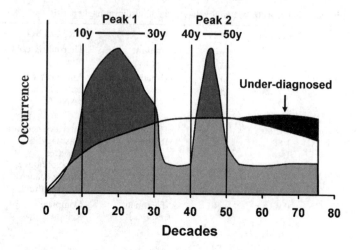

Figure 1 Age distribution of autoimmune hepatitis at presentation. Original concept of bimodal age distribution between 10 years (y) and 30 years and 40 years and 50 years has been replaced by the concept that disease occurrence affects all age groups, including infants, and may be underdiagnosed in the elderly (arrow)

of concurrent immune diseases (34% versus 17%, $p = 0.05$) and HLA DR4 (49% versus 24%, $p = 0.007$). They also have a higher occurrence of non-*DRB1*0401* alleles than men (15% versus 0%, $p = 0.02$). Clinical and laboratory indices of disease severity, frequency of cirrhosis at presentation, and responses to corticosteroid therapy, however, are similar between men and women[19].

ACUTE ONSET

Forty per cent of patients with autoimmune hepatitis have an acute, rarely fulminant presentation[20,21], and serum aminotransferase levels can exceed 1000 IU/dl[22]. The acute presentation may reflect pre-existing subclinical disease that is unmasked by a spontaneous exacerbation[23] or *de-novo* disease that resembles an acute viral or toxic hepatitis[24–27]. Patients with exacerbated chronic disease typically have clinical (ascites, aesophageal varices, or spider angiomas), laboratory (thrombocytopenia or hypoalbuminaemia), and histological changes (fibrosis or cirrhosis) that suggest established disease[22]. Patients with *de-novo* acute autoimmune hepatitis satisfy the codified international criteria for the diagnosis (Table 1), but they must be suspected so that the criteria can be applied.

Autoimmune hepatitis must be included in the differential diagnosis of all patients with an acute or fulminant presentation, and in these cases liver biopsy assessment is justified. Centrilobular or zone 3 inflammation has been described in some patients with acute disease[26–30], and sequential liver tissue examinations have demonstrated transition from the centrilobular changes to the classical features of portal interface hepatitis[29,30]. The centrilobular pattern may be an early acute form of autoimmune hepatitis, and its presence should not dissuade the diagnosis.

SYMPTOMS AND PHYSICAL FINDINGS

Fatigue is the most common symptom (86%), and hepatomegaly is the most common physical finding (78%)[31] (Table 3). Thirty-four per cent of patients may be asymptomatic at initial consultation, and they are most commonly men with lower serum aminotransferase and immunoglobulin levels than symptomatic patients[32]. Histological features are similar between asymptomatic and symptomatic patients, and both groups respond well to glucocorticoids. Most asymptomatic patients become symptomatic during follow-up, and differences between the asymptomatic and symptomatic state may reflect variations in disease activity and patient tolerance.

Table 3   Clinical manifestations of autoimmune hepatitis

	Occurrence (%)
Symptoms	
Fatigue	86
Upper abdominal discomfort	48
Anorexia	30
Polymyalgias	30
Cushingoid features	19
Fever ($\leqslant 40°C$)	18
None (at presentation)	14–34
Physical findings	
Hepatomegaly	78
Jaundice	69
Spider angiomas	58
Splenomegaly	32
Ascites	20
Encephalopathy	14
Concurrent immune diseases	38

REGIONAL AND ETHNIC DIFFERENCES

Autoimmune hepatitis was originally reported in white patients from Australia, North America, and northern Europe, and descriptions of clinical manifestations and natural history derived from these populations, as did treatment strategies based on clinical trials. Autoimmune hepatitis is now recognized to have a global distribution, and the clinical manifestations, severity, and outcome of the disease may vary by region and race.

Cirrhosis is present at accession more commonly in black North American patients with autoimmune hepatitis than in white North American patients (85% versus 38%), and hepatic synthetic function is decreased more frequently[33]. Both groups respond similarly to corticosteroids, but black North American patients are younger at presentation than white counterparts. These findings suggest that black North Americans have a more aggressive disease than white North Americans and that their higher frequency of advanced disease reflects intrinsic disease behaviour rather than delays in diagnosis or difficulties in accessing medical care.

Alaskan natives have a higher frequency of acute icteric disease, asymptomatic illness, and advanced fibrosis at presentation than white counterparts[34]. Japanese patients typically have mild, late-onset disease that can respond to non-steroidal medication such as ursodeoxycholic acid[35,36]. South American patients in Brazil and Argentina are younger than white North American patients, and they have more severe laboratory derangements[18]. African, Asian and Arab patients have an earlier age of disease onset than white northern European counterparts, and they have a higher frequency of cholestatic laboratory findings, greater occurrence of biliary changes on histological examination, and poorer initial response to standard therapy[37].

Indigenous aetiological agents may naturally select patients with genetic predispositions that favour their propagation and cause the disease[38]. Other autoimmune promoters may be linked to each region-specific susceptibility factor, and they may further modify the expression and outcome of the disease to a degree that imparts an ethnic or geographical distinction. The diagnosis of autoimmune hepatitis should be considered in all patients with chronic hepatitis of undetermined cause regardless of ethnic origin, and clinical deviations from the classical Caucasoid disease should be accommodated.

LABORATORY MANIFESTATIONS

The laboratory features of autoimmune hepatitis reflect its hepatitic nature (Table 4). The predominant abnormalities are the serum aminotransferase levels, which can range to over 1000 U/dl and mimic a severe acute hepatitis[22,24,27] (Figure 2). Most patients have a substantial hypergammaglobulinaemia, and the immunoglobulin G level is the predominant globulin fraction which is elevated. An increased serum alkaline phosphatase level is common, but 79% of patients have serum alkaline phosphatase abnormalities less than twice the upper limit of normal[7,39]. Only 21% of patients have serum alkaline phosphatase levels that exceed two-fold normal, and none with classical disease has a serum alkaline phosphatase level that exceeds four-fold normal. A disproportionately increased serum alkaline phosphatase level is an important clue to an alternative diagnosis.

Table 4 Laboratory manifestations of autoimmune hepatitis

	Occurrence (%)
Laboratory features	
Aspartate aminotransferase elevation	100
Hypergammaglobulinaemia	92
Increased immunoglobulin G level	91
Hyperbilirubinaemia	83
Alkaline phosphatase ⩾ 2-fold normal	21
Serological markers	
SMA, ANA, or anti-LKM1	100
Perinuclear antineutrophil cytoplasm	92 (type 1 only)
Anti-asialoglycoprotein receptor	82
Anti-actin	74
Anti-chromatin	42 (ANA+ only)
Anti-liver cytosol 1	32 (type 2 only)
Anti-Saccharomyces cerevisiae	28
Anti-soluble liver antigen/liver-pancreas	11–17

Autoantibodies in italics are non-standard markers, and they have not been incorporated in routine diagnostic algorithms. SMA, smooth muscle antibodies; ANA, antinuclear antibodies; anti-LKM1, antibodies to liver/kidney microsome type 1.

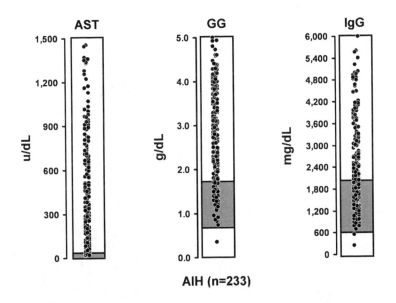

AIH (n=233)

Figure 2 Laboratory features of autoimmune hepatitis at presentation. Serum aspartate aminotransferase (AST) levels, gamma-globulin (GG), and immunoglobulin G (IgG) levels are shown. Normal ranges are depicted by the light bar at the base of each column. Original data

Table 5 Autoantibodies associated with autoimmune hepatitis

	Target (s)	Clinical value
Standard repertoire		
Antinuclear antibodies	Centromere, ribonucleoproteins	Diagnosis of type 1
Smooth muscle antibodies	Actin, tubulin, vimentin, desmin, skeletin	Diagnosis of type 1
Anti-liver kidney microsome type 1	CYP2D6	Diagnosis of type 2
Supplemental repertoire		
Perinuclear antineutrophil cytoplasmic antibodies	Possible nuclear membrane lamina	Diagnosis of type 1, cryptogenic hepatitis
IgA antibodies to endomysium hepatitis	Endomysium monkey oesophagus	Coeliac disease, cryptogenic
Non-standard repertoire		
Antibodies to actin	Microfilaments	Diagnosis of type 1
Antibodies to soluble liver antigen/liver pancreas	Ribonucleoprotein complex	Relapse, cryptogenic hepatitis
Anti-asialoglycoprotein receptor	Asialoglycoprotein receptor	Histologic activity, relapse
Anti-chromatin	Chromatin	Relapse
Anti-liver cytosol type 1 failure	Formiminotransferase cyclodeaminase	Diagnosis of type 2, treatment

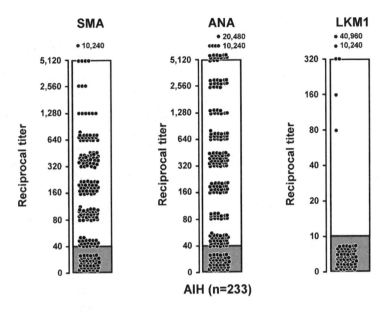

Figure 3 Serological features of autoimmune hepatitis at presentation. Serum titres of smooth muscle antibodies (SMA), antinuclear antibodies (ANA), and antibodies to liver/kidney microsome type 1 (LKM1) are shown. Normal ranges are depicted by the light bar at the base of each column. Original data

CONVENTIONAL SEROLOGICAL FINDINGS

Smooth muscle antibodies (SMA), antinuclear antibodies (ANA), and antibodies to liver kidney microsome type 1 (anti-LKM1) constitute the conventional battery of serological markers for autoimmune hepatitis[1-3] (Table 5). Smooth muscle antibodies and ANA can be present in any titre[40] (Figure 3). Antibodies to liver kidney microsome type 1 are detected in only 4% of adult patients with autoimmune hepatitis in the United States, and they can occur alone or in conjunction with SMA and ANA[41]. Low titres should not be discounted as unimportant in patients with other classical findings, especially in children[1,2].

Forty-three per cent of patients with autoimmune hepatitis have concurrent ANA and SMA; 44% have SMA only; and 13% have ANA only[40]. The expression of SMA and ANA can be simultaneous; one autoantibody can be detected before the other; one antibody can disappear and be replaced by the other; or neither may be detected at presentation but develop later in the course of the disease. The conventional autoantibodies behave variably and unpredictably, and they are not by themselves pathogenic or prognostic.

NON-STANDARD SEROLOGICAL FINDINGS

Non-standard autoantibodies have been characterized, and they may in time be assimilated into conventional diagnostic algorithms[42,43] (Tables 4 and 5). These autoantibodies may have diagnostic as well as prognostic value, and their application may supplement the standard markers. Perinuclear antineutrophil cytoplasmic antibodies (pANCA) have been useful in classifying patients seronegative for the conventional antibodies, and immunoglobulin A (IgA) antibodies to endomysium (EMA) may have a similar value[43] (Table 5). Forty-two per cent of patients with coeliac disease have abnormal liver tests, and they may be mistakenly classified as cryptogenic hepatitis[44]. Furthermore, coeliac disease (often asymptomatic) occurs in 3% of patients with autoimmune hepatitis, and these patients warrant appropriate treatment[45] (Table 5). Endomysial antibodies are more predictive of coeliac disease in autoimmune hepatitis than IgA antibodies to tissue transglutaminase which can be stimulated by hepatic inflammation and fibrogenesis[46].

Antibodies to SLA/LP, actin (anti-actin), chromatin (anti-chromatin), and LC1 have been associated with severe disease or poor treatment response, and they may prove valuable as prognostic tools[47] (Table 5). Furthermore, these autoantibodies may be associated with genetic propensities that also affect outcome. Antibodies to SLA/LP and actin have been associated with HLA DR3[47–50], and there may be other serological expressions with a similar genetic association that augur a poor prognosis.

The major clinical limitation of the non-standard autoantibodies has been their low individual occurrence in patients with autoimmune hepatitis[47]. Antibodies to SLA/LP have been associated with severe disease and a propensity to relapse after corticosteroid withdrawal, but they are present in only 9–54% of individuals with the disease[49–52]. Similarly, anti-actins identify patients with a higher frequency of treatment failure and death from liver failure or requirement for liver transplantation than seronegative patients, but they are restricted to patients with SMA[48]. Antibodies to chromatin are associated with higher serum levels of γ-globulin and immunoglobulin G, and greater occurrence of relapse after drug withdrawal than seronegative counterparts, but they are found in only 39% of patients[53,54]. Lastly, anti-LC1 occur in patients with severe liver inflammation and rapid progression to cirrhosis, but they are detected in no more than 32% of patients with anti-LKM1[55,56]. The variable and infrequent expression of these important antibodies in individuals with the disease limit the value of any one determination in assessing prognosis, especially since absence of the prognostic marker does not preclude a poor outcome.

CONCURRENT IMMUNE DISEASES

Concurrent immune diseases are common, and they underscore the heightened autoimmune propensity of the host[20,31]. Autoimmune thyroiditis, ulcerative colitis, and Graves' disease are the most common findings, but virtually any extrahepatic immune disease may be present (Figure 4). Coeliac disease is

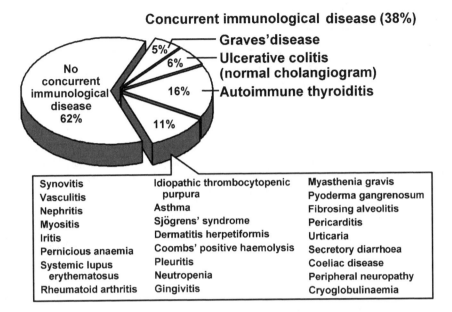

Concurrent immunological disease (38%)

Graves'disease
Ulcerative colitis (normal cholangiogram)
Autoimmune thyroiditis

No concurrent immunological disease 62%

5%
6%
16%
11%

Synovitis	Idiopathic thrombocytopenic	Myasthenia gravis
Vasculitis	purpura	Pyoderma gangrenosum
Nephritis	Asthma	Fibrosing alveolitis
Myositis	Sjögrens' syndrome	Pericarditis
Iritis	Dermatitis herpetiformis	Urticaria
Pernicious anaemia	Coombs' positive haemolysis	Secretory diarrhoea
Systemic lupus	Pleuritis	Coeliac disease
erythematosus	Neutropenia	Peripheral neuropathy
Rheumatoid arthritis	Gingivitis	Cryoglobulinaemia

Figure 4 Concurrent immune diseases in autoimmune hepatitis. Original data

important to recognize because it is typically asymptomatic and treatable[44,45]. Autoimmune hepatitis may also be present as a background disorder in patients with systemic autoimmune diseases, such as systemic lupus erythematosus and rheumatoid arthritis. Immune diseases may develop at any time during the course of autoimmune hepatitis, and continuous monitoring for their emergence is warranted.

CLINICAL SUBTYPES

Three types of autoimmune hepatitis have been proposed based on serological markers[57] (Table 6). These types have not been established as valid clinical entities, and the designations have not been endorsed by the International Autoimmune Hepatitis Group. Nevertheless, they have been used as clinical descriptors, and the different types illustrate the diverse manifestations of autoimmune hepatitis.

Type 1 autoimmune hepatitis is the most common form worldwide in Caucasoid adults, and it is characterized by the presence of ANA and/or SMA[57] (Table 6). Eighty per cent of adults with autoimmune hepatitis in the United States have this type. Over 70% of patients with type 1 autoimmune hepatitis are women,

Table 6 Subclassifications of autoimmune hepatitis based on autoantibodies

Clinical features	Type 1	Type 2	Type 3
Signature autoantibodies	SMA ANA	LKM1	SLA/LP
Autoantigen	Unknown	CYD2D6	50 kDa protein tRNP[(Ser)Sec]
Age (years)	Infancy to old age	Pediatric (2–14)	Adults (30–50)
Women (%)	78	89	90
Immune diseases (%)	38	34	58
Typical concurrent immune diseases	Thyroiditis, Graves' disease, ulcerative colitis	Thyroiditis, vitiligo, type 1 diabetes, APECED	Same as type 1
HLA associations	B8, DR3, DR4	B14, DR3, C4A-Q0, DR7	DR3
Allelic risk factors	DRB1*0301 DRB1*0401	DRB1*07	DRB1*0301
Steroid-responsive	+++	++	+++

LKM1, antibodies to liver/kidney microsome type 1; SLA/LP, antibodies to soluble liver antigen/liver pancreas; pANCA, perinuclear anti-neutrophil cytoplasmic antibodies; tRNP[(ser)sec], transfer ribonucleoprotein complex involved in serine (Ser) metabolism; APECED, polyendocrinopathy–candidiasis–ectodermal dystrophy.

and 48% are less than 40 years old[20]. An abrupt onset of symptoms is common, and a fulminant presentation is possible. It is important to recognize this propensity since the disease may be mistaken for an acute viral or toxic hepatitis, and the institution of potentially life-saving corticosteroid therapy may be delayed. Thirty-eight per cent of patients have concurrent immune diseases, mainly autoimmune thyroiditis, and 25% have cirrhosis already established at the time of presentation. This latter finding indicates that type 1 autoimmune hepatitis can have an indolent, subclinical, aggressive stage.

Type 2 autoimmune hepatitis is characterized by the presence of antibodies to liver/kidney microsome type 1[57,58] (Table 6). This disease occurs mainly in children, but 20% of patients with type 2 disease in Europe are adults. Concurrent immune diseases are common, especially type 1 diabetes mellitus, vitiligo and autoimmune thyroiditis. As in type 1 disease, an acute, even fulminant, presentation is possible and important to recognize early[21]. Type 2 autoimmune hepatitis is rare in the United States, occurring in only 4% of adult patients[41].

Type 2 autoimmune hepatitis can occur in patients with autoimmune-polyendocrinopathy-candidiasis-ectodermal dystrophy (APECED)[59]. This syndrome includes multiple endocrine organ failure, mucocutaneous candidiasis, ectodermal dystrophy, and autoimmune hepatitis in various syndromic combinations. The disease is autosomal recessive, and the mutation is on chromosome 21q22.3[60]. The altered gene product is an autoimmune regulator that modulates the negative selection of autoreactive immunocytes by the thymus. CYP1A2 and 2A6 have been implicated as the autoantigens associated with the liver disease.

Type 3 autoimmune hepatitis is the least established form of the disease, and it is characterized by the presence of anti-SLA/LP[57,61,62] (Table 6). The target autoantigen is a 50 kDa cytosolic protein which is probably a transfer ribonucleoprotein complex responsible for incorporating selenocysteine into peptide chains[63–65]. Patients with anti-SLA/LP commonly have other antibodies, including antinuclear antibodies, smooth muscle antibodies, and anti-LKM1[50,51,66,67]. Only 26% of patients have anti-SLA/LP as their sole marker[61]. Patients with anti-SLA/LP are indistinguishable from patients without anti-SLA/LP in age, gender distribution, frequency and nature of concurrent immune diseases, frequency of ANA and SMA, and response to corticosteroid therapy[51,67].

The subclassification of autoimmune hepatitis by autoantibody type has been justified by the presumptions that autoantibodies reflect distinctive pathogenic mechanisms, subgroups defined by autoantibody expression are different clinical entities, and classification is valuable in directing treatment. None of these presumptions is established. The subgroups do not predict prognosis; nor do patients classified by their serological manifestations respond differently to conventional corticosteroid therapy. The major clinical value of subclassification may be to communicate clinical phenotypes, and its major investigative value may be to facilitate identification of underlying pathogenic mechanisms in serologically homogeneous populations. Each of these potential values, however, is jeopardized by the lack of mutual exclusivity between the antibodies and their phenotypes and the regional and ethnic variations in the occurrence of the serological markers.

'AUTOANTIBODY-NEGATIVE AUTOIMMUNE HEPATITIS'

Thirteen per cent of adult white patients in the United States with features of autoimmune hepatitis lack conventional autoantibodies, and they may be designated as cryptogenic chronic hepatitis[68–70]. Studies have indicated that many patients in this cryptogenic category have autoimmune hepatitis that has escaped detection by conventional autoantibody testing[52,68–70], and perhaps a more appropriate designation for these patients is 'autoantibody-negative autoimmune hepatitis'.

Patients who satisfy criteria for autoimmune hepatitis but lack the conventional autoantibodies are indistinguishable from patients with classical type 1 autoimmune hepatitis by HLA phenotype and responses to corticosteroid therapy, including frequencies of remission, relapse, treatment failure, and death from hepatic failure[68,69]. Sequential serological testing will disclose the late appearance of smooth muscle and antinuclear antibodies in 25% of these patients[40]. Other autoantibodies such as pANCA, anti-SLA/LP, and IgA endomysial antibodies may characterize others[51,52]. A small group of patients will remain truly autoantibody-negative, and they may have a different disease or an autoimmune condition that continues to escape serological classification. Patients with the classical features of autoimmune hepatitis should not be denied the potential life-saving benefits of corticosteroid therapy simply because they lack serological markers. Their designation as 'autoantibody-negative autoimmune hepatitis' rather than cryptogenic chronic hepatitis may protect against this possibility.

GENETIC MARKERS

Genetic factors influence the occurrence, clinical expression and treatment outcome of autoimmune hepatitis[71–75], but HLA typing has not been advocated as a diagnostic tool[1,2]. HLA DR3 is the main susceptibility factor in white northern Europeans and North Americans, and HLA DR4 is a secondary but independent risk factor for the disease[76]. Eighty-five per cent of white patients with type 1 autoimmune hepatitis from the United States and northern Europe have HLA DR3, DR4 or DR3-DR4.

High-resolution DNA-based techniques have indicated that the principal susceptibility factors for type 1 autoimmune hepatitis reside on the *DRB1* gene, and they are the alleles, *DRB1*0301* and *DRB1*0401*[77,78]. Patients with *DRB1*0301* are younger than patients with *DRB1*0401*, and they fail corticosteroid therapy more often, die of liver failure or require liver transplantation more commonly, and have a significantly greater frequency of an adverse treatment outcome than patients with *DRB1*0401*[73]. The HLA phenotype may also influence autoantibody expression[48–50], and both the antibody profile and the HLA phenotype probably say more about the pathogenic mechanisms of the disease, its aetiological factors, treatment requirements, and outcome than clinicians can now appreciate[79].

Different ethnic groups have different susceptibility alleles[35,80–82], and *DRB1*07* has been associated with the expression of anti-LKM1[83,84]. The susceptibility alleles may be clues to region-specific indigenous aetiological agents[38], and they may ultimately allow categorization of autoimmune hepatitis by aetiological factor or pertinent pathogenic mechanism[85]. Tests that reflect the cause of the disease or its intrinsic autoimmune nature would be the ultimate diagnostic tools.

In summary, autoimmune hepatitis affects all ages and diverse ethnic groups. Diagnostic criteria have been codified, and they must be applied. Patients may have acute, even fulminant, presentations, and patients seronegative for conventional autoantibodies should be included within the diagnosis. Serological profiles define clinical subtypes, but they are not mutually exclusive and they do not alter therapy.

References

1. Johnson PJ, McFarlane IG, Alvarez F et al. Meeting Report. International Autoimmune Hepatitis Group. Hepatology. 1993;18:998–1005.
2. Alvarez F, Berg PA, Bianchi FB et al International Autoimmune Hepatitis Group Report: Review of criteria for diagnosis of autoimmune hepatitis. J Hepatol. 1999;31:929–38.
3. Czaja AJ, Freese DK. Diagnosis and treatment of autoimmune hepatitis. Hepatology. 2002; 36:479–97.
4. Ludwig J, Czaja AJ, Dickson ER, LaRusso NF, Wiesner RH. Manifestations of non-suppurative cholangitis in chronic hepatobiliary disease: morphologic spectrum, clinical correlations and terminology. Liver. 1984;4:105–16.
5. Czaja AJ, Carpenter HA. Autoimmune hepatitis with incidental histologic features of bile duct injury. Hepatology. 2001;34:659–65.
6. Czaja AJ. Chronic active hepatitis: the challenge for a new nomenclature. Ann Intern Med. 1993;119:510–17.
7. Czaja AJ. The variant forms of autoimmune hepatitis. Ann Intern Med. 1996;125:588–98.
8. Czaja AJ. Frequency and nature of the variant syndromes of autoimmune liver disease. Hepatology. 1998;28:360–5.
9. Czaja AJ, Carpenter HA, Santrach PJ, Moore SB. Autoimmune cholangitis within the spectrum of autoimmune liver disease. Hepatology. 2000;31:1231–8.
10. Talwalkar JA, Keach JC, Angulo P, Lindor KD. Overlap of autoimmune hepatitis and primary biliary cirrhosis: an evaluation of a modified scoring system. Am J Gastroenterol. 2002;97:1191–7.
11. Czaja AJ, Carpenter HA. Validation of a scoring system for the diagnosis of autoimmune hepatitis. Dig Dis Sci. 1996;41:305–14.
12. Boberg KM, Fausa O, Haaland T et al. Features of autoimmune hepatitis in primary sclerosing cholangitis: an evaluation of 114 primary sclerosing cholangitis patients according to a scoring system for the diagnosis of autoimmune hepatitis. Hepatology. 1996;23: 1369–76.
13. McFarlane IG. The relationship between autoimmune markers and different clinical syndromes in autoimmune hepatitis. Gut. 1998;42:599–602.
14. McFarlane IG. Autoimmune hepatitis: clinical manifestations and diagnostic criteria. Can J Gastroenterol. 2001;15:107–13.
15. McFarlane IG. Autoimmune hepatitis: diagnostic criteria, subclassifications, and clinical features. Clin Liver Dis. 2002;6:605–21.
16. Schramm C, Kanzler S, Meyer zum Buschenfelde K-H, Galle PR, Lohse AW. Autoimmune hepatitis in the elderly. Am J Gastroenterol. 2001;96:1587–91.
17. Wang KK, Czaja AJ. Prognosis of corticosteroid-treated hepatitis B surface antigen-negative chronic active hepatitis in postmenopausal women: a retrospective analysis. Gastroenterology. 1989;97:1288–93.
18. Czaja AJ, Souto EO, Bittencourt PL et al. Clinical distinctions and pathogenic implications of type 1 autoimmune hepatitis in Brazil and the United States. J Hepatol. 2002;37:302–8.
19. Czaja AJ, Donaldson PT. Gender effects and synergisms with histocompatibility leukocyte antigens in type 1 autoimmune hepatitis. Am J Gastroenterol. 2002;97:2051–7.
20. Czaja AJ, Davis GL, Ludwig J, Baggenstoss AH, Taswell HF. Autoimmune features as determinants of prognosis in steroid-treated chronic active hepatitis of uncertain etiology. Gastroenterology. 1983;85:713–17.
21. Porta G, Da Costa Gayotto LC, Alvarez F. Anti-liver-kidney microsome antibody-positive autoimmune hepatitis presenting as fulminant liver failure. J Pediatr Gastroenterol Nutr. 1990;11:138–40.
22. Davis GL, Czaja AJ, Baggenstoss AH, Taswell HF. Prognostic and therapeutic implications of extreme serum aminotransferase elevation in chronic active hepatitis. Mayo Clin Proc. 1982;57:303–9.
23. Burgart LJ, Batts KP, Ludwig J, Nikias GA, Czaja AJ. Recent onset autoimmune hepatitis: biopsy findings and clinical correlations. Am J Surg Pathol. 1995;19:699–708.
24. Amontree, JS, Stuart, TD, Bredfeldt, JE. Autoimmune chronic active hepatitis masquerading as acute hepatitis. J Clin Gastroenterol. 1989;11:303–7.
25. Nikias GA, Batts KP, Czaja AJ. The nature and prognostic implications of autoimmune hepatitis with an acute presentation. J Hepatol. 1994;21:866–71.

26. Okano N, Yamamoto K, Sakaguchi K et al. Clinicopathological features of acute-onset autoimmune hepatitis. Hepatol Res. 2003;25:263–70.

27. Kessler WR, Cummings OW, Eckert G, Chalasani N, Lumeng L, Kwo P. Fulminant hepatic failure as the initial presentation of acute autoimmune hepatitis. Clin Gastroenterol Hepatol. 2004;2:625–31.

28. Pratt DS, Fawaz KA, Rabson A, Dellelis R, Kaplan MM. A novel histological lesion in glucocorticoid-responsive chronic hepatitis. Gastroenterology. 1997;113:664–8.

29. Singh R, Nair S, Farr G, Mason A, Perrillo R. Acute autoimmune hepatitis with centizonal liver disease: case report and review of the literature. Am J Gastroenterol. 2002;97:2670–3.

30. Misdraji J, Thiim M, Graeme-Cook FM. Autoimmune hepatitis with centrilobular necrosis. Am J Surg Pathol. 2004;28:471–8.

31. Czaja AJ. Autoimmune hepatitis. In: Feldman M, Friedman LS, Sleisenger MH, editors. Sleisenger and Fordtran's Gastrointestinal and Liver Disease, 7th edn. Philadelphia: Saunders, 2002:1462–73.

32. Kogan J, Safadi R, Ashur Y, Shouval D, Ilan Y. Prognosis of symptomatic versus asymptomatic autoimmune hepatitis. A study of 68 patients. J Clin Gastroenterol. 2002; 35:75–81.

33. Lim KN, Casanova RL, Boyer TD, Bruno CJ. Autoimmune hepatitis in African Americans: presenting features and responses to therapy. Am J Gastroenterol. 2001;96:3390–4.

34. Hurlburt KJ, McMahon BJ, Deubner H, Hsu-Trawinski B, Williams JL, Kowdley KV. Prevalence of autoimmune hepatitis in Alaska natives. Am J Gastroenterol. 2002;97:2402–7.

35. Seki T, Ota M, Furuta S et al. HLA class II molecules and autoimmune hepatitis susceptibility in Japanese patients. Gastroenterology. 1992;103:1041–7.

36. Nakamura K, Yoneda M, Yokohama S et al. Efficacy of ursodeoxycholic acid in Japanese patients with type 1 autoimmune hepatitis. J Gastroenterol Hepatol. 1998;13:490–5.

37. Zolfino T, Heneghan MA, Norris S, Harrison PM, Portmann BC, McFarlane IG. Characteristics of autoimmune hepatitis in patients who are not of European Caucasoid ethnic origin. Gut. 2002;50:713–17.

38. Fainboim L, Velasco VCC, Marcos CY et al. Protracted, but not acute, hepatitis A virus infection is strongly associated with HLA-*DRB1*1301*, a marker for pediatric autoimmune hepatitis. Hepatology. 2001;33:1512–17.

39. Kenny RP, Czaja AJ, Ludwig J, Dickson ER. Frequency and significance of antimitochondrial antibodies in severe chronic active hepatitis. Dig Dis Sci. 1986;31:705–11.

40. Czaja AJ. Behavior and significance of autoantibodies in type 1 autoimmune hepatitis. J Hepatol. 1999;30:394–401.

41. Czaja AJ, Manns MP, Homburger HA. Frequency and significance of antibodies to liver/kidney microsome type 1 in adults with chronic active hepatitis. Gastroenterology. 1992; 103:1290–5.

42. Czaja AJ, Homburger HA. Autoantibodies in liver disease. Gastroenterology. 2001;120: 239–49.

43. Czaja AJ, Norman GL. Autoantibodies in the diagnosis and management of liver disease. J Clin Gastroenterol. 2003;37:315–29.

44. Bardella MT, Fraquelli M, Quatrini M, Molteni N, Bianchi P, Conti D. Prevalence of hypertransaminasemia in adult celiac patients and effect of gluten-free diet. Hepatology. 1995;22:833–6.

45. Volta U, De Franceschi L, Molinaro N et al. Frequency and significance of anti-gliadin and anti-endomysial antibodies in autoimmune hepatitis. Dig Dis Sci. 1998;43:2190–5.

46. Czaja AJ, Shums Z, Donaldson PT, Norman GL. Frequency and significance of antibodies to *Saccharomyces cerevisiae* in autoimmune hepatitis. Dig Dis Sci. 2004;49:611–18.

47. Czaja AJ, Shums Z, Norman GL. Nonstandard antibodies as prognostic markers in autoimmune hepatitis. Autoimmunity. 2004;37:195–201.

48. Czaja AJ, Cassani F, Cataleta M, Valentini P, Bianchi FB. Frequency and significance of antibodies to actin in type 1 autoimmune hepatitis. Hepatology. 1996;24:1068–73.

49. Czaja AJ, Donaldson PT, Lohse AW. Antibodies to soluble liver antigen/liver pancreas and HLA risk factors in type 1 autoimmune hepatitis. Am J Gastroenterol. 2002;97:413–19.

50. Ma Y, Okamoto M, Thomas MG et al. Antibodies to conformational epitopes of soluble liver antigen define a severe form of autoimmune liver disease. Hepatology. 2002;35:658–64.

51. Czaja AJ, Carpenter HA, Manns MP. Antibodies to soluble liver antigen, P450IID6, and mitochondrial complexes in chronic hepatitis. Gastroenterology. 1993;105:1522–8.
52. Baeres M, Herkel J, Czaja AJ et al. Establishment of standardized SLA/LP immunoassays: specificity for autoimmune hepatitis, worldwide occurrence, and clinical characteristics. Gut. 2002;51:259–64.
53. Li L, Chen M, Huang DY, Nishioka M. Frequency and significance of antibodies to chromatin in autoimmune hepatitis type 1. J Gastroenterol Hepatol. 2000;15:1176–82.
54. Czaja AJ, Shums Z, Binder WL, Lewis SJ, Nelson VJ, Norman GL. Frequency and significance of antibodies to chromatin in autoimmune hepatitis. Dig Dis Sci. 2003;48: 1658–64.
55. Martini E, Abuaf N, Cavalli F, Durand V, Johanet C, Homberg J-C. Antibody to liver cytosol (anti-LC1) in patients with autoimmune chronic active hepatitis type 2. Hepatology. 1988;8:1662–6.
56. Abuaf N, Johanet C, Chretien P et al. Characterization of the liver cytosol antigen type 1 reacting with autoantibodies in chronic active hepatitis. Hepatology. 1992;16:892–8.
57. Czaja AJ, Manns MP. The validity and importance of subtypes of autoimmune hepatitis: a point of view. Am J Gastroenterol. 1995;90:1206–11.
58. Homberg J-C, Abuaf N, Bernard O et al. Chronic active hepatitis associated with antiliver/ kidney microsome antibody type 1: a second type of 'autoimmune' hepatitis. Hepatology. 1987;7:1333–9.
59. Clemente MG, Meloni A, Obermayer-Staub P, Frau F, Manns MP, De Virgiliis S. Two cytochromes P450 are major hepatocellular autoantigens in autoimmune polyglandular syndrome type 1. Gastroenterology. 1998;114:324–8.
60. Aaltonen J, Borses P, Sandkuijl L, Perheentupa J, Peltonen L. An autosomal locus causing autoimmune disease: autoimmune polyglandular disease type 1 assigned to chromosome 21. Nature Genet. 1994;8:83–7.
61. Manns M, Gerken G, Kyriatsoulis A, Staritz M, Meyer zum Buschenfelde KH. Characterization of a new subgroup of autoimmune chronic active hepatitis by autoantibodies against a soluble liver antigen. Lancet. 1987;1:292–4.
62. Stechemesser E, Klein R, Berg PA. Characterization and clinical relevance of liver-pancreas antibodies in autoimmune hepatitis. Hepatology. 1993;18:1–9.
63. Wies I, Brunner S, Henninger J, Herkel J, Meyer zum Buschenfelde KH, Lohse AW. Identification of target antigen for SLA/LP autoantibodies in autoimmune hepatitis. Lancet. 2000;355:1510–15.
64. Costa M, Rodriques-Sanchez JL, Czaja AJ, Gelpi C. Isolation and characterization of cDNA encoding the antigenic protein of the human tRNA[(Ser)Sec] complex recognized by autoantibodies from patients with type 1 autoimmune hepatitis. Clin Exp Immunol. 2000; 121:364–74.
65. Torres-Collado AX, Czaja AJ, Gelpi C. Anti-tRNP[(SER)SEC]/SLA/LP autoantibodies. Comparative study using in-house ELISA with recombinant 48.8 kDa protein, immunoblot and analysis on immunoprecipitated RNAs. Liver Int. 2005 (In press).
66. Vitozzi S, Djilali-Saiah I, Lapierre P, Alvarez F. Anti-soluble liver antigen/liver-pancreas (SLA/LP) antibodies in pediatric patients with autoimmune hepatitis. Autoimmunity. 2002; 35:485–92.
67. Kanzler S, Weidemann C, Gerken G et al. Clinical significance of autoantibodies to soluble liver antigen in autoimmune hepatitis. J Hepatol. 1999;31:635–40.
68. Czaja AJ, Hay JE, Rakela J. Clinical features and prognostic implications of severe corticosteroid-treated cryptogenic chronic active hepatitis. Mayo Clin Proc. 1990;65:23–30.
69. Czaja AJ, Carpenter HA, Santrach PJ, Moore SB, Homburger HA. The nature and prognosis of severe cryptogenic chronic active hepatitis. Gastroenterology. 1993;104: 1755–61.
70. Kaymakoglu S, Cakaloglu Y, Demir K et al. Is severe cryptogenic chronic hepatitis similar to autoimmune hepatitis?. J Hepatol. 1998;28:78–83.
71. Czaja AJ, Rakela J, Hay JE, Moore SB. Clinical and prognostic implications of human leukocyte antigen B8 in corticosteroid-treated severe autoimmune chronic active hepatitis. Gastroenterology. 1990;98:1587–93.
72. Czaja AJ, Carpenter HA, Santrach PJ, Moore SB. Significance of HLA DR4 in type 1 autoimmune hepatitis. Gastroenterology. 1993;105:1502–7.

73. Czaja AJ, Strettell MDJ, Thomson LJ et al. Associations between alleles of the major histocompatibility complex and type 1 autoimmune hepatitis. Hepatology. 1997;25:317–23.
74. Czaja AJ, Donaldson PT. Genetic susceptibilities for immune expression and liver cell injury in autoimmune hepatitis. Immunol Rev. 2000;174:250–9.
75. Czaja AJ, Doherty DG, Donaldson PT. Genetic bases of autoimmune hepatitis. Dig Dis Sci. 2002;47:2139–50.
76. Donaldson PT, Doherty DG, Hayllar KM, McFarlane IG, Johnson PJ, Williams R. Susceptibility to autoimmune chronic active hepatitis: human leukocyte antigens DR4 and A1-B8-DR3 are independent risk factors. Hepatology. 1991;13:701–6.
77. Doherty DG, Donaldson PT, Underhill JA et al. Allelic sequence variation in the HLA class II genes and proteins in patients with autoimmune hepatitis. Hepatology. 1994;19: 609–15.
78. Strettell MDJ, Donaldson PT, Thomson LJ et al. Allelic basis for HLA-encoded susceptibility to type 1 autoimmune hepatitis. Gastroenterology 1997;112:2028–35.
79. Czaja AJ. Understanding the pathogenesis of autoimmune hepatitis. Am J Gastroenterol. 2001;96:1224–31.
80. Vazquez-Garcia MN, Alaez C, Olivo A et al. MHC class II sequences of susceptibility and protection in Mexicans with autoimmune hepatitis. J Hepatol. 1998;28:985–90.
81. Pando M, Larriba J, Fernandez GC et al. Pediatric and adult forms of type 1 autoimmune hepatitis in Argentina: evidence for differential genetic predisposition. Hepatology. 1999; 30:1374–80.
82. Goldberg AC, Bittencourt PL, Mougin B et al. Analysis of HLA haplotypes in autoimmune hepatitis type 1: identifying the major susceptibility locus. Hum Immunol. 2001;62:165–9.
83. Czaja AJ, Kruger M, Santrach PJ, Moore SB, Manns MP. Genetic distinctions between types 1 and 2 autoimmune hepatitis. Am J Gastroenterol. 1997;92:2197–200.
84. Bittencourt PL, Goldberg AC, Cancado ELR et al. Genetic heterogeneity in susceptibility to autoimmune hepatitis types 1 and 2. Am J Gastroenterol. 1999;94:1906–13.
85. Donaldson PT. Genetics in autoimmune hepatitis. Semin Liver Dis. 2002;22:353–64.

16
Diagnosis of primary biliary liver diseases, overlap syndromes and changing diagnoses

J. HEATHCOTE

INTRODUCTION

Primary diseases of the biliary tract covered in this chapter are primary biliary cirrhosis (PBC) and primary sclerosing cholangitis (PSC). Congenital diseases of the biliary tract, e.g. Caroli's disease, hypoplastic duct syndromes, biliary atresia and choledochal cysts will not be covered in this chapter. Overlap syndromes refer to chronic liver diseases which appear to have features of more than one autoimmune liver disease. Sometimes one liver disease is dominant, and other times two full-blown autoimmune liver diseases are present concurrently. Occasionally it would appear that an individual with one autoimmune disease subsequently loses the typical features of that disease and gains the features of another. Such a change in diagnosis is rare. All of these liver diseases to be discussed below probably have an autoimmune (i.e. genetically controlled) basis for their aetiology with different precipitating factors; thus it is not surprising that overlapping features may be observed.

DIAGNOSIS OF PBC

The recognition in 1965 that antimitochondrial antibodies (AMA) were a serological hallmark for PBC greatly facilitated our ability to make a definitive diagnosis in this disease[1]. Without this hallmark it was necessary to delineate the biliary tree, as well as perform a liver biopsy, to confirm the diagnosis. The three criteria required to make a 'definite' diagnosis of PBC include AMA positivity, elevation in serum alkaline phosphatase and a liver biopsy showing the typical bile duct destruction with or without portal tract granulomas. It has been suggested that a diagnosis of 'probable' PBC can be made if any two of these three aforementioned features are present, suggesting that a diagnosis can be made without performing a liver biopsy[2].

DILEMMAS: DIAGNOSIS OF PBC

Do all individuals who test AMA-positive have PBC?

Mitchison et al.[3] described the liver histology in 29 individuals who had been found to test AMA-positive in serum, but who were symptom-free and had normal liver biochemical tests. Liver histology in these 29 individuals showed that all but three had findings typical of, or compatible with, PBC, albeit at a very early stage, none of whom had cirrhosis. Metcalf et al.[4] reported a 10-year follow-up on these 29 individuals, five of whom had died from non-hepatic causes. The remaining 24 developed elevation in serum alkaline phosphatase and some developed typical symptoms of PBC, namely fatigue and/or pruritus, in follow-up. The median age of these AMA-positive individuals was 67 years, very much older than the usual age at presentation of symptomatic PBC. This study would suggest that indeed detectable AMA in serum means there is underlying PBC. However, AMA may on occasion be detected in individuals with other liver diseases, e.g. autoimmune hepatitis (AIH)[5], PSC[6] and perhaps even in association with other non-hepatic chronic diseases[7]. The work of Mitchison and Metcalf suggests that, when AMA are present in otherwise asymptomatic individuals, examination of liver histology usually shows typical bile duct lesions. This may not be the case in AMA-positive AIH[5].

AMA-negative PBC

There are now several reports of patients who have the typical symptoms, cholestatic liver biochemistry, immunological findings and liver biopsy appearances of PBC but who consistently have undetectable AMA in serum, even when using the most highly sensitive techniques[8,9]. Long-term follow-up studies indicate that the natural history of AMA-negative PBC is no different from that of AMA-positive PBC. However, it is probably unwise to make a diagnosis of AMA-negative PBC just on the basis of an elevated ALP and typical liver biopsy appearances without radiologically delineating the biliary tree and excluding PSC. The majority of individuals thought to have AMA-negative PBC test positive, often in high titre for ANA. It is because ANA may also be present in serum in PSC that it is necessary to delineate the biliary tract.

ANA is considered a hallmark for AIH, and when these cases were first described they were described as having 'immune cholangitis' – a form of AIH – and treatment with corticosteroids was recommended[10].

ANA-positivity in PBC

ANA may be detected in 6% of the general population, this rate increases with age so that in individuals over the age of 60, 14% test ANA-positive[11]. At any age approximately a half of those with definite PBC test ANA-positive, generally in high titre. Several different patterns of nuclear staining may be seen. In a recent study by Muratori et al.[12], 40/83 (48%) of AMA-positive PBC tested positive for ANA; this figure was 85% in 13 patients with AMA-negative PBC. The pattern was more likely to be multiple nuclear dot (38%) in the

AMA-negative (this figure was 12% in AMA-positive patients). In a study by Worman et al.[13], the authors suggest that both sp100 and gp[1]210 are very specific for PBC, but their control group did not include patients with other causes of vanishing bile duct syndromes.

Granulomatous destruction of interlobular bile ducts: differential diagnosis

Interlobular bile ducts seen on liver histology normally have the appearance of a 'string of pearls' in that the biliary epithelial cells are similarly sized and the nuclei are centred. It is not unusual to notice mild distortion of interlobular bile ducts in patients with any form of chronic liver disease, best described in chronic hepatitis C[14] and in AIH[5]. However, in PSC and PBC, bile duct loss is present in addition to ductal distortion. The vanishing bile duct syndrome may be found in many other conditions, e.g. following drug-induced liver injury (most commonly following antibiotic-induced cholestatic hepatitis)[15] with malignancies, and with the many causes for congenital and idiopathic hypoplastic duct syndromes[16]. Interlobular bile duct destruction with invasion by granulomas is nearly always due to PBC; rarely it is caused by hepatic sarcoidosis[17]. The granulomas of sarcoidosis tend to extend into the parenchyma and may have a fibrinous appearance. Simultaneous PBC and sarcoidosis is well described[18]!

The bile duct loss of PSC is not always accompanied by the histological hallmark, namely 'onion skin' fibrosis around the ducts, so when this pattern is absent it may be hard to distinguish small duct PSC[19] from the many causes of vanishing bile duct syndrome, including PBC, particularly when AMA testing is negative.

Symptoms: are they disease-specific?

The major symptoms of PBC are pruritus and jaundice. Fatigue does not correlate with the severity of liver disease and because of its subjectivity, and the fact that this symptom is quite prevalent in society as a whole, it cannot be considered a symptom which is specific for chronic cholestatic liver disease (PBC or PSC). Pruritus in the presence of chronic liver disease is not confined to biliary diseases; this symptom may be present in individuals with chronic hepatitis C[20], or in any person with chronic liver disease on exogenous oestrogen therapy!

Hyperbilirubinaemia in its conjugated form is a feature of severe intrahepatic or extrahepatic cholestasis. Confusion may occur in individuals with a diagnosis of PBC who have superimposed Gilbert syndrome. As this affects 7% of the population, superimposed Gilbert syndrome is not rare, and may be suspected when the bilirubin is very much higher than would be suggested appropriate by the other tests of liver function. If the mixed picture of conjugated and unconjugated hyperbilirubinaemia fails to be recognized, the patient may be referred for liver transplantation a lot sooner than is necessary. Oestrogen therapy will promote a conjugated hyperbilirubinaemia.

It is usually not difficult to make a confident diagnosis of PBC. Now that this disease has recently been described in teenagers[21], this mandates AMA testing in a much wider age range of individuals with chronic liver disease. Asymptomatic PBC is more likely to be present in an older individual and a new diagnosis of PBC may be made, even in the eighth decade. AMA may become undetectable upon the introduction of therapy with ursodeoxycholic acid; however, in those individuals who test consistently AMA-negative, even prior to UDCA treatment, it may be much harder to make a definite diagnosis of PBC. The diagnosis of AMA-negative PBC should not be made without both examination of the biliary tree and a liver biopsy.

DIAGNOSIS OF PSC

Currently the gold standard for the diagnosis of PSC is the finding of the typical stricturing and beading of the biliary tree seen either with endoscopic retrograde cholangiography (ERCP) or magnetic resonance cholangiography (MRC). However, this pattern of damage to the biliary tree likely takes several years to develop, meaning there are many individuals who have PSC but who appear to have a normal biliary tract on radiological visualization. ERCP was introduced in the mid-1970s; it was then very helpful in defining the pattern of this chronic liver disease. Many patients prior to this definitive test had been misdiagnosed on liver biopsy as having other causes of the vanishing bile duct syndrome, particularly AMA-negative PBC[22]. Whereas ERCP is extremely valuable in facilitating the diagnosis of PSC when the typical findings are observed, it is not so valuable when used to follow progression of disease, because of the difficulty in ensuring that exactly the same conditions exist at the time of the procedure, e.g. pressure applied and volume of dye introduced into the biliary tract. Similarly exogenous factors may influence the interpretation of MRC; thus both ERCP and MRC have their shortcomings. Recent studies suggest, however, that their accuracy is probably comparable when abnormalities are present, but neither can be used to identify early PSC[23].

Liver histology is particularly unreliable in PSC, both in prompting the diagnosis and assessing the prognosis. First, unless a large liver biopsy specimen is provided, with at least 10 portal tracts present, the diagnosis of any cause of vanishing bile duct syndrome is unreliable; but in the case of PSC there may be very patchy involvement of the biliary system so there is potentially even greater sampling error than for other liver diseases. To be confident about the diagnosis of cirrhosis, both the left and right lobes of the liver need to be biopsied.

DILEMMAS: DIAGNOSIS OF PSC

Superimposed AIH

Particularly in children given a diagnosis of PSC, the liver biopsy may suggest that the diagnosis is AIH. Similarly in children given a primary diagnosis of

AIH, examination of the biliary tree may suggest additional biliary tract disease. In children and young adults, concurrent AIH with PSC is present 50% of the time[24,25].

Vanishing bile duct syndrome with a normal biliary tree; is it all small duct PSC?

Sometimes work-up of an individual who is AMA-negative may show a normal biliary tree, identified using radiological techniques, but liver biopsy identifies bile duct loss. It is thus important to decide whether the biliary tract disease is small duct PSC[26] or AMA-negative PBC[8] or even idiopathic ductopenia[27]. An algorithm needs to be developed which determines how to clearly separate these three diagnostic categories. Long-term follow-up studies in small duct PSC have been conducted, including repeat radiological examination of the biliary tree, and progression to large duct PSC occurs in a small percentage, but this does not help make the diagnosis at the initial presentation! Our current state of knowledge does not indicate that autoantibody screening at the time of presentation in such individuals with a vanishing bile duct syndrome helps a great deal in narrowing down the differential diagnosis, because the wrong liver control groups have been tested to date.

Absence of markers of cholestasis in PSC

In patients being followed for inflammatory bowel disease, liver biochemical tests are often used to screen for those who may have additional PSC. Whereas the finding of an elevated serum alkaline phosphatase may identify an individual at risk for PSC, PSC may be present in the face of a normal ALP, although it is unusual for the gammaglutamyl transpeptidase (GGT) to be normal[24]. Thus GGT is probably a more reliable screening test than ALP for this disease. PSC may develop after a patient has undergone pancolectomy for inflammatory bowel disease, and in a third of patients with PSC no colitis is present at initial diagnosis, although this may develop subsequently. In children with both AIH and PSC the liver biochemistry does not help distinguish whether or not concurrent AIH or PSC is present[25].

OVERLAP SYNDROMES

Overlap syndromes in hepatology refer to individuals who appear to have more than one autoimmune liver disease present simultaneously or have a single autoimmune liver disease with features of another.

AIH and PSC

As mentioned above, this particular overlap appears to be most common in children and young adults, but may occur in older patients. Two paediatric studies, one retrospective[24] and the other prospective[25], indicate that children either given a primary diagnosis of PSC or a primary diagnosis of AIH have at least a 50% chance of having both diseases simultaneously; thus it is essential that the two diagnoses always be considered with either primary diagnosis. Both these studies indicated that there were no symptoms, biochemical findings, autoantibody features, serological tests or liver histological findings that could reliably indicate whether a child had just the one autoimmune disorder or both. This means that in children given the primary diagnosis of PSC using ERCP or MRC, a liver biopsy is essential to exclude superimposed AIH and vice-versa in that all children given a diagnosis of AIH need radiological examination of the biliary tree. The features of the biliary disease in children with AIH appear to be a little different from the beading and stricturing of classical PSC. More often the common bile duct is just irregular, even somewhat dilated but not strictured. Unfortunately there are no follow-up data published to indicate whether or not these features remain stable or progress to a pattern more typical of PSC. A retrospective study in children with PSC would suggest that the typical beading and stricturing of classical PSC may be seen in children, even in those as young as 6 months. This suggests that perhaps the findings of Gregorio et al.[25] described as autoimmune sclerosing cholangitis (ASC) are not exactly the same as what is thought of as PSC. Although inflammatory bowel disease is more likely in children with ASC, complicated by AIH, they report that children with simply AIH may have inflammatory bowel disease too.

PBC with overlapping features of AIH

It was suggested by Chazoullieres et al.[28] that patients with otherwise overt PBC may have additional features of AIH. The criteria these authors used were the presence of two of the following; either an ALT > 5-fold elevated, or the presence of detectable SMA and/or an IgG > 2-fold elevated or the presence of interface hepatitis of moderate–severe severity. Their data indicated that 9% of patients with PBC have these features. The outcome of individuals with PBC and a lymphoplasmacytic interface hepatitis was also worse, and the authors suggested additional corticosteroid therapy should be given to such individuals[29]. In another study of patients with PBC it was determined that, even in those individuals with these overlapping features of AIH, the outcome was no different following introduction of UDCA whether or not these features were present[30]. Thus it is debatable whether the addition of steroid therapy is truly warranted. Other investigators applied the revised AIH score to patients with PBC; none had a score of 15 or greater (definite AIH), although 19% did have a score that suggested a probable AIH[31]. Hence it appears not unusual for patients with PBC to have overlapping features of AIH – whether this means they have an overlap syndrome, or that the spectrum of PBC is broader than is considered appropriate by some is a 'hot topic' for debate!

Individuals with AIH who test positive for AMA

In almost every case series of AIH reported, there are a few individuals who test positive for AMA[5]. When this was first recognized it was felt that sometimes the mitochondrial antibodies had been misread, and they were really microsomal antibodies typical of type 2 AIH[32]. Since the introduction of more specific tests using the PDE2 antigen, either using ELISA or Western blotting, however, it has been confirmed that some patients with all the features of AIH test positive for AMA[33]. Whether these individuals have AMA-positive AIH or an overlap of AIH and PBC, or perhaps even AIH with early PBC, is impossible to tell without long-term follow-up studies; few have been reported to date.

CHANGING DIAGNOSES

PBC to AIH

There have been a number of case reports of patients with all the typical features of PBC; these are namely AMA-positivity, elevated alkaline phosphatase and liver biopsy features, who have responded well to UDCA therapy. However, in subsequent follow-up they have suddenly developed a more hepatitic picture with high levels of serum aminotransferases and with a change in liver histology, a loss of AMA-positivity and a gain of ANA-positivity. They then require the addition of corticosteroid therapy to control their liver disease[34,35]. These are clear-cut cases of changing diagnoses, but they are rare.

AIH to PSC

There has been a recent case series of six individuals given a primary diagnosis of AIH with all the typical biochemical and histological findings (with no evidence of any bile duct injury) who several years later, when poor response to immunosuppressive therapy was observed, were reinvestigated and were found to have the radiological findings typical of PSC[36]. Three of these six had undergone ERCP at the time of their original diagnosis and in all three the ERCP was normal; thus the diagnosis clearly changed from that of AIH to that of PSC. Four of these six individuals were young, i.e. this overlap of AIH and PSC seems to predominate in the younger patient. There were no identifiable features at time of initial presentation that would have suggested that these individuals would progress or had superimposed PSC.

None of the studies reported above indicates what triggers the autoimmune liver disease. There are many genetic studies that indicate a preponderance of certain HLA patterns for the different autoimmune liver diseases. It is likely that, in the setting of a certain genetic background, particularly of genes which determine the immune response, a number of exogenous and even endogenous factors may promote the development of autoimmune liver disease. It is not entirely surprising that, over the course of time with a background that predisposes a heightened autoimmune response, overlapping features of the different autoimmune liver diseases may be seen either simultaneously or consecutively.

References

1. Walker JG, Doniach D, Roitt M et al Serological tests in diagnosis of primary biliary cirrhosis. Lancet. 1965;1:827.

2. Zein CO Angulo P, Lindor KD. When is liver biopsy needed in the diagnosis of primary biliary cirrhosis. Clin Gastroenterol Hepatol. 2003;1:89–95.

3. Mitchison HC, Bassendine MF, Hendrick A et al. Positive antimitochondrial antibody but normal alkaline phosphatase: is this primary biliary cirrhosis? Hepatology. 1986;6:1279–84.

4. Metcalf JV, Mitchison HC, Palmer JM, Jones DE, Bassendine MF, James OF. Natural history of early primary biliary cirrhosis. Lancet. 1996;348:1399–402.

5. Czaja AJ, Muratori P, Muratori L, Carpenter HA, Bianchi FB. Diagnostic and therapeutic implications of bile duct injury in autoimmune hepatitis. Liver Int. 2004;24:322–9.

6. Burak KW, Urbanski SJ, Swain MG. A case of co-existing primary biliary cirrhosis and primary sclerosing cholangitis: a new overlap of autoimmune liver diseases. Dig Dis Sci. 2001;46:2043–7.

7. Diederichsen H, Pallisgaard G. Mitochondrial antibodies without antinuclear antibodies in non-hepatic diseases. Acta Med Scand. 1977;202:97–101.

8. Michieletti P, Wanless IR, Katz A et al. Antimitochondrial antibody negative primary biliary cirrhosis: a distinct syndrome of autoimmune cholangitis. Gut. 1994;35:260–5.

9. Invernizzi P, Crosignani A, Battezzati PM et al. Comparison of the clinical features and clinical course of antimitochondrial antibody-positive and -negative primary biliary cirrhosis. Hepatology. 1997;25:1090–5.

10. Brunner G, Klinge O. A chronic destructive non-suppurative cholangitis-like disease picture with antinuclear antibodies (immunocholangitis). Dtsch Med Wochenschr. 1987;112:1454–8.

11. Hooper B, Whittingham S, Mathews JD, Mackay IR, Curnow DH. Autoimmunity in a rural community. Clin Exp Immunol. 1972;12:79–87.

12. Muratori P, Muratori L, Ferrari R et al. Characterization and clinical impact of antinuclear antibodies in primary biliary cirrhosis. Am J Gastroenterol. 2003;98:431–7.

13. Worman HJ, Courvalin JC. Antinuclear antibodies specific for primary biliary cirrhosis. Autoimmun Rev. 2003;2:211–17.

14. Kaji K, Nakanuma Y, Tsuneyama K, Van de Water J, Gershwin ME. Hepatic bile duct injury in chronic hepatitis C: histopathologic and immunohistochemical studies. Mod Pathol. 1994;7:937–45.

15. Al-Traif I, Lilly L, Wanless IR, Heathcote J. Cholestatic liver disease with ductopenia (vanishing bile duct syndrome) following clindamycin and trimethoprim-sulfamethoxazole administration. Am J Gastroenterol. 1994;89:1230–4.

16. Burak KW, Pearson DC, Swain MG, Kelly J, Urbanski SJ, Bridges RJ. Familial idiopathic ductopenia: a report of five cases in three generations. J Hepatol. 2000;32:159–63.

17. Nakanuma Y, Kouda W, Herada K, Hiramatsu K. Hepatic sarcoidosis with vanishing bile duct syndrome, cirrhosis, and portal phlebosclerosis. Report of an autopsy case. J Clin Gastroenterol. 2001;32:181–4.

18. Keeffe EB. Sarcoidosis and primary biliary cirrhosis. Literature review and illustrative case. Am J Med. 1987;83:977–80.

19. Angulo P, Maor-Kendler Y, Lindor KD. Small duct primary sclerosing cholangitis: a long term follow up study. Hepatology. 2002;35:1494–500.

20. Kumar KS, Saboorian MH, Lee WM. Cholestatic presentation of chronic hepatitis C. Dig Dis Sci. 2001;46:2066–73.

21. Dahlan Y, Smith L, Simmonds D et al. Pediatric onset primary biliary cirrhosis. Gastroenterology. 2003;125:1476–9.

22. Chapman RWG, Marborgh BA, Rhodes JM et al. Primary sclerosing cholangitis: a review of its clinical features, cholangiography and hepatic histology. Gut. 1980;21:870–7.

23. Talwalkar JA, Angulo P, Johnson CD, Petersen BT, Lindor KD. Cost-minimization analysis of MRC versus ERCP for the diagnosis of primary sclerosing cholangitis. Hepatology. 2004;40:39–45.

24. Wilchanski M, Chait P, Wade JA et al. Primary sclerosing cholangitis in 32 children: clinical, laboratory and radiographic features, with survival analysis. Hepatology. 1995;22:1415–22.

25. Gregorio GV, Portmann B, Karana J et al. Autoimmune hepatitis/sclerosing cholangitis overlap syndrome in childhood: a 16 year prospective study. Hepatology. 2001;33:544–53.
26. Bjornsson E, Boberg KM, Cullen S et al. Patients with small duct primary sclerosing cholangitis have a favourable long term prognosis. Gut. 2002;51:731–5.
27. Ludwig J. Idiopathic adulthood ductopenia: an update. Mayo Clin Proc. 1998;73;285–91.
28. Chazoullieres O, Wendum D, Serfaty L, Montembault S, Rosmorduc O, Poupon R. Primary biliary cirrhosis–autoimmune hepatitis overlap syndrome: clinical features and response to therapy. Hepatology. 1998;2002:296–301.
29. Corpechot C, Carrat F, Poupon R, Poupon RE. Primary biliary cirrhosis: incidence and predictive factors of cirrhosis development in ursodiol-treated patients. Gastroenterology. 2002;122:652–8.
30. Joshi S, Cauch-Dudek K, Wanles IR et al. Primary biliary cirrhosis with additional features of autoimmune hepatitis: response to therapy with ursodeoxycholic acid. Hepatology. 2002;35:409–13.
31. Talwalkar JA, Keach JC, Angulo P, Lindor KD. Overlap of autoimmune hepatitis and primary biliary cirrhosis: an evaluation of a modified scoring system. Am J Gastroenterol. 2002;97:1191–7.
32. Kenny RP, Czaja AJ, Ludwig J, Dickson ER. Frequency and significance of automitochondrial antibodies in severe chronic active hepatitis. Dig Dis Sci. 1986;31:705–11.
33. Joshi S, Guindi M, Heathcote J. The significance of anti-mitochondrial antibodies (AMA) in patients with classical autoimmune hepatitis: twenty-eight years of follow-up. Gastroenterology. 2004;126(Suppl. 2): abstract M1198.
34. Colombato LA, Alvarez F, Cote J et al. Autoimmune cholangiopathy: the result of consecutive primary biliary cirrhosis and autoimmune hepatitis? Gastroenterology. 1994;107:1839–43.
35. Weyman RL, Voigt M. Consecutive occurrence of primary biliary cirrhosis and autoimmune hepatitis: a case report and review of the literature. Am J Gastroenterol. 2001;96:585–7.
36. Abdo A, Bain VG, Kichian K, Lee SS. Evolution of autoimmune hepatitis to primary sclerosing cholangitis: a sequential syndrome. Hepatology. 2002;36:1393–9.

Section VI
Treatment and prognosis I

Chair: E.L. KRAWITT and A. STIEHL

17
Standard treatment of autoimmune hepatitis

S. LÜTH, E. BAYER, C. SCHRAMM and A. W. LOHSE

INTRODUCTION

Appropriate immunosuppressive therapy in autoimmune hepatitis is life-saving, while untreated disease has a poor prognosis in most cases[1]. The severity of autoimmune hepatitis, especially in young patients, was recognized by those first describing the disease entity, and therefore it was quite early in the history of the disease when Ian Mackay and co-workers described the use of immunosuppression (the first chronic liver disease which could be treated successfully)[2,3]. The favourable results led other investigators to initiate randomized controlled trials to demonstrate the efficacy of corticosteroid therapy in patients with autoimmune hepatitis (Figure 1a). These were some of the earliest randomized controlled trials performed, and autoimmune hepatitis is clearly the first liver disease in which the effectiveness of a therapeutic intervention has been convincingly demonstrated by this approach. Three independent trials (at the Royal Free Hospital in London, the Mayo Clinic in Rochester and in Copenhagen) proved the advantage of corticosteroid therapy over placebo[4-7]. At the same time these trials demonstrated the poor prognosis of patients with untreated disease with more than 60% of the placebo patients having died within the first 5 years[7]. Even though the natural history of patients diagnosed today may on average be more favourable, because many patients are diagnosed in an asymptomatic state, the general need for immunosuppression in patients with autoimmune hepatitis is beyond doubt, especially in patients with clinically symptomatic disease.

Despite these clear-cut studies there is no general consensus on the standard treatment of autoimmune hepatitis. Increasing evidence suggests that optimal therapy needs to be individualized according to the acuteness of the disease, presence of cirrhosis, degree of impairment of liver function, presence of comorbid conditions, risk factors for treatment-induced side-effects etc. Thus, general guidelines can be given, but individual adaptation needs to be undertaken. The treatment of autoimmune hepatitis should therefore be guided by expertise, and best be monitored by specialized centres. The excellent survival of patients treated in such a fashion is depicted in Figure 1B.

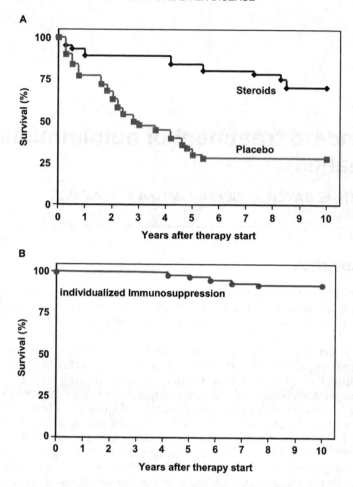

Figure 1 A: Survival rate in steroid-treated and placebo-treated patients in the Royal Free Hospital (London) trial[5,7], **B**: survival rate in patients treated by an individualized immunosuppressive approach in Mainz in 2001[8]

REMISSION INDUCTION

The drugs of choice for induction of remission are corticosteroids, preferably prednisolone, while the drug of choice for maintenance of remission is azathioprine[9,10].

The exact treatment schedule is controversial, and probably needs adjustment according to the individual severity of the disease[11,12]. In the majority of patients treatment should be started with about 1 mg/kg body weight prednisolone per day until there is a treatment response[13]. In patients with mild disease half this dose may be sufficient. In patients with severe disease, starting with 100 mg prednisolone given intravenously may improve the response.

With a starting dose of 1 mg/kg per day a biochemical response is usually observed within days. As soon as improvement has been documented the dose can be reduced fairly rapidly, usually in steps of 10 mg weekly dose reductions. However, relapse at lower doses is common, so tapering of steroids should be slowed down from a daily dosage of about 20 mg onwards. Our usual protocol is to then give 15 mg/day for 2 weeks, then 12.5 mg/day for 2 weeks and then 10 mg/day. In patients with a slow or incomplete response even slower tapering, starting at 25 mg prednisolone per day, may be warranted. The dose of 10 mg/day should not be lowered unless genuine remission is achieved with normal levels for AST and ALT.

Azathioprine should be added to the treatment as soon as the diagnosis of AIH is confirmed, because its immunosuppressive effect takes several weeks to fully develop[14]. In patients in whom the diagnosis is uncertain, steroid monotherapy should be maintained until the diagnosis is confirmed or refuted. Once the diagnosis is confirmed (even if this is only after a relapse following cessation of steroid therapy), azathioprine should be added to corticosteroids at a dose of 1 mg/kg per day. Again, the degree of disease activity determines the dosage: patients with very active disease should be given higher doses up to 2 mg/kg per day[10]. While corticosteroids are tapered early, azathioprine doses are maintained at least until steroids are reduced below 10 mg prednisolone/day.

A typical stepwise approach in a patient (70–80 kg body weight) with initally very active disease is shown in Table 1. We would call this the 'European approach' in comparison to the 'North American approach', mainly propagated by the Mayo Clinic group, and largely based on the treatment protocol used by Summerskill and colleagues in the original Mayo Clinic treatment trial[6]. This approach starts with 60 mg of prednisolone per day given for 1 week, followed by 40 mg/day for 1 week, and 30 mg/day for 2 weeks followed by 20 mg/day until remission is achieved[15]. Alternatively, azathioprine is given

Table 1 Stepwise reduction of steroids and combined azathioprine immunosuppressive therapy in a 70–80 kg patient

	Corticosteroid (mg/day)	Azathioprine* (mg/day)
Week 1	75	100
Week 2	60	100
Week 3	50	100
Week 4	40	100
Week 5	30	100
Week 6	25	100
Week 7	20	100
Week 8 + 9	15	100
Week 10 + 11	12.5	100
Week 12	10	100

*Azathioprine should be added as soon as the diagnosis of AIH is considered definite.

immediately at a low dose of 50 mg/day and the prednisolone dose halved to 30 mg/day for the first week, 20 mg/day for the second week, 15 mg/day for 2 weeks and 10 mg/day thereafter until remission is achieved.

In the absence of comparative studies it is difficult to reliably judge the advantages and disadvantages of these different approaches. It appears important that in patients presenting with acute hepatitis liver destruction is stopped quickly; therefore these patients are probably best served by starting treatment with high doses. Our personal experience favours the initially higher steroid approach for two major reasons:

1. Remission induction is rapid and reliable, thus preventing further liver tissue destruction. Using this 'European approach' we have been able to induce remission (i.e. normal transaminases) within the first 6 months in more than 90% of our patients (more than 80% within the first 3 months), and less than 2% of patients had transaminases above twice the upper limit of normal at this time point[8].

2. The more rapid normalization allows a faster subsequent tapering of steroids below the Cushing dose; thus this is likely to reduce the overall steroid side-effects. High steroid doses for up to 3 months tend to cause almost exclusively reversible side-effects, while steroid doses of 10 mg/day prednisolone or more given, for more than 6 months, are more likely to result in irreversible damage such as osteoporosis or cataract development.

Response to treatment is usually prompt and often impressive. Non-response or poor response to corticosteroid therapy is extremely uncommon, and should lead to re-evaluation of the diagnosis. In icteric patients poor response may be due to insufficient absorption of orally administered steroids, so that in these patients intravenous treatment is a good option until serum bilirubin starts to fall substantially.

MAINTENANCE THERAPY

There is a gradual progression from remission induction therapy to remission maintenance therapy, and in up to 10% of the patients complete remission is never reached without risking inadequate side-effects of high-dose immuno-suppression. In these patients maintenance regimens should also be started after 6 months.

The drug of choice for maintenance therapy is azathioprine. The King's College group in particular demonstrated the effectiveness of azathioprine in the maintenance of remission in two important studies[9,10]. The exact dose of azathioprine probably has to be adapted individually. The usual treatment regimen uses 1–1.5 mg/kg body weight per day. This dose is sufficient for the majority of patients during the first year of treatment, when they are still receiving low-dose prednisolone at the same time. After 1 year, as long as transaminases and IgG are normal, steroids can be slowly tapered out, but up to 50% of patients will experience relapse at this time point. This has led some authors to recommend switching to a higher dose of azathioprine (2 mg/kg) at the time of steroid withdrawal. The King's college group have described a stable remission rate of around 80% using this regimen[10]; however, in their azathioprine maintenance trial using this regimen, four patients (out of 72) died from a solid cancer, one from sepsis and one from lymphoma during the study period. The worry is that high-dose azathioprine might be associated with a higher risk of tumour development, and should therefore be avoided[10]. On the other hand, data from patients with inflammatory bowel disease suggest a much lower cancer risk. An alternative approach would be to continue low-dose steroids (usually 5 mg prednisolone/day) and not to increase azathioprine above 1–1.5 mg/kg per day. It is doubtful if this difference of opinion will ever be solved by a randomized trial; therefore observational studies will need to guide us in the foreseeable future. The excellent survival of our patients with the more individualized approach described above, and the lack of an increased tumour incidence with this regimen, encourages us to continue this approach. Thus we taper steroids after 1 year of remission by slowly decreasing the daily dose in steps of 2.5 mg every 3 months. When transaminases or IgG levels begin to rise again, steroids are re-introduced, with a short (about 4 weeks) phase of increased steroid dose (usually starting around 20 mg daily). Transaminases need to be checked every 4–6 weeks during a phase of treatment reduction, in order to detect and treat flares early so that the need for higher steroid doses does not arise.

The role of liver biopsies during the follow-up of patients with autoimmune hepatitis is controversial, too. In patients with complete normalization of both transaminases and IgG or gamma-globulins, treatment can be guided by the laboratory response alone, as almost all of these patients will at the most show minimal activity histologically. This is different in patients with elevated transaminases. In patients with transaminases elevated mildly, i.e. less than twice the upper limit of normal, 50% still show moderate to severe activity on liver biopsy. These patients should probably be treated with higher doses of immunosuppression. Patients with transaminases above this level are very likely to have at least moderate histological disease activity, and therefore do not require a liver biopsy to justify an intensification of treatment. The justification for an increased immunosuppression in the presence of moderate or severe histological disease activity comes from two observations: first, some studies report the development of cirrhosis in as many as 40% of AIH patients during therapy, when the treatment aim is set at transaminase levels below twice normal[16,17]; secondly, we have been able to demonstrate marked over-expression of transforming growth factor β (TGF-β) in areas of the liver where inflammatory infiltrates could be found[18]. Since TGF-β is thought to be the most important profibrogenic cytokine in the liver, it is reasonable to assume that inflammatory lesions associated with high TGF-β expression are responsible for continued progression of liver fibrosis in these AIH patients. Therefore, immunosuppression should aim at minimal or no inflammatory activity in the liver in order to prevent cirrhosis development or progression[19].

SIDE-EFFECTS OF TREATMENT

Despite the very good response of AIH to immunosuppression, and the excellent long-term prognosis, risks and side-effects of treatment need to be considered. Side-effects, and in particular the fear of side-effects, are the main reason for non-compliance. In view of the importance of adequately dosed immunosuppression in AIH, the potential side-effects and their prevention or management should be explained to patients repeatedly (Table 2).

Table 2   Treatment side-effects

Short-term high-dose steroid effects
Cutaneous side-effects
Electrolyte abnormalities
Arterial hypertension
Hyperglycaemia (diabetes)
Pancreatitis
Psychiatric side-effects
Haematopoietic side-effects
Long-term steroid side-effects
Osteoporosis
Aseptic joint necrosis
Myopathy
Gastrointestinal side-effects (intestinal ulcers, diverticular perforation, hepatic steatosis, hyperlipidaemia)
Adrenal insufficiency
Ophthalmological side-effects (cataract, glaucoma)
Aazathioprine intolerance
Long-term azathioprine toxicity
Infections
Malignomas?

Short-term side-effects of high-dose steroids, as are required for remission induction, occur to some extent in every patient, and therefore need to be explained to and discussed with the patient. The degree and character of these short-term side-effects differ enormously between patients, and are therefore only partly dose-related. The most notable, and for many patients worrying, side-effects of steroids are due to their anabolic effects. Moonface appearance can start to develop as soon as 2 weeks after initiating treatment, but is quickly reversible as soon as steroids can be tapered below 10 mg/day. Weight gain is an indirect effect due to the increase in appetite induced by steroids. This gain in appetite can be difficult to manage for the patient, as steroids often induce a craving for food. However, explaining these effects can greatly reduce weight gain: for example, patients should be advised to always have some low-calorie food with them, which will prevent them from running to the next shop to buy a bar of chocolate, if hunger overcomes them.

The second most important transient side-effects during high-dose treatment are due to the psychotropic effects of steroids. Both mania and depression can be induced by steroids, and frank psychosis can occasionally occur. The patient and the family need to be aware of this potential side-effect and be advised to contact their doctor immediately. The problem can usually be solved by more rapid tapering of steroids as best as can be justified in view of the activity and stage of the liver disease. The occasional patient, however, may require in-patient psychiatric treatment. Milder psychotropic effects are very common, and the majority of patients experience a somewhat elevated mood that is quite pleasant to most of them.

Corticosteroids increase blood pressure in a dose-dependent fashion within 24 h after treatment initiation. Systolic pressure increases approximately 15 mmHg at a dose of 20–50 mg prednisolone[20], but falls as the dose is tapered.

Osteopenia and osteoporosis are important risks of long-term steroid therapy, and need to be prevented and/or, if present, treated appropriately. As long as patients are receiving more than 7.5 mg prednisolone a day, vitamin D and calcium should be supplemented. Secondly, upon remission, prednisolone dosage should be decreased to about 5 mg/day. Thirdly, in about 50% of patients steroids can be withdrawn completely, and this should be tried after a minimum of 1 year immunosuppressive therapy. If steroids cannot be withdrawn, vitamin D (500–1000 IU per day) and calcium (0.5–1 g per day) should probably be given in parallel. Measurement of bone density can help to guide osteoporosis prophylaxis and treatment. In a study of our own patients with AIH, we found that only patients with other major risk factors for osteoporosis were at risk of osteopenia, most importantly smoking, lack of calcium in their diet, and lack of physical exercise. The few patients who do have progressive osteopenia or overt osteoporosis should probably receive bisphosphonates.

Cataract development is another side-effect of long-term steroid treatment, and there is no effective prophylaxis. Furthermore, steroids may cause or contribute to glaucoma development. For these two reasons yearly checks by an ophthalmologist seem advisable in order to intervene appropriately, when these complications arise.

Azathioprine toxicity also falls into two categories: primary intolerance and secondary toxicity. Primary intolerance is of an allergic nature and occurs in about 3% of patients. The clinical picture can be quite dramatic with fever, nausea and vomiting and general malaise. When this occurs shortly after starting azathioprine the diagnosis is obvious. In some patients the intolerance is less obvious with milder symptoms or a more delayed presentation of symptoms. In all these patients intolerance should be considered and tested. This can best be done by stopping azathioprine treatment, which results in improvement of symptoms within a few days. In view of the importance of azathioprine in the long-term treatment of the disease, the diagnosis of intolerance should be confirmed, which can be accomplished by careful re-exposure in all those in whom initial symptoms were less clear-cut, or in whom a transient gastroenteritis or other infection might have accounted for the symptoms. Upon re-exposure patients truly intolerant will experience renewed symptoms usually very promptly, often within the first day or two. As this is an idiosyncratic effect, life-long azathioprine intolerance should be explained to the patient and marked in the medical records. For these patients, alternative maintenance treatment needs to be given. Most of these patients can be safely managed by steroid monotherapy (with osteoporosis prophylaxis), while in others alternative drugs such as mycophenolate mofetil, methotrexate or cyclophosphamide can be considered.

Azathioprine toxicity may also develop in the long term, and can be observed by a further 3–5% of patients. Of particular note are hepatotoxicity and bone-marrow damage. Hepatotoxicity is uncommon, but it is an important differential diagnosis in all those patients showing a rise of transaminases on maintenance therapy, in particular if transaminases rise without a parallel rise in IgG or gamma-globulins. A liver biopsy can reliably differentiate between the two entities. Alternatively, assessing the response to a transient increase in steroid dose can also help to differentiate, as azathioprine toxicity will not improve in response to steroids. Bone-marrow toxicity is not common as long as treatment doses are below 2 mg/kg, but it should still be watched for by regular blood counts (initially monthly, later on every 3 months). Mild leukopenia is frequently observed and should not lead to dose reduction. Only in progressive leukopenia should azathioprine be stopped, as the bone-marrow damage is often irreversible, or may reverse very slowly. Azathioprine is metabolized in part by thiopurine methyl transferase (TPMT), and the activity of this enzyme is lacking in about 0.3% of the population. However, severe bone-marrow toxicity does not seem to be reliably predicted by the assessment of TPMT activity and therefore testing cannot be advised on a routine basis.

TREATMENT WITHDRAWAL

Patients not reaching remission should probably be kept on their immunosuppression indefinitely, as tapering the dose or withdrawing treatment is almost certain to lead to reactivation. As reactivation may be subclinical, regular monitoring of these patients is important. Patients attaining clinical and laboratory remission may be considered for a trial of treatment withdrawal. However, it needs to be considered that relapse is common (50–80% of patients) within 1–2 years after treatment withdrawal. Relapse can be predicted by two important parameters[9]:

1. Duration of treatment (and remission) prior to withdrawal.

2. Histological signs of disease activity.

Treatment should probably be a minimum of 3–4 years prior to a trial of treatment withdrawal, as otherwise the relapse rate is very high[9]. Presumably treatment period is a surrogate marker, and duration of stable remission is likely to be the more important predictor, but this has not been studied systematically.

Relapse prediction by liver biopsy is reliable enough to justify a liver biopsy prior to a trial of treatment withdrawal. Anything above minimal disease activity (HAI scores of 4 or more) should be considered a contraindication for treatment withdrawal. Absence of histological disease activity in turn is the best predictor of successful treatment withdrawal.

Treatment withdrawal should probably not be abrupt, but should be undertaken in slow steps, probably best reducing the dose every 2–3 months. In case of continued steroid treatment this is also required in order to avoid a steroid withdrawal syndrome, but similarly in azathioprine-treated patients a slow reduction is advisable, as sometimes reactivation of disease activity can already be observed upon this dose reduction, and remission can then be achieved more easily.

In principle, late relapse can occur at any time in the life of an AIH patient, as the genetic predisposition for the disease remains present throughout life. On the other hand, relapse after more than 2 years of treatment-free stable remission is uncommon. Control of transaminases every 6 months seems adequate for the first 5 years after treatment withdrawal (after more frequent check-ups during the first 2 years), and yearly checks thereafter. It is advisable not to simply wait for a clinical relapse, as AIH may take an insidious course and cirrhosis development may progress in an asymptomatic patient, which could have been prevented by timely diagnosis and treatment of a relapse.

In summary, therefore, treatment of autoimmune hepatitis is exceptionally successful and rewarding, but requires skill and patience. Treatment regimens need to be individually adapted, and side-effects prevented by careful dose adjustment. Regular follow-up is necessary. With very few exceptions the vast majority of AIH patients nowadays can have a normal life expectancy and a very good quality of life. Liver transplantation can be avoided in almost all cases except those presenting with fulminant hepatitis and a few patients presenting already with advanced cirrhosis.

References

1. Krawitt EL. Autoimmune hepatitis. N Engl J Med. 1996;334:897–903.
2. Mackay IR, Weiden S, Ungar B. Treatment of active chronic hepatitis and lupoid hepatitis with 6-mercaptopurine and azathioprine. Lancet. 1964;42:899–902.
3. O'Brien EN, Goble AJ, Mackay IR. Plasma-transaminase activity as an index of the effectiveness of cortisone in chronic hepatitis. Lancet. 1958;1:1245–9.
4. Copenhagen Study Group for Liver Diseases. Effect of prednisone on the survival of patients with cirrhosis of the liver. Lancet. 1969;1:119–21.
5. Cook GC, Mulligan R, Sherlock S. Controlled prospective trial of corticosteroid therapy in active chronic hepatitis. Q J Med. 1971;40:159–85.
6. Summerskill WHJ, Korman MG, Ammon HV, Baggenstoss AH. Prednisone for chronic active liver disease: dose titration, standard dose, and combination with azathioprine compared. Gut. 1975;16:876–83.
7. Kirk AP, Jain S, Pocock S, Thomas HC, Sherlock S. Late results of the Royal Free Hospital prospective controlled trial of prednisolone therapy in hepatitis B surface antigen negative chronic active hepatitis. Gut. 1980;21:78–83.
8. Kanzler S, Loehr H, Gerken G, Galle PR, Lohse AW. Long-term management and prognosis of autoimmune hepatitis (AIH): a single center experience. Z Gastroenterol 2001;39:339–48.
9. Stellon AJ, Keating JJ, Johnson PJ, McFarlane IG, Williams R. Maintenance of remission in autoimmune chronic active hepatitis with azathioprine after corticosteroid withdrawal. Hepatology. 1988;8:781–4.
10. Johnson PJ, McFarlane IG, Williams R. Azathioprine for long term maintenance of remission in autoimmune hepatitis. N Engl J Med. 1995;333:958–63.
11. Czaja AJ, Freese DK. Diagnosis and treatment of autoimmune hepatitis. Hepatology. 2002; 36:479–97.

12. Heneghan MA, McFarlane IG. Current and novel immunosuppressive therapy for autoimmune hepatitis. Hepatology. 2002;35:7–13.
13. Meyer zum Büschenfelde KH, Lohse AW. Autoimmune hepatitis. N Engl J Med. 1995;333: 1004–5.
14. Johnson PJ. Immunosuppressive drug mechanisms. In: McFarlane IG, Williams R, editors. Molecular Basis of Autoimmune Hepatitis. Austin: Landes, 1996:177–91.
15. Czaja AJ. Autoimmune hepatitis. In: M Feldman, B Scharschmidt, MH Sleisenger, JS Fordtran, editors. Sleisenger and Fordtran's Gastrointestinal and Liver Disease, 6th edn. Philadelphia: WB Saunders, 1998:1265.
16. Czaja AJ, Ludwig J, Baggenstoss AH et al. Corticosteroid-treated chronic active hepatitis in remission. Uncertain prognosis of chronic persistent hepatitis. N Engl J Med. 1981;304: 5–9.
17. Davis GL, Czaja AJ, Ludwig J. Development and prognosis of histologic cirrhosis in corticosteroid-treated HBsAg-negative chronic active hepatitis. Gastroenterology. 1984;87: 1222–7.
18. Bayer EM, Herr W, Kanzler S et al. Transforming growth factor-beta1 in autoimmune hepatitis: correlation of liver tissue expression and serum levels with disease activity. J Hepatol. 1998;28:803–11.
19. Czaja AJ, Carpenter HA. Progressive fibrosis during corticosteroid therapy of autoimmune hepatitis. Hepatology. 2004;39:1631–8.
20. Kelly JJ, Mangos J, Williamson PM et al. Cortisol and hypertension. Clin Exp Pharmacol Physiol Suppl. 1998;25:S51–6.

18
Novel approaches to the treatment of autoimmune hepatitis

T. O. LANKISCH, C. P. STRASSBURG and M. P. MANNS

INTRODUCTION

Autoimmune hepatitis (AIH) is an immune-mediated chronic inflammation of the liver of unknown aetiology potentially leading to hepatic parenchyma destruction when left untreated[1]. Absolute treatment indications are amino-transferase levels of at least 5–10-fold of normal; histological evidence of bridging or multilobular necrosis; or severe hepatic or extrahepatic clinical symptoms such as fatigue, upper right quadrant pain as well as immune-mediated symptoms. Mild clinical symptoms and less severe laboratory or histological abnormalities remain relative treatment indications.

Induction of remission is usually achieved by standard therapy which is based on prednisone or prednisolone monotherapy or a lower steroid dose in combination with azathioprine. These two treatment regimes are equally effective and a biochemical, histological as well as clinical remission is achieved in 87% of cases after 3 years of therapy[2] (Figure 1). Azathioprine alone is not sufficient for the induction of remission. Conventional treatment is continued until remission, treatment failure, drug toxicity, or an incomplete response result. Remission is defined as a complete biochemical and histological resolution of inflammation as well as the disappearance of clinical symptoms. All in all, remission is achieved in only 15% of patients after withdrawal of immunosuppressive therapy. Therefore, maintenance of remission is frequently performed with prednisone alone or in combination with azathioprine.

However, other therapeutic approaches have emerged in recent years, mainly because of severe side-effects of standard therapy or recurrent relapse or non-response. Resistance to treatment occurs in 10–15% and steroid toxicity is present in about 5–10%[1,2]. Main steroid side-effects are Cushing's syndrome, weight gain, glucose intolerance, hypertension, osteoporosis, cataracts or depression, among others (Table 1). Furthermore, in some patients there are contraindications for steroid or azathioprine treatment.

In the past decade there have been some novel approaches to the treatment of AIH, mainly derived from the transplantation field where new immunosuppressive agents were successfully administered to prevent rejection: calcineurin

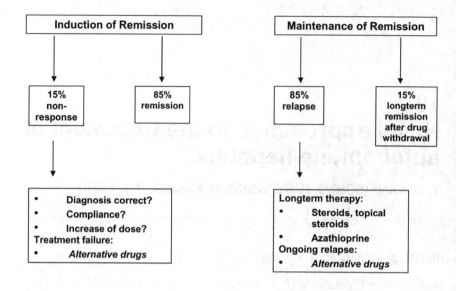

Figure 1 A suggested approach to the use of alternative treatment

inhibitors such as cyclosporin A (CsA) and tacrolimus, as well as antimetabolites such as mycophenolate mofetil, cyclophosphamide and methotrexate, were used in difficult-to-treat patients to replace standard therapy. Furthermore, the second-generation steroid budesonide plays an increasing role in novel treatment strategies. Approaches other than immunosuppression were also tried, with the administration of ursodeoxycholic acid as a potential immunomodulative agent. Although some drug regimes may be promising they all have in common that the published results are based on only small numbers of patients.

NOVEL TREATMENT STRATEGIES

Cyclosporin A

Cyclosporin A is a peptide produced by the fungus *Tolypocladium inflatum*, which binds cyclophilin; this complex inhibits calcineurin, resulting in a down-regulation of interleukin-2 and T cell function.

CSA has been administered predominantly in children for the induction of remission as well as for the maintenance of remission. The beneficial treatment of AIH with CsA was first described in 1985 in an adult with autoimmune hepatitis type 1 (AIH-1), followed by a few case reports demonstrating that CsA may be effective, but only shown in single individuals[3]. In a multicentre study, 32 children with AIH were treated with CsA (trough level 200–250 ng/ml) for 6 months, followed by a low-dose standard combination therapy[4]. Two patients were withdrawn, one for non-compliance and the other for liver failure not responding to CsA. The majority of the children had AIH-1 (28 versus 4), 25 patients had normalized aminotransferase levels after 6 months, and all after 1 year of treatment. Interestingly, there was no difference between AIH-1 and AIH-2. The adverse effects of CsA were described as being mild.

In 1999 the beneficial use of CsA (initial trough level 200–250 ng/ml) has also been shown in children and adolescents with AIH-2 in a retrospective study[5]. Clinical and biochemical remission were achieved in eight children using CsA as a primary immunosuppression and in five children with standard treatment failure. Furthermore, the addition of CsA to standard combination therapy in two children with acute liver failure was successful in one patient. In two patients CsA was discontinued because of renal dysfunction, but no other serious side-effects associated to CsA were seen in others.

In concordance with these studies similar results were also seen in 15 adolescents and adults with predominantly AIH-1 in a 26-week course of CsA treatment[6].

Based on these studies it seems that CsA can be a beneficial therapeutic agent in patients with AIH, and remains the best-investigated alternative drug in the treatment of AIH. However, CsA has mainly been used as treatment in studies with small patient numbers, predominantly in children, and has not yet been compared with conventional treatments in a randomized fashion. Severe side-effects may also occur only after long-term treatment, which has not so far been taken into consideration in AIH patients. Clinicians have to be especially aware of renal insufficiency, infection, neoplasm, hypertension and neurotoxicity.

Tacrolimus

Tacrolimus (FK506) is produced by *Streptomyces tsukubaensis* and inhibits the synthesis of cytokines and the expression of the IL-2 receptor thereby suppressing T cell function. Its immunosuppressive activity is 10–200 times greater than that of CsA. Experience with tacrolimus in patients with AIH is limited, and based on three studies demonstrating a possible role for the induction of remission and maintenance of remission in combination with standard therapy. In an uncontrolled trial with 21 adults with autoimmune chronic active hepatitis, application of 3 mg twice daily for 1 year resulted in an improvement of aminotransferase and bilirubin levels, but blood urea nitrogen and serum creatinine slightly increased as a consequence of tacrolimus application[7].

Similar results were shown in a published abstract which described six out of seven patients with severe new-onset AIH successfully treated with a lower dose of tacrolimus in combination with prednisolone, leading to biochemical improvement[8]. More recently, 11 adults with steroid-refractory AIH were treated with low-dose tacrolimus (mean serum level 3 ng/ml) and followed up for 16 months in a retrospective study[9]. All patients were on steroids and 10 patients additionally received azathioprine. The application of tacrolimus achieved biochemical remission in 10 patients. In nine patients the alternative treatment allowed tapering steroids off, but azathioprine remained as a maintenance therapy (100 mg/day) in nine patients. A follow-up liver biopsy

Table 1 Site of action and major side-effects of immunosuppressive agents

Drug	Modes of action	Side-effects	Administration for induction of remission	Administration for maintenance of remission
Steroids	Inhibition of pro-inflammatory transcription factors/cytokines	Cushing syndrome, osteopenia, diabetes, glaucoma, hypertension, depression	+ monotherapy + combination with azathioprine	+ combination with azathioprine
Azathioprine	DNA damage, inhibition of *de-novo* purine synthesis	Myelotoxicity, pancreatitis, infection, nausea	− monotherapy + combination with steroids	+ monotherapy + combination therapy
Cyclosporin A	Downregulation of IL-2 and T-cell function.	Nephrotoxicity, hypertension, neurotoxicity, nausea, diarrhoea	+ Alvarez et al.[4], Malekzedah et al.[6]	+ Debray et al.[5]
Tacrolimus	Inhibition of IL-2/-3/IFN-γ, of IL-2-receptor expression and of cytotoxic T-cell function	Nephrotoxicity, neurotoxicity, hypertension, pancytopenia	+ with steroids: Heneghan et al.[8] Aqel et al.[9]	+ in combination with standard therapy: Aqel et al.[9]
Budesonide	Local inhibition of transcription factors/cytokines (high first-pass metabolism)	In the absence of cirrhosis or portosystemic shunts side-effects are rare and resemble those of systemic steroids	+ Schüler et al.[11]	− Czaja et al.[13]
Deflazacort	Inhibition of proinflammatory transcription factors/cytokines	Side-effects are rare and resemble those of systemic steroids		+ Rebollo et al.[14]

Table 1 (continued)

Drug	Modes of action	Side-effects	Administration for induction of remission	Administration for maintenance of remission
Mycophenolate mofetil	Inhibition of DNA synthesis and of B-/T-cell proliferation	Infections, GI bleeding, nausea/diarrhoea, leukopenia		+ with steroids: Richardson et al.[15]
Cyclo-phosphamide	Alkylating/DNA-damaging agent	Myelotoxicity, cystitis, nausea/diarrhoea	+ with steroids: Kanzler et al.[16]	+ with steroids: Kanzler et al.[16]
Methotrexate	Antifolate, inhibition of dihydrolate eductase and therefore cell proliferation	Myelotoxicity, GI infection, bleeding, cystitis	+ with steroids: Burak et al.[17]	+ with steroids: Burak et al.[17]

was available in seven patients, with all showing a reduction of inflammation activity and improvement of the fibrosis score. One 80-year-old patient developed severe side-effects (tremors, hypertension and generalized oedema) and tacrolimus was withdrawn. The most common adverse effect was headache seen in four patients.

Budesonide and deflazacort

Another approach is the replacement of conventional corticosteroids. Budesonide is a synthetic glucocorticoid, which is derived from 16α-hydroxyprednisolone and is a widespread drug used for allergic pulmonary disease and inflammatory bowel disease. Its affinity to the glucocorticoid receptor is about 15 times higher than that of prednisolone. Budesonide undergoes a high first-pass metabolism in the liver, leading to a reduced systemic bioavailability of 10%.

In a study with 13 AIH patients budesonide decreased aminotransferase levels to normal limits and was well tolerated[10]. No individual experienced steroid side-effects or marked signs of Cushing's syndrome. Cirrhotic patients had a significantly stronger decrease of plasma cortisol than non-cirrhotic patients, but the mean reference value remained above the lowest reference value.

In our experience budesonide, either as monotherapy or in combination with azathioprine, achieved complete remission in four of nine cases[11]. Another three patients showed incomplete remission. Histological improvement could be observed in two patients, while one patient remained unchanged and four showed progressive disease. Side-effects were managed with dose reduction or improved after decrease of hepatic inflammation. Thus, in cases with AIH and severe inflammation or cirrhosis, budesonide is metabolized to a lower degree. This may be explained by a reduced liver function and occurrence of portosystemic shunts, whereas a reduced hepatic metabolism may have a greater influence on systemic budesonide side-effects than portosystemic shunts[12]. Overall, the data indicate that remission of AIH can be achieved when prednisolone is replaced by budesonide. However, budesonide has not so far shown an advantage in long-term maintenance therapy[13]. The change to budesonide in 10 patients on an immunosuppressive maintenance therapy was associated with a low frequency of remission, occurrence of treatment failure, and side-effects; therefore, further studies should evaluate safety, efficacy and tolerance of budesonide in larger controlled trials for maintenance therapy of AIH. An international controlled multicentre trial is currently investigating the efficacy of budesonide in treatment of naive patients in comparison with prednisone in combination with azathioprine (see Figure 1).

Another synthetic glucocorticoid which has been evaluated in AIH is deflazacort. This is an oxazoline derivative of prednisolone with anti-inflammatory and immunosuppressive activity. Clinical experience is limited to patients already in remission. In the maintenance therapy of 15 patients, deflazacort (7.5 mg) could replace prednisone (5 mg) without relapse or major side-effects[14]. Additional studies are necessary to further evaluate the efficacy in larger patient cohorts.

Mycophenolate mofetil

Mycophenolate mofetil is an ester prodrug of mycophenolic acid that inhibits inosine monophosphate dehydrogenase, resulting in a depletion of guanosine nucleotides and a reduction of *de-novo* synthesis lymphocyte DNA. Mycophenolate mofetil has been challenged in AIH patients in only a single study, which reported on seven adults with AIH-1 who were intolerant or not responding to azathioprine[15]. Five of seven patients reached normalization of aminotransferases after 3 months of therapy with 1 g mycophenolate mofetil twice daily in addition to prednisolone. Histological inflammatory activity was significantly reduced after 7 months and the steroid dose was reduced from a median of 20 mg per day to 2 mg per day. One patient required dose reduction because of a fall in white cell blood count.

Cyclophosphamide

Cyclophosphamide (1–1.5 mg/kg daily) is an alkylating chemotherapeutic drug that leads to DNA damage. In a small study with three patients cyclophosphamide (1–1.5 mg/kg per day) in combination with a tapering dose of corticosteroids, was able to induce remission and to maintain remission with 50 mg of cyclophosphamide every other day[16]. Neither severe side-effects nor relapses were observed in a follow-up of a cumulative observation period of more than 12 years; however, due to its potentially severe haematological side-effects, continued application of cyclophosphamide remains a highly experimental treatment option.

Methotrexate

There is a single case report describing a 52-year-old woman who did not respond to steroids and was intolerant of azathioprine[17]. This patient refused CsA and thus was treated successfully with 7.5 mg methotrexate per week, resulting in normalization of liver enzymes, improvement of liver histology and maintained remission, with a steroid-sparing effect.

Ursodesoxycholic acid

Another new treatment approach is the administration of ursodesoxycholic acid (UDCA), often used in patients with primary biliary cirrhosis. UDCA is presumed to alter HLA class I antigen expression on cell membranes and to suppress immunoglobulin production[18]. In Japanese patients with mild AIH, 600 mg/day for 2 years resulted in clinical and biochemical improvement, but not in a reduction of fibrosis[19]. In contrast, no benefit was observed in North American individuals who received 13–15 mg/kg daily[20]. Thus, the role of UDCA as primary AIH therapy or in combination with immunosuppressants remains unclear, and needs to be determined in larger studies.

Table 2 Multicentre study investigating the efficacy of budesonide in naive patients (BUC 38, AIH, sponsored by Falk Pharma GmbH, Freiburg)

medication	segment A 6 months, double-arm, randomized, doubleblind			segment B 6 months, single-arm, open-labeled
	months 1 W1 W2 W3 W4	months 2 W5 W6 W7 W8	months 3-6	months 7-12
budesonide azathioprine	budesonide-group 3mg ◄— 3mg 3x/d, at biochemical response 3mg 2x/d ——► ◄—— 1-2mg/kg daily ——►			
prednisone (mg/d) prednisone (mg/d) budesonide azathioprine	prednisone-group 40 40 40 40 40 40 30 25 - - - - ◄——	30 25 20 15 20 15 10 10 - - - - 1-2 mg/kg daily	10 10 -	3mg 3x/d or 3mg 2x/d ——►

Future therapeutic approaches

Increasing insights in the molecular pathomechanisms of AIH may open new strategies for the development of novel immunosuppressive agents. Drugs have been established that disrupt intracellular signalling pathways involved in AIH such as apoptosis or transcription of proinflammatory cytokines. In this respect FTY-720, a novel immunosuppressive agent that exerts its effect via induction of lymphocyte apoptosis, may be a promising candidate[21]. Cytokine interactions can be manipulated by administering recombinant species or can be inhibited by the use of monoclonal antibodies. Targeting of various cytokines might therefore be effective in the down-regulation of the immune response. T cell vaccination that eliminates disease-specific T cell clones might be another promising strategy since it has been shown, in an experimental AIH model, not only to prevent disease but also to decrease disease activity as effectively as steroids. Soluble CTLA-4 that competes with CD4 helper T cells for interaction with antigen-presenting cells could also down-regulate the autoimmune response[22–24]. Finally, gene therapy may represent a further option in the future that is capable of counterbalancing the overproduction of proinflammatory cytokines and the progression to fibrosis, in addition to promoting liver regeneration in the course of AIH[25].

SUMMARY

The administration of alternative drugs may be useful in patients with AIH who do not tolerate side-effects or do not respond to standard therapy. However, in some patients a thorough history may reveal non-compliance, or further clinical investigations may detect other causes of liver disease such as primary sclerosing cholangitis or primary biliary cirrhosis (Table 2). Once these issues are excluded, clinicians should think of alternative immunosuppressive agents. Their therapeutic goal includes the achievement of long-term remission in difficult-to-treat patients. For most new immunosuppressants the published results are promising; however, long-term outcome and side-effects still remain unknown, and larger studies are required to provide more evidence for their risks and benefits that would justify their use as an alternative strategy to standard therapy. Thus, the alternative treatment remains highly individualized in each patient and should be performed in control studies only.

References

1. Manns MP, Strassburg CP. Autoimmune hepatitis: clinical challenges. Gastroenterology. 2001;120:1502–17.
2. Heneghan MA, McFarlane IG. Current and novel immunosuppressive therapy for autoimmune hepatitis. Hepatology. 2002;35:7–13.
3. Mistilis SP, Vickers CR, Darroch MH, McCarthy SW. Cyclosporin, a new treatment for autoimmune chronic active hepatitis. Med J Aust. 1985;11;143:463–5.
4. Alvarez F, Ciocca M, Canero-Velasco C et al. Short-term cyclosporine induces a remission of autoimmune hepatitis in children. J Hepatol. 1999;30:222–7.
5. Debray D, Maggiore G, Girardet JP, Mallet E, Bernard O. Efficacy of cyclosporin A in children with type 2 autoimmune hepatitis. J Pediatr. 1999;135:111–14.
6. Malekzadeh R, Nasseri-Moghaddam S, Kaviani MJ, Taheri H, Kamalian N, Sotoudeh M. Cyclosporin A is a promising alternative to corticosteroids in autoimmune hepatitis. Dig Dis Sci. 2001;46:1321–7.
7. Van Thiel DH, Wright H, Carroll P et al. Tacrolimus: a potential new treatment for autoimmune chronic active hepatitis: results of an open-label preliminary trial. Am J Gastroenterol. 1995;90:771–6.
8. Heneghan MA, Rizzi P, McFarlane IG, Portmann B, Harrsion PM. Low dose tacrolimus as treatment of severe autoimmune hepatitis: potential role in remission induction. Gut. 1999; 44(Suppl. 1):A61.
9. Aqel BA, Machicao V, Rosser B, Satyanarayana R, Harnois DM, Dickson RC. Efficacy of tacrolimus in the treatment of steroid refractory autoimmune hepatitis. J Clin Gastroenterol. 2004;38:805–9.
10. Danielsson A, Prytz H. Oral budesonide for treatment of autoimmune chronic active hepatitis. Aliment Pharmacol Ther. 1994;8:585–90.
11. Schüler A, Möllmann HW, Manns MP. Treatment of autoimmune hepatitis with budesonide. Hepatology. 1995;22:488A.
12. Geier A, Gartung C, Dietrich CG, Wasmuth HE, Reinartz P, Matern S. Side effects of budesonide in liver cirrhosis due to chronic autoimmune hepatitis: influence of hepatic metabolism versus portosystemic shunts on a patient complicated with HCC. World J Gastroenterol. 2003;9:2681–5.
13. Czaja AJ, Lindor KD. Failure of budesonide in a pilot study of treatment-dependent autoimmune hepatitis. Gastroenterology. 2000;119:1312–6.
14. Rebollo BJ, Cifuentes MC, Pinar MA et al. Deflazacort for long-term maintenance of remission in type I autoimmune hepatitis. Rev Esp Enferm Dig. 1999;91:630–8.
15. Richardson PD, James PD, Ryder SD. Mycophenolate mofetil for maintenance of remission in autoimmune hepatitis in patients resistant to or intolerant of azathioprine. J Hepatol. 2000;33:371–5.

16. Kanzler S, Gerken G, Dienes HP, Meyer zum Büschnefelde KH, Lohse AW. Cyclophosphamide as alternative immunosuppressive therapy for autoimmune hepatitis – report of three cases. Z Gastroenterol. 1997;35:571–8.
17. Burak KW, Urbanski SJ, Swain MG. Successful treatment of refractory type 1 autoimmune hepatitis with methotrexate. J Hepatol. 1998;29:990–3.
18. Calmus Y, Gane P, Rouger P, Poupon R. Hepatic expression of class I and class II major histocompatibility complex molecules in primary biliary cirrhosis: effect of ursodeoxycholic acid. Hepatology. 1990;11:12–15.
19. Nakamura K, Yoneda M, Yokohama S et al. Efficacy of ursodeoxycholic acid in Japanese patients with type 1 autoimmune hepatitis. J Gastroenterol Hepatol. 1998;13:490–5.
20. Czaja AJ, Carpenter HA, Lindor KD. Ursodeoxycholic acid as adjunctive therapy for problematic type 1 autoimmune hepatitis: a randomized placebo-controlled treatment trial. Hepatology. 1999;30:1381–6.
21. Czaja AJ, Manns MP, McFarlane IG, Hoofnagle JH. Autoimmune hepatitis: the investigational and clinical challenges. Hepatology. 2000;31:1194–200.
22. Lohse AW, Dienes HP, Meyer zum Büschenfelde KH. Suppression of murine experimental autoimmune hepatitis by T-cell vaccination or immunosuppression. Hepatology. 1998;27:1536–43.
23. Thompson CB, Allison JP. The emerging role of CTLA-4 as an immune attenuator. Immunity. 1997;7:445–50.
24. Schwartz RS. The new immunology – the end of immunosuppressive drug therapy? N Engl J Med. 1999;340:1754–56.
25. Touhy VK, Mathisen PM. Gene therapy for autoimmune diseases. In: Manns MP, Paumgartner G, Leuschner U, editors. Immunology and Liver. Dordrecht: Kluwer Academic Publishers, 2000:376–85.

19
Standard treatment for primary sclerosing cholangitis and overlap autoimmune hepatitis/primary sclerosing cholangitis

E. SCHRUMPF and K. M. BOBERG

INTRODUCTION

In primary sclerosing cholangitis (PSC) drugs are usually ineffective in slowing disease progression and affecting prognosis[1]. Due to this circumstance, and the relatively high prevalence of PSC, PSC has become the number one indication for hepatic transplantation in the Nordic countries[2]. Nevertheless, several factors must be taken into consideration from the time of diagnosis until evaluation for hepatic transplantation (Table 1).

Table 1 PSC: therapeutic flow chart

Immunosuppression
Endoscopic treatment?
Cholangiocarcinoma?
Transplantation?
UDCA
Other treatment

IMMUNOSUPPRESSION

No randomized study has shown an effect of immunosuppressive therapy in PSC[1]. However, some selected patients seem to respond to corticosteroids[3]. In a retrospective study we collected data from 135 consecutive PSC patients, of whom 47 had received corticosteroid treatment whereas 88 were untreated. Among the treated patients 20 had some kind of response with improvement in symptoms and blood tests. Only three patients had a complete response as evaluated from the response criteria after steroid treatment in autoimmune hepatitis (AIH). Several smaller studies from other centres have shown similar results: that a small subgroup of PSC patients responds to corticosteroids. Overlap AIH/PSC has been suggested a separate entity, but has not yet been well defined. The clinical impression so far is that this group of overlap patients comprises patients most likely to respond to such treatment.

In our study it was shown that elevated levels of bilirubin and alanine aminotransferase and low levels of alkaline phosphatase, were prognostic factors for treatment response[3]. As patients who will respond to immunosuppression are difficult to identify before treatment, such treatment must be tried in more patients than in the small fraction of patients who in the end turn out to benefit from immunosuppression. Therefore, the risk of side-effects from this treatment must be carefully considered.

Nevertheless, we advise that PSC patients with traits of AIH should always be evaluated for a short-term test period with corticosteroids.

ENDOSCOPIC TREATMENT

For several reasons many centres were previously reluctant to try endoscopic treatment in PSC, due to the risk of complications such as pancreatitis, biliary duct perforation, and sepsis. Many also were reluctant to try endoscopic treatment of one major dominant stenosis as the majority of patients often also had several other stenoses. During the past decade many groups have, however, shown a clear benefit from endoscopic treatment. Stiehl and co-workers[4] have gained more experience than most other groups; 106 patients were followed for a long time period and around 50% of these had endoscopic treatment, most with regular balloon dilations; in some cases even short-term stents were applied. The risk of complications was acceptable and a significant effect on survival was obtained, as predicted with the Mayo multicenter survival model. Although these patients were given ursodeoxycholic acid (UDCA) simultaneously, it seems reasonable to conclude that the effect obtained was due to the endoscopic treatment[5].

Due to the potential risks with this treatment, it is advised that the interventions are carried out by colleagues with experience in treating PSC patients and with expertise in endoscopic treatment modalities. First, selection of patients for the treatment should be made carefully. In particular one should not overlook the development of cholangiocarcinoma which, however, is extremely difficult to diagnose[6]. Secondly, even in patients with major extrahepatic stenoses, it may in some cases be advisable not to apply endoscopic treatment if, for example, disseminated intrahepatic changes are also seen. If such treatment is indicated, the method of treatment, however, has not been defined. Should treatment be with balloon dilation alone, or should short-time stenting also be applied? How frequently should endoscopic treatment be repeated? Even though some important questions are still unanswered, we regard endoscopic treatment as an important treatment modality in patients with PSC.

CHOLANGIOCARCINOMA IN PSC?

PSC is definitely a premalignant disease: the risk of cholangiocarcinoma development is particularly high[6,7]. If cholangiocarcinoma has already developed possible treatment modalities are limited[8]. Some patients, however, can have radical surgery which therefore should always be considered[9]. Among palliative treatment modalities, biliary stenting for drainage is a possible option; other treatment modalities are being explored, but have not yet been proven effective[8].

In the Nordic series 255 PSC patients were accepted for liver transplantation during 1990–2001[10]. In one-third of these patients malignant disease was suspected but never demonstrated before transplantation. Nevertheless, 20% of the PSC patients accepted for liver transplantation turned out to have malignant disease. Predictors of hepatobiliary malignancy in these series were recent diagnoses of PSC, no UDCA treatment, clinical suspicion of malignancy and previous colorectal cancer. The 1-, 3- and 5-year patient survival rates after transplantation for patients with PSC and cholangiocarcinoma were 65%, 35% and 35%, respectively.

On this basis it is concluded that satisfactory results can be obtained in liver transplantation in PSC patients where cholangiocarcinoma development is demonstrated peroperatively or during close examination of the explanted liver.

LIVER TRANSPLANTATION IN PSC

PSC is a common indication for liver transplantation, but evaluation of patients and timing of liver transplantation remain as major problems[2,11]. As for survival after transplantation, results are as good in PSC as in other end-stage liver disease. During recent years there has been increasing interest in the development of recurrent disease after transplantation of PSC patients; the magnitude of this problem remains to be decided[12].

OTHER MEDICAL TREATMENT IN PSC

In addition to immunosupression UDCA has been most frequently tried in PSC[3]. All studies performed have shown an effect on biochemical tests, whereas a symptomatic effect is usually not seen, and there is usually an effect on liver histology. Effects on survival have either not been evaluated or not been demonstrated. Recently experimental support has been found for a chemopreventive effect of UDCA on the development of cholangiocarcinoma[10,11] and on the development of colorectal cancer in patients with ulcerative colitis and PSC[13]. This highly interesting chemopreventive effect of UDCA must of course be confirmed in more studies before any conclusions can be drawn, but in our opinion this may turn out to be the most interesting effect of UDCA in PSC.

Other treatment modalities include symptomatic treatment of pruritus, treatment of steatorrhoea and of fat-soluble vitamin deficiency and osteopenic bone disease which is often seen in PSC – particularly in the advanced stage of liver disease[1].

SUMMARY AND THE FUTURE

There are no medical treatment modalities which universally delay disease progression in PSC. Immunosuppression may have an effect in a small number of PSC patients. Endoscopic treatment is efficient and should be at hand in centres treating these patients. The development of cholangiocarcinoma is a major problem, and should always be considered. Transplantation is the only treatment option in advanced cases of PSC, and is not contraindicated in early-stage cholangiocarcinoma.

As there is such a great phenotypic variation in PSC: small duct PSC – overlap AIH/PSC, PSC with and without inflammatory bowel disease, asymptomatic and symptomatic PSC – and such a great variation in the clinical course, we suspect that there also is a variation in aetiology and pathogenesis, possibly linked to different genotypes. This may mean that PSC patients respond differently to different treatment modalities. The great challenge for the future may therefore be to tailor treatment according to phenotypes and/or genotypes.

References

1. Boberg KM, Schrumpf E. Treatment of primary sclerosing cholangitis. In: Krawitt EL, Wiesner RH, Nishioka M, editors. Autoimmune Liver Diseases. Amsterdam: Elsevier, 1998:529–51.
2. Bjoro K, Friman S, Hockerstedt K et al. Liver transplantation in the Nordic countries 1982–1998: changes of indications and improving results. Scand J Gastroenterol. 1999;34: 714–22.
3. Boberg KM, Egeland T, Schrumpf E. Long-term effect of corticosteroid treatment in primary sclerosing cholangitis. Scand J Gastroenterol. 2003;38:991–5.
4. Stiehl A, Rudolph G, Klöters-Plachky, Sauer P, Walker S. Development of dominant bile duct stenoses in patients with primary sclerosing cholangitis treated with ursodeoxycholic acid: outcome after endoscopic treatment. J Hepatol. 2002;36:151–6.
5. Schrumpf E, Boberg KM. Endoscopic treatment for primary sclerosing cholangitis? (Editorial). J Hepatol. 2002;36:278–9.
6. Ponsioen CY, Vroutenraets SM, Van Milligen de Wit AW et al. Value of brush cytology for dominant strictures in primary sclerosing cholangitis. Endoscopy. 1999;31:305–9.
7. Bergquist A, Ekbom A, Olsson R et al. Hepatic and extrahepatic malignancies in primary sclerosing cholangitis. J Hepatol. 2002;36:321–7.
8. Boberg KM, Schrumpf E. Diagnosis and treatment of cholangiocarcinoma. Curr Gastroenterol Rep. 2004;6:52–9.
9. Chamberlain RS, Blumgart LH. Hilar cholangiocarcinoma: a review and commentary. Ann Surg Oncol. 2000;7:55–66.
10. Brandsæter B, Isoniemi H, Broomé U et al. Liver transplantation for primary sclerosing cholangitis: predictors and consequensces of hepatobiliary malignancy. J Hepatol. 2004;40: 815–22.
11. Brandsæter B, Broomé U, Isoniemi H et al. Liver transplantation for primary sclerosing cholangitis in the Nordic countries: outcome after acceptance to the waiting list. Liver Transplant. 2003;9:961–9.

Prescribing Information
(Please refer to full SPC before prescribing)

Presentations: *Ursofalk capsules* each containing 250mg ursodeoxycholic acid (UDCA); *Ursofalk Suspension* containing 250 mg UDCA per 5ml. **Indications:** Treatment of primary biliary cirrhosis and for the dissolution of radiolucent gallstones in patients with a functioning gall bladder. **Dosage:** *Adults and elderly: Primary biliary cirrhosis:* 10–15 mg UDCA/kg/day in 2 to 4 divided doses. *Gallstones: Adults:* 8–12 mg UDCA/kg/day in 2 divided doses. If doses are unequal, the larger dose should be taken late evening. Treatment should be continued until two successive cholecystograms and/or ultrasound investigations 4–12 weeks apart fail to demonstrate gallstones. *Elderly:* as for adults, but with relevant precautions. *Children:* Dose should be related to body weight. **Contra-indications:** Ursofalk is not suitable for the dissolution of radio-opaque gallstones or in patients with a non-functioning gall bladder. **Warnings/Precautions:** A product of this class has been found to be carcinogenic in animals. The relevance of this finding to the clinical use of Ursofalk has not been established. **Interactions:** Cholestyramine, charcoal, colestipol and certain antacids (e.g. aluminium hydroxide) bind bile acids *in vitro* and could have a similar effect *in vivo*, interfering with the absorption of Ursofalk. Drugs which increase cholesterol elimination in bile such as oestrogenic hormones, oestrogen-rich contraceptive agents and certain blood cholesterol lowering agents such as clofibrate should not be taken with Ursofalk. UDCA may increase the absorption of cyclosporin in transplantation patients. **Use in pregnancy:** Ursofalk should not be used in pregnancy. When treating women of child bearing potential non-hormonal or low oestrogen oral contraceptive measures are recommended. **Undesirable effects:** Diarrhoea may occur rarely. **Legal category:** POM. **Basic NHS cost:** *Ursofalk capsules:* 60-capsule pack — £31.10, 100-capsule pack £32.85; *Ursofalk Suspension* 250 ml bottle – £28.50. **Product licence holder:** Dr. Falk Pharma UK Limited, Unit K, Bourne End Business Park, Cores End Road, Bourne End, Bucks SL8 5AS, United Kingdom. **Product licence number:** *Ursofalk capsules* –14658/0001; *Ursofalk Suspension* – 14658/0008. **Date of preparation:** December 2004.

References:

1. Matsuzaki Y *et al.* Improvement of biliary enzyme levels and itching as a result of long-term administration of ursodeoxycholic acid in primary biliary cirrhosis. Am Gastroenterol 1999; **85:** 15-23.

2. Parés A *et al.* Long-term effects of ursodeoxycholic acid in primary biliary cirrhosis: results of a double-blind controlled multicentric trial. Hepatol 2000; **32:** 561-6.

3. Bergasa NV. Pruritus and fatigue in primary biliary cirrhosis. Clin Liver Dis 2003; **7:** 879-900.

Date of preparation: July 2005   DrF 05/0013

Dr Falk Pharma UK Limited
Bourne End Business Park
Cores End Road
Bucks SL8 5AS

Ursofalk®
ursodeoxycholic acid

Evidence and licence

- Ursofalk was the first formulation of ursodeoxycholic acid licensed for primary biliary cirrhosis (PBC)

- Ursofalk may ease pruritus and improve fatigue,[1,2] both of which symptoms have a marked negative impact on quality of life[3]

- Capsules and suspension formulations

For the treatment of primary biliary cirrhosis (PBC)

10-15mg per kg per day

in 2-4 divided doses

For the dissolution of gallstones

8-12mg per kg per day

in 2 divided doses

12. Bjøro K, Schrumpf E. Liver transplantation for primary sclerosing cholangitis (Review). J Hepatol. 2004;40:570–7.
13. Pardi DS, Lofthus EV Jr, Kremers WK, Lindor KD. Ursodeoxycholic acid as a chemopreventive agent in patients with ulcerative colitis and primary sclerosing cholangitis. Gastroenterology. 2003;124:889–93.

20
Novel approaches to the treatment of primary sclerosing cholangitis

R. W. CHAPMAN

INTRODUCTION

Primary sclerosing cholangitis (PSC) is a chronic cholestatic liver disease characterized by a progressive obliterating fibrosis of the intrahepatic and extrahepatic bile ducts. The disease in symptomatic patients often progresses to secondary biliary cirrhosis, and premature death from liver failure, hepato-biliary and colon cancer[1]. PSC is closely associated with inflammatory bowel disease (IBD), particularly ulcerative colitis (UC) which is found in approximately two-thirds of patients with PSC of northern European origin. PSC is the most common hepatobiliary disease associated with UC. The prevalence in UC populations of PSC is between 2% and 6%. There are also associations between PSC with other immune-mediated diseases such as coeliac disease and rheumatoid arthritis.

Although the aetiopathogenesis of PSC is poorly understood it is thought to be immune-mediated. The detection of a disease specific autoantibody (ANCA)[2], the association with the HLA-B8-DR3 haplotype found in other autoimmune diseases, the relationship between HLA status and prognosis[3] and the presence of an organ-specific T cell infiltrate in PSC[4] provide indirect evidence for the role of genetic and immune mechanisms in the aetiology of PSC. The mechanisms underlying the loss of immune tolerance which allow such 'autoimmune' disease processes to occur have not as yet been defined.

All these studies provide indirect evidence that PSC is an autoimmune disease possibly involving an exaggerated cell-mediated immune response with immunological damage targeted at the biliary epithelial cells. The putative antigens that may set off this immune response have not yet been identified.

An alternative hypothesis has been proposed; viz. that bacterial cell products ascending in portal venous blood are taken up by hepatic macrophages which in turn set up an immune response leading to peribiliary fibrosis in immuno-genetically susceptible hosts.

NATURAL HISTORY AND PROGNOSIS

Unlike primary biliary cirrhosis (PBC), the natural history of PSC is very difficult to predict in an individual patient as the rate of progression of the disease can be highly variable. This clinical variability makes the timing of medical and endoscopic therapy and liver transplantation difficult when compared with other chronic liver disease. Even when the diagnosis is made in the asymptomatic phase PSC can be a progressive disease with the insidious development of biliary cirrhosis and the complications arising from this. The median survival of patients, including those who are symptomatic at presentation, appears to be about 12 years, although a study from Sweden has suggested a better prognosis in asymptomatic patients with a median survival of 21 years[5].

Prognostic indices or models of the natural history of the PSC have proved useful in studying large populations, but are generally of less use in predicting the course of an individual patient. This is largely because no markers or risk factors can predict which patients will develop cholangiocarcinoma or progressive jaundice due a dominant biliary stricture, both of which may lead to premature death. Acute bacterial cholangitis is uncommon and usually follows instrumental biliary intervention. It tends also to occur in endstage disease where multiple strictures and biliary sludging can lead to the formation of brown pigment stones intensifying the degree of biliary obstruction[6]. The development of cholangiocarcinoma carries a poor prognosis and precludes orthotopic liver transplantation (OLT). In a recent series cholangiocarcinoma has been reported in 14–27% of patients with PSC[7,8] The annual risk of developing hepatobiliary cancer in PSC has been estimated at 1. 5% per year[7].

TREATMENT

At present there is no curative treatment for PSC. Treatments either serve to manage the general and specific complications of the disease or more specifically to retard and reverse the rate of disease progression. The optimal therapy which successfully improves symptoms, delays progression towards liver failure and OLT, and prevents the onset of cholangiocarcinoma remains elusive.

Management of complications

As PSC slowly progresses to biliary cirrhosis and portal hypertension. complications may arise from chronic cholestasis or endstage liver failure (as in PBC and other liver diseases) or complications specific to PSC such as biliary strictures, biliary sludge and the development of cholangiocarcinoma. The general management of these complications will not be discussed in this chapter. Endoscopic therapy with balloon dilation and/or biliary stenting is usually reserved for complications such as main duct stricturing causing jaundice, when good short-term relief of symptoms can be achieved. However Stiehl et al.[8] have reported the technique of prospective regular dilation of biliary strictures leading to increased survival rates when compared with predicted survival calculated from the Mayo multicentre survival model. These impressive results need to be confirmed by other studies.

Specific medical therapy – the prevention of disease progression

In both PBC and PSC the primary site of inflammation and damage is the biliary epithelium. When severely damaged or destroyed the bile ducts do not have the capacity to regenerate like hepatocytes, which are the primary target for injury in various parenchymal liver diseases. Given the finite number of bile ducts in the liver the natural history of PSC, like PBC, is that of progressive loss of functioning intrahepatic bile ducts (ductopenia). This ductopenia leads to a progressive and irreversible failure of hepatic biliary excretion. To delay and reverse this process physicians have tried a variety of agents but in PSC, in contrast to PBC, few prospective randomized controlled trials have been performed.

Corticosteroids

It is surprising that there have been no long-term studies of the effect of corticosteroid therapy on histological progression and survival in PSC, especially as the disease may be immune-mediated. This may reflect concerns about the long-term side-effect profile of corticosteroids. Systemic and topical corticosteroid therapy has been evaluated in a number of small often uncontrolled trials. In an uncontrolled pilot study 10 patients with PSC, selected because they had elevated aminotransferases, were given oral prednisolone and the majority responded with improvement in their biochemistry[10]. In a subsequent study Lindor et al. [11] were unable to confirm these optimistic results. They treated 12 patients with a combination of low-dose prednisone (10 mg daily) and Colchicine (0. 6 mg twice daily). The clinical course of the treated patients was compared with a control group, but the study was not randomized. After 2 years no significant differences in the biochemistry and liver histology were detected between the two groups. In this study treatment did not alter the rate of disease progression or improve survival. The absence of a beneficial response and the suspicion that corticosteroid therapy enhanced cortical bone loss, and hence the risk of developing compression fractures of the spine even in young male patients, led the authors to advise against empirical corticosteroid therapy in these patients. This conclusion was strengthened by the observation that spontaneous fractures in the post-liver transplantation occur almost exclusively in PSC patients who are already osteopenic at the time of transplantation[12].

Topical corticosteroids are usually administered through a nasobiliary drain left in situ following ERCP. Three anecdotal studies[13-15] have reported benefit. The only controlled trial of nasobiliary lavage with corticosteroids from the Royal Free Hospital[16] showed no benefit when compared with a placebo group. Although the numbers were small the bile of all the treated patients became rapidly colonized with enteric bacteria, and a higher incidence of bacterial cholangitis was recorded in the treatment group. Recently a complete therapeutic response to steroids was found retrospectively in only 4.5% of a large group of PSC patients from Norway. Reponse was associated with marked elevations in serum transaminases in a small subgroup of patients[17].

More recent clinical trials have studied the possible benefit of budesonide, a second generation corticosteroid with a high first-pass metabolism and minimal systemic availability. Unfortunately preliminary results, both alone[18] and in combination with ursodeoxycholic acid[19], have been disappointing.

A recent Cochrane review has concluded that there is no direct evidence to suggest that either oral or topical corticosteroids are beneficial in PSC. Indeed when PSC patients with coexistent UC are given courses of corticosteroids to treat their UC this treatment appears to have little influence on the behaviour of their liver disease. It may be difficult to justify a trial using corticosteroids as monotherapy, but a large controlled trial could clarify their role in combination with a choleretic agent. Potentially serious side-effects may be reduced by new agents such as biphosphonates which prevent cortical bone loss.

Methotrexate

After demonstrating a promising response to low-dose oral pulse methotrexate in an open study[20] involving 10 PSC patients without evidence of portal hypertension, Knox and Kaplan[21] performed a prospective double-blind, randomized control trial comparing oral pulse methotrexate at a dose of 15 mg per week with a well-matched placebo group. Twelve patients with PSC were entered into each group. Although each patient was monitored with both liver biopsy and ERCP (at baseline and yearly) and biochemical tests the only significant change was a fall in the serum alkaline phosphatase by 31% in those receiving methotrexate.

There was no significant improvement in liver histology, or any differences in outcome of the two groups with regard to treatment failure or death. In a pilot study Lindor et al.[22] found that methotrexate given in combination with ursodeoxycholic acid (UDCA) to 19 PSC patients was associated with toxicity (alopecia, pulmonary complications) but no further improvement in liver biochemistries compared with UDCA given alone to a matched group of nine patients.

Other immunosuppressants

Despite the evidence that PSC may be an immune-mediated disease there have been few randomized controlled trials of immunosuppressive agents containing sufficient numbers of patients with early disease. Immunosuppression is unlikely to be effective in patients with advanced liver disease and irreversible bile duct loss, and this may account for the disappointing results so far seen in PSC with these agents. No control trials of azathioprine in PSC have been reported. In one case report[23], two patients improved clinically on azathioprine, but in another[24] the patient deteriorated. The use of cyclosporin in PSC has been evaluated in a randomized controlled trial from the Mayo Clinic[25] involving 34 patients with PSC and in the majority coexistent UC. Treatment with cyclosporin may help the symptoms of UC[26], but had no effect on the course or prognosis of PSC. Follow-up liver histology after 2 years of treatment revealed progression in 9/10 of the placebo group but only 11/20 of the cyclosporin-treated group[25]. This was not reflected by any beneficial effect on the biochemical tests. The prevalence of side-effects was low; serious renal complications were not reported.

Tacrolimus (FK506) an immunosuppressive macrolide antibiotic, has been used to treat 10 patients with PSC in an open study[27]. After 1 year of treatment with a twice-daily oral regime all patients experienced an improvement in their liver biochemical tests. For example the median serum bilirubin level was reduced by 75% and the serum alkaline phosphatase was reduced by 70%. No major adverse events were reported in this initial study in PSC. A randomized controlled trial is required to confirm these encouraging preliminary results.

The hepatobiliary injury which occurs in rats with experimental bacterial overgrowth is said to result from peptidoglycan-polysaccharide-mediated activation of Kupffer cells which in turn release cytokines such as tumour necrosis factor (TNF-α). In rats the liver injury can be prevented by pentoxifylline. In an open pilot study, 20 patients with PSC were treated with pentoxyfylline 400 mg q.i.d. for 1 year. In this dose pentoxyfylline did not improve symptoms or liver tests[28]. Negative results were also obtained in a open pilot study of 10 PSC patients using etanercept, an anti-TNF antibody administered twice-weekly subcutaneously for 1 year[29]. However, the anti-TNF antibody infiximab, which has been shown to be efficacious in the treatment of Crohn's disease, in marked contrast to entanercept, has surprisingly not been studied in the treatment of PSC.

Antifibrogenic agents

In the light of initial reports which suggested a positive trend of the antifibrogenic agent colchicine on survival in PBC and other types of cirrhosis, a randomized trial from Sweden[30] compared colchicine in a dose of 1 mg daily by mouth in 44 patients with PSC with a matched placebo group of 40 patients. At 3-year follow-up there were no differences in clinical symptoms, serum biochemistry, liver histology or survival between the two groups. The absence in this study of any proven effect of colchicine on disease progression, outcome or survival is in keeping with more recent long-term studies of colchicine in PBC and other chronic liver diseases which have failed to confirm the initial reported survival benefits.

Ursodeoxycholic acid (UDCA)

This hydrophilic bile acid has become widely used in the treatment of cholestatic liver of all causes. UDCA appears to exert a number of effects, all of which may be beneficial in chronic cholestasis: a choleretic effect by increasing bile flow; a direct cytoprotective effect; an indirect cytoprotective effect by displacement of the more hepatotoxic endogenous hydrophobic bile acids from the bile acid pool; an immunomodulatory effect and finally an inhibitory effect on apoptosis.

Table 1 Controlled studies of ursodeoxycholic acid in PSC

Reference	No. of patients	Study type	Dose	Study duration	LFT improved				Symptoms improved	Liver histology improved
					Alk P	γ-GT	Bili	AST		
Beuers et al., 1992[34]	14	DBPC	13–15 mg/kg daily	12 months	Yes	Yes	Yes	Yes	No	Yes
Lo et al., 1992[31]	23	DBPC	10–15 mg/kg/daily	24 months	Trend	Trend	No	Trend	No	No
Stiehl et al., 1994[35]	20	DBPC, Uncontrolled	750 mg daily	Controlled for 3 months; uncontrolled up to 4 years	Yes	Yes	No	Yes	No	Yes
Mitchell et al., 2001[40]	26	DBPC	20–25 mg/kg daily	24 months	Yes	Yes	No	No	No	Yes
van Hoogstraten et al., 1998[46]	48	DB	10 mg/kg daily in single (group 1) or three (group 2) doses	24 months	Yes	Yes	No	Yes	No	n.a.
Lindor et al., 1997[36]	105	DBPC	13–15 mg/kg daily	Mean 2.9 years	Yes	Yes	Yes	Yes	No	No

DB, double-blinded trial; PC, placebo-controlled trial; Alk P, alkaline phosphatase; γ-GT, γ-glutamyltranspeptidase; Bili, bilirubin; AST, aspartate transaminase; LFT, liver function tests; n.a., data not available.

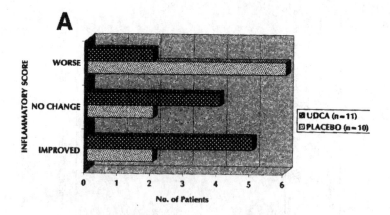

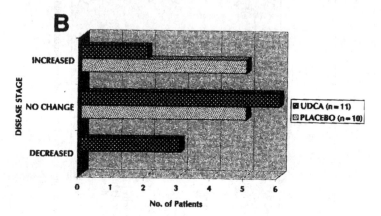

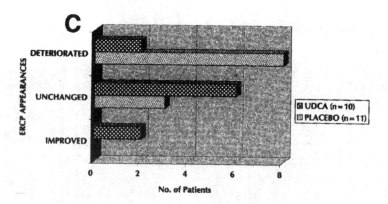

Figure 1 Comparison of (**A**) changes in portal inflammatory score, (**B**) disease progression by staging, and cholangiographic assessment (**C**) in patients treated with high-dose UDCA versus placebo over the 2-year trial (reproduced from ref. 41)

Using a labelled bile acid analogue Jazrawi et al.[31] demonstrated a defect in hepatic bile acid excretion but not in uptake in patients with PBC and PSC resulting in bile acid retention. They observed an improvement of hepatic excretory function with UDCA in patients with PBC but only a trend towards improvement in the small number of patients with PSC. Not only is hepatic bile acid excretion affected by UDCA, but so is ileal reabsorption of endogenous bile acids. The net result is enrichment of the bile acid pool with UDCA. Hydrophobic bile acids are more toxic than UDCA, which can protect and stabilize membranes.

Studies have demonstrated that long-term treatment with UDCA decreases aberrant expression of HLA class I on hepatocytes and reduces levels of soluble cell adhesion molecules (sICAM) in PBC patients. *In-vitro* studies have shown that UDCA may alter cytokine production by human peripheral mononuclear cells. In PSC, one study has shown that UDCA has been shown to decrease aberrant HLA DR expression on bile ducts[32]. However, a more recent study could not demonstrate any alteration in expression of either HLA class I and II or ICAM-1 on either biliary epithelial cells or hepatocytes[33]. The body of evidence suggests that UDCA does have some modulatory effects on immune function, but how important these are remains unclear.

Numerous studies have attempted to address the clinical efficacy of UDCA treatment in PSC (Table 1). The majority have been uncontrolled studies in small numbers of patients. In a pilot study O'Brien et al.[34] treated 12 patients with UDCA on an open basis over 30 months. They documented improvement in fatigue, pruritus and diarrhoea, and significant improvement of all liver biochemical tests, particularly alkaline phosphatase, during the two UDCA treatment periods. Symptoms and liver biochemistry relapsed during a 6-month withdrawal period between treatment phases. During UDCA treatment the amount of cholic acid declined slightly but the levels of other relatively hydrophobic bile acids did not change significantly.

In the first prospective randomized double-blind controlled trial of UDCA in PSC Beuers et al.[35] compared, over a 12-month period, six patients who received UDCA 13–15 mg per kg body weight with eight patients who received placebo. The majority of patients had early disease (Ludwig classification stages I and II). After 6 months a significant reduction in alkaline phosphatase and aminotransferases was achieved in the treatment group. A significant fall in bilirubin was only noted after 12 months. Using a multiparametric score the UDCA-treated group showed significant improvement in their liver histology, mainly attributed to decreased portal and parenchymal inflammation. Unfortunately treatment did not ameliorate their symptoms. UDCA-induced diarrhoea was the only important side-effect, requiring one patient to withdraw.

Similar results were obtained by Stiehl et al.[36], who randomized 20 patients to either 750 mg daily of UDCA or placebo.

However, in a larger prospective randomized placebo-controlled trial of UDCA in PSC by Lindor et al.[37], no benefit could be demonstrated. In this trial 105 patients were randomized to treatment with UDCA in conventional doses, *viz.* 13–15 mg/kg body weight daily, or placebo and followed up for up to 6 years (mean 2.9 years). Treatment with UDCA had no effect upon the time until treatment failure defined as death, liver transplantation, the development of cirrhosis, quadrupling of bilirubin, marked relapse of symptoms or the development of signs of chronic liver disease. Furthermore the significant improvement in liver biochemical tests seen in the treated group was not reflected by any beneficial changes in liver histology. On the contrary there was a suggestion that the liver histology of patients on UDCA showed a greater tendency to progress towards fibrosis. However, this could also be explained by sampling variability between serial liver biopsies[38].

The failure of standard doses of UDCA to provide clinical benefit led our group to consider the use of higher doses. Our rationale is that, with increasing cholestasis, there is decreasing enrichment of the bile acid pool with UDCA and higher doses are required to achieve the same level of enrichment[39]. Furthermore the in-vitro immunomodulatory effects of UDCA are enhanced with increasing UDCA concentrations[40].

In a pilot study we evaluated 26 patients with PSC who were randomized to either high-dose (20–25 mg/kg) UDCA or placebo[41] for 2 years (Figure 1). High-dose UDCA had no effect on symptoms but, as expected, there was a significant improvement in liver biochemistry. More importantly we found a significant reduction in cholangiographic appearances and liver fibrosis. In the treatment group bile acid saturation with UDCA >70% confirmed patient compliance. No significant side-effects were reported, in particular no worsening of colitis was seen.

These encouraging results were confirmed by an open study in 30 patients with PSC treated for 1 year[42]. When compared with historical controls a significant improvement in projected survival using the Mayo risk score was observed with a high dose but not with the conventional dose (13–15 mg/kg per day) of UDCA. In the light of these promising results a large controlled trial of a moderately high dose of UDCA (17–22 mg/kg) has been completed in Scandinavia. The provisional results have shown no significant differences between the two groups, although there was a strong trend in favour of improved survival in the UDCA group. Moreover, the study was probably underpowered to show a positive result, as the endpoints of death or transplant were lower than expected in both groups (Olsson and Broome, 2004, personal communication).

It is established that patients with UC with PSC have a higher rate of colonic dysplasia and cancer than patients with PSC alone[43]. Recent studies have suggested that treatment with UDCA reduces the rate of colonic dysplasia and cancer[44,45]. Whether UDCA reduces the high rate of cholangiocarcinoma in PSC remains to be established, although UDCA appeared to be protective against the development of cholangicarcinoma in a study in PSC transplant patients from the Nordic countries[46].

MISCELLANEOUS TREATMENTS

In keeping with ulcerative colitis, there is a strong inverse relationship between PSC and cigarette smoking. This led Angulo et al.[47] to test the hypothesis that oral nicotine might have a beneficial effect in PSC. Eight non-smoking patients with PSC were treated with nicotine 6 mg q.i.d. for up to 1 year. Side-effects were high, requiring cessation in three patients, and no beneficial effects were seen.

COMBINED THERAPY

In an important pilot study the potential of combination therapy was explored by Schramm et al.[48], who treated 15 patients with PSC. All patients received low-dose UDCA (500–750 mg daily), prednisolone 1 mg/kg daily and azathioprine 1–1.5 mg/kg daily. After a median follow-up period of 41 months, all patients had a significant improvement in liver function tests. Seven patients had been previously treated with UDCA but liver enzymes improved only after immunosuppressive therapy was added. More importantly, six of 10 with follow-up biopsies showed histological improvement, and significant radiological deterioration was seen in only 1 of 10 patients who had endoscopic retrograde cholangiography.

In a prospective trial Stiehl et al.[8,49] studied the survival of 106 patients with PSC for up to 13 years, treated with 750 mg UDCA daily and by endoscopic balloon dilation of major dominant stenoses whenever necessary. Some of the patients developed dominant strictures during the trial, and UDCA did not prevent such stricture formation. This combined approach of UDCA and endoscopic intervention significantly improved survival compared with predicted survival rates. This was an uncontrolled study and it is therefore difficult to ascertain whether UDCA or endoscopic therapy, if either, prolonged survival, although the results are promising.

CONCLUSION

There is no established effective medical treatment for PSC. However, promising recent studies suggest that high-dose UDCA may have a role in at least slowing disease progression and reducing the rate of colonic dysplasia and cancer, although larger long-term studies are awaited. The chemoprotective effect of UDCA on the colon and possibly the bile ducts probably means that all PSC patients with inflammatory bowel disease should be treated with UDCA. Randomized controlled trials of combination therapy in early PSC are needed, possibly high-dose UDCA in combination with immunosuppressant agents, and/or antibiotics. With the identification of T cell subsets involved in PSC, and the cytokines they produce, it may be possible to use particular recombinant cytokines or antibodies to specific cytokines such as anti-TNF antibody (infliximab) to manipulate the immune response in PSC and alter disease progression. Greater insight into the pathogenetic mechanisms involved in PSC would enable therapy to be targeted more specifically at the area of initial damage, namely the biliary epithelium.

Liver transplantation remains the mainstay of treatment for patients with endstage disease[46,50].

References

1. Chapman RW, Arborgh BA, Rhodes JM et al. Primary sclerosing cholangitis: a review of its clinical features, cholangiography, and hepatic histology. Gut. 1980;21:870–7.
2. Lo SK, Fleming KA, Chapman RW. Prevalence of anti-neutrophil antibody in primary sclerosing cholangitis and ulcerative colitis using an alkaline phosphatase technique. Gut. 1992;33:1370–5.
3. Chapman R. Does HLA status influence prognosis in primary sclerosing cholangitis? Gastroenterology. 1995;108:937–40.
4. Martins E, Graham AK, Chapman RW, Fleming KA. Elevation of gamma delta T lymphocytes in peripheral blood and livers of patients with primary sclerosing cholangitis and other autoimmune liver diseases. Hepatology. 1996;23:988–93.
5. Broome U, Olsson R, Loof L et al. Natural history and prognostic variables in 305 Swedish patients with primary sclerosing cholangitis. Gut. 1996;38:610–15.
6. Pokorny CS, McCaughan GW, Gallagher ND, Selby WS. Sclerosing cholangitis and biliary tract calculi – primary or secondary? Gut. 1992;33:1376–80.
7. Bergquist A, Ekbom A, Olsson R et al. Hepatic and extra hepatic malignancies in primary sclerosing cholangitis. J Hepatol. 2002;36:321–7.
8. Stiehl A, Rudolph G, Kloters-Plachy P, Sauer P, Walker S. Development of dominant bile duct stenoses in patients with primary sclerosing cholangitis treated with ursodeoxycholic acid: outcome after endoscopic treatment. J Hepatol. 2002;36:151–6.
9. Sivak M Jr, Farmer RG, Lalli AF. Sclerosing cholangitis: its increasing frequency of recognition and association with inflammatory bowel disease. J Clin Gastroenterol. 1981; 3:261–6.
10. Burgert SL, Brown BP, Kirkpatrick RB, LaBrecque DR. Positive corticosteroid response in early primary sclerosing cholangitis. Gastroenterology. 1984;86:1037 (abstract).
11. Lindor KD, Wiesner RH, Colwell LJ et al. The combination of prednisone and colchicine in patients with primary sclerosing cholangitis. Am J Gastroenterol. 1991;86:57–61.
12. Porayko MK, Wiesner RH, Hay JE et al. Bone disease in liver transplant recipients:incidence, timing and risk factors. Transplant Proc. 1991;23:1462–5.
13. Grijm R, Huibregtse K, Bartelsman J et al. Therapeutic investigations in primary sclerosing cholangitis. Dig Dis Sci. 1986;31:792–8.
14. Jeffrey GP, Reed WD, Laurence BH, Shilkin KB. Primary sclerosing cholangitis: clinical and immunopathological review of 21 cases. J Gastroenterol Hepatol. 1990;5:135–40.
15. Craig PI, Williams SJ, Hatfield ARW, Ng M, Cotton PB. Endoscopic management of primary sclerosing cholangitis. Gut. 1990;31:1182a (abstract).
16. Allison MC, Burroughs AK, Noone P, Summerfield JA. Biliary lavage with corticosteroids in primary sclerosing cholangitis. A clinical, cholangiographic and bacteriological study. J Hepatol. 1986;3:118–22.
17. Boberg KM, Egeland T, Schrumpf E. Long term corticosteroid treatment in PSC. Scand J Gastroenterol. 2003;38:991–5.
18. Angulo P, Batts KP, Jorgensen A, Lindor KD. Budesonide in the treatment of primary sclerosing cholangitis: a pilot study. Hepatology. 1999;30:477A (abstract).
19. van Hoogstraten HJF, Vieggar FP, Boland GI et al. Budesonide or prednisone in combination with ursodeoxycholic acid in primary sclerosing cholangitis: a randomized double-blind pilot study. Am J Gastroenterol. 2000;95:2015–22.
20. Knox TA, Kaplan MM. Treatment of primary sclerosing cholangitis with oral methotrexate. Am J Gastroenterol. 1991;86:546–52.
21. Knox TA, Kaplan MM. A double-blind controlled trial of oral-pulse methotrexate therapy in the treatment of primary sclerosing cholangitis. Gastroenterology. 1994;106:494–9.
22. Lindor KD, Jorgensen RA, Anderson ML et al. Ursodeoxycholic acid and methotrexate for primary sclerosing cholangitis: a pilot study. Am J Gastroenterol. 1996;91:511–15.
23. Javett SL. Azathioprine in primary sclerosing cholangitis. Lancet. 1971;i:810–11.
24. Wagner A. Azathioprine treatment in primary sclerosing cholangitis. Lancet. 1971;2:663–4.
25. Wiesner RH, Steiner B, LaRusso NF, Lindor KD, Baldus WP. A controlled clinical trial evaluating cyclosporine in the treatment of primary sclerosing cholangitis. Hepatology. 1991;14:63A (abstract).

26. Sandborn WJ, Wiesner RH, Tremaine WJ, Larusso NF. Ulcerative colitis disease activity following treatment of associated primary sclerosing cholangitis with cyclosporin. Gut. 1993;34:242–6.

27. Van-Thiel DH, Carroll P, Abu-Elmagd K et al. Tacrolimus (FK 506), a treatment for primary sclerosing cholangitis: results of an open-label preliminary trial. Am J Gastroenterol. 1995;90:455–9.

28. Harucha AE, Jorgensen R, Lichtman SN, La Russo NF, Lindor KD. A pilot study of pentoxifylline for the treatment of primary sclerosing cholangitis. Am J Gastroenterol. 2000;95:2338–42.

29. Epstein MP, Kaplan MM. Pilot study of etanercept in the treatment of PSC. Dig Dis Sci. 2004;49:1–4.

30. Olsson R, Broome U, Danielsson A et al. Colchicine treatment of primary sclerosing cholangitis. Gastroenterology. 1995;108:1199–203.

31. Jazrawi RP, de-Caestecker JS, Goggin PM et al. Kinetics of hepatic bile acid handling in cholestatic liver disease: effect of ursodeoxycholic acid. Gastroenterology. 1994;106:134–42.

32. Lo SK, Hermann R, Chapman RW et al. Ursodeoxycholic acid in primary sclerosing cholangitis: a double blind controlled trial. Hepatology. 1992:16A.

33. van Milligen de Wit AW, Kuiper H, Camoglio L et al. Does ursodeoxycholic acid mediate immunomodulatory and anti inflammatory effects in patients with primary sclerosing cholangitis? Eur J Gastroenterol Hepatol. 1999;11:129–36.

34. O'Brien CB, Senior JR, Arora-Mirchandani R, Batta AK, Salen G. Ursodeoxycholic acid for the treatment of primary sclerosing cholangitis: a 30-month pilot study. Hepatology. 1991;14:838.

35. Beuers U, Spengler U, Kruis W et al. Ursodeoxycholic acid for treatment of primary sclerosing cholangitis: a placebo-controlled trial. Hepatology. 1992;16:707–14.

36. Stiehl A, Walker S, Stiehl L et al. Effect of ursodeoxycholic acid on liver and bile duct disease in primary sclerosing cholangitis. A 3-year pilot study with a placebo-controlled study period. J Hepatol. 1994;20:57–64.

37. Lindor KD, The Mayo PSC/UDCA Study Group. Ursodiol for primary sclerosing cholangitis. N Engl J Med. 1997:336:691–5.

38. Olsson R, Hagerstrand I, Broome U et al. Sampling variability of percutaneous liver biopsy in primary sclerosing cholangitis. J Clin Pathol. 1995;48:933–5.

39. Rost D, Rudolph G, Kloeters-Plachky P, Stiehl A. The effect of high-dose ursodeoxycholic acid on its biliary enrichment in primary sclerosing cholangitis. J Hepatol. 2004;40:695–8.

40. Hirano F, Tanaka H, Makino Y, Okamoto K, Makino I. Effects of ursodeoxycholic acid and chenodeoxycholic acid on major histocompatibility complex class I gene expression. J Gastoenterol. 1996;31:55–60.

41. Mitchell SA, Bansi DS, Hunt N, von Bergmann K, Fleming KA, Chapman RW. A preliminary trial of high dose ursodeoxycholic acid in primary sclerosing cholangitis. Gastroenterology. 2001;122.

42. Harnois DM, Angulo P, Jorgensen RA, La Russo NF, Lindor KD. High-dose ursodeoxycholic acid as a therapy for patients with primary sclerosing cholangitis. Am J Gastroenterol. 2001;96:1558–66.

43. Broome U, Lofberg R, Veress B et al. Primary sclerosing cholangitis and ulcerative colitis:evidence for increased neoplastic potential. Hepatology. 1995;22:1404–8.

44. Tung BY, Emond MJ, Haggitt RC et al. Ursodiol use is associated with lower prevalence of colonic neoplasia in patients with ulcerative colitis and primary sclerosing cholangitis. Ann Intern Med, 2001:134:89–95.

45. Pardi DS, Loftus EV, Kremers WK, Leach J, Lindor KD. Ursodeoxycholic acid as a chemoprevetive agent in patients with ulcerative colitis and primary sclerosing cholangitis. Gastroenterology. 2003;124:889–93

46. Brandsaeter B, Broome U, Isoniemi H et al. Liver transplantation for primary sclerosing cholangitis in the Nordic countries:outcome after acceptance to the waiting list. Liver Transplant. 2003;9:961–9.

47. Angulo P, Bharucha AE, Jorgensen RA et al. Oral nicotine in treatment of primary sclerosing cholangitis: a pilot study. Dig Dis Sci. 1999;44:602–5.

48. Schramm C, Schirmacher P, Helmreich-Becker I et al. Combined therapy with azathioprine and prednisolone and ursodiol in patients with primary sclerosing cholangitis. A case series. Ann Intern Med. 1999;131:94–946.
49. Stiehl A, Rudolph G, Sauer P, Benz C et al Efficacy of ursodeoxycholic acid treatment and endoscopic dilatation of major duct stenoses in primary sclerosing cholangitis. A 8 years prospective study. J Hepatol. 1997;26;56–61.
50. Gow PJ, Chapman RW. Liver transplantation for primary sclerosing cholangitis. Liver. 2000:20:97–103.
51. Van Hoogstraten HJ, Wolfhagen FJ, Van de Meeberg PC, van Buuren HR, van Berge H, Schalm SW. Ursodeoxycholic acid therapy for primary sclerosing cholangitis: results of a 2 year randomized controlled trial to evaluate single versus multiple doses. J Hepatol. 1998; 29:417–23.

Section VII
Treatment and prognosis II

Chair: J. HEATHCOTE and R. WILLIAMS

21
Novel approaches to the treatment of primary biliary cirrhosis

U. LEUSCHNER

INTRODUCTION

In earlier chapters it has been shown that treatment of patients with auto-immune liver diseases with cyclosporin A, tacrolimus, mycophenolate mofetil or pranlukast had some positive results, but that the number of studies and the number of included patients was rather small. Since this is the same regarding PBC, these data will not be discussed.

Ursodeoxycholic acid (UDCA) in a dose of 13–15 mg/kg body weight per day is the treatment of choice for primary biliary cirrhosis (PBC). In 16 randomized controlled trials with 1422 patients it has been shown that UDCA significantly reduced the risk of ascites/oedema, jaundice, values of bilirubin, liver enzymes and the incidence of liver transplantation. In addition UDCA did not show any side-effects. On the other hand, up to now it could not be demonstrated that UDCA reduces mortality or prolongs life expectancy[1–3] (Table 1).

This result, namely that UDCA is unable to improve survival rate, was to be expected, because in the above-mentioned 16 randomized controlled trials 50% of the patients included were in a late stage of the disease (stage III/IV). As with other liver diseases and their treatment regimens the therapeutic effect of the recommended drug or treatment measures decreases with the progression of the disease. This has been shown for interferon and ribavirin in the treatment of chronic hepatitis C, for steroids in patients with autoimmune hepatitis, for phlebotomy in haemochromatosis or in alcoholic cirrhosis, where fibrosis can be reverted to a minor degree but cirrhosis of the liver not at all, although the patient stopped drinking. According to data from the literature, and taking into account the natural course of PBC[4,5], the optimal treatment time for PBC patients lies between year 8 (first abnormal liver function tests: diagnosis established) and 13 after the first detection of antimitochondrial antibodies; this means in the progressive, inflammatory stage of the disease and not when fibrosis has already developed or the disease has progressed to cirrhosis. When compensated or decompensated cirrhosis is present the therapeutic effect of UDCA is less prominent and the drug finally becomes ineffective. Therefore the

Table 1 Effects of ursodeoxycholic acid for primary biliary cirrhosis versus placebo (data until April 2002)

- Randomized controlled trials: 16
- Only trials using adequate randomization
- UDCA at any dose versus placebo
- Patient no: 1422
- Results: UDCA significantly (< 0.05) reduced:
  Ascites
  Jaundice
  Bilirubin
  Liver enzymes
  Incidence of LTX (< 0.04)
  No significant adverse effect
  No effect on mortality $\pm$ LTX

(Ref. 2)

cited Cochrane Analysis[2] is unsuited to answer the question of whether UDCA prolongs life expectancy or not.

On the other hand, many investigations have shown[6–9], that using so-called prognostic markers, UDCA not only reduces the incidence of liver transplantation but obviously is also able to prolong life expectancy[10–12]. Prognostic markers included in the Mayo Risk Score are bilirubin, prothrombin time, albumin, and age of the patient. Other accepted prognostic markers not included in the Mayo Risk Score are piecemeal necroses, histological progression and fibrosis. A multivariate analysis indicated that UDCA is an independent factor associated with lack of progression of PBC; therefore prolongation of survival is most likely[13–15]. Even though prognostic models are not precise, they can improve characterization of patient groups and can assist in clinical decision making.

WHY DOES PBC NOT RESPOND TO GLUCOCORTICOIDS?

PBC is a cholestatic liver disease and an autoimmune disease. It presents with characteristic autoantibodies, an augmentation of IgM, with concomitant autoimmune diseases, immunocompetent cells in liver infiltrates and it overlaps with autoimmune hepatitis. Amazingly PBC seems not to respond to prednisone monotherapy or treatment with other immunosuppressants, and is not cured by UDCA. This raises the question of whether a combination therapy with UDCA and the immunosuppressants prednisone or budesonide would be superior to monotherapies. Perhaps, after having modulated cholestasis using UDCA, immunosuppressants could regain their specific anti-inflammatory and immunosuppressive properties.

As has been shown previously[16,17] in PBC patients and patients with obstructive cholestasis endotoxin concentrations in serum increase significantly. Endotoxins stimulate macrophages, hepatocytes and lymphocytes to express proinflammatory cytokines such as IL-1, IL-6 and TNF-α, which in the hypothalamus induce the synthesis of corticotropin-releasing hormone (CRH) (Figure 1)[18–20]. CRH and arginine vasopressin (AVP) stimulate the secretion of

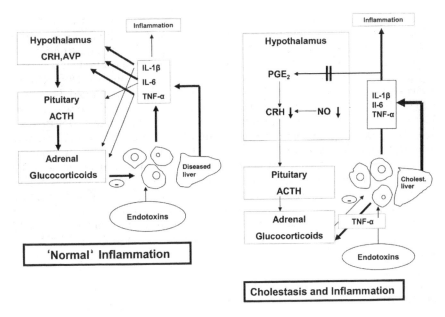

Figure 1 Hypothalamic–pituitary–adrenal axis (HPA) and immune system interactions. CRH, corticotropin-releasing hormone; LPS, endotoxins; PGE, prostaglandin E; NO, nitric oxide; AVP, arginine vasopressin

corticotropin in the anterior pituitary gland, which activates the synthesis and secretion of glucocorticoids from the adrenal glands (hypothalamic–pituitary–adrenal axis: HPA axis). This means that corticotropin is the key regulator of glucocorticoid release by the adrenal glands. Glucocorticoids as well-known anti-inflammatory and immunosuppressive compounds are able to suppress several functions of activated immunocompetent cells, influence their differentiation and proliferation and suppress the expression of IL-6 and IL-1β[21,22].

The HPA axis has been poorly studied in cholestatic liver diseases, and only a few data are available in patients with PBC. From these few investigations we learned that in cholestasis the proinflammatory cytokines obviously stimulate CRH production in the hypothalamus to a lesser degree than in non-cholestatic diseases[23,24], which means that the synthesis of glucocorticoids in the adrenal glands and their secretion into the circulation will be impaired. On the other hand, in cholestatic liver diseases cytokines, especially TNF-α, directly stimulate adrenal glucocorticoid production, as has been shown in the hypophysectomized cholestatic rat in contrast to the non-cholestatic rat[25]. This means that in cholestatic liver disease there is a dysregulation with an enhanced adrenal steroid production rather than a hypocorticosteroidism caused by a reduced stimulation of the HPA axis.

Since in cholestatic rats a number of central abnormalities in response to systemic interleukins could be demonstrated[26], one, two, three or even several dysregulations of the HPA axis could be major defects that inhibit or completely abolish the therapeutic effect of orally or intravenously administered glucocorticoids or other immunosuppressants in PBC. In immune-mediated inflammations a clinically relevant dysregulation of the immunoneuroendocrine system could be due either to a short- or long-term adaptation of either the HPA axis or immunocompetent cells, or the development of resistance or tolerance to glucocorticoids (Table 2)[27]. As mentioned, in animal experiments chronic inflammation and cholestasis were associated with a mild rather than a severe hypercortisolism with low CRH expression[28]. Further, an increase of the hypothalamic concentration of substance P, a potent inhibitor of CRH synthesis, was observed[29,30] which could diminish the augmented TNF-α-mediated corticoid secretion by the adrenal glands. Further it has been shown that simultaneous expression of IL-2 and IL-4 induces resistance to steroids in T lymphocytes by decreasing the affinity of glucocorticoid receptors on the cell surface[31,32], and that the conversion of cortisol to less active or inactive metabolites reduces the sensitivity of immunocompetent cells to glucocorticoids[33]. Cytokines and other mediators that influence the HPA axis are summarized in Table 3[27].

Table 2 Adaptation of the HPA-axis to immune-mediated inflammations and resistance to glucocorticoids

Adaption
- Animals with chronic inflammation have a mild rather than severe hypercorticosterolism with low CRH expression
- Hypothalamic elevation of substance P inhibits CRH secretion in inflammation

Resistance
- IL-2 and IL-4 together induce resistance in T-cells by decreasing the affinity of glucocorticoid receptors
- Conversion of cortisol into less active or inactive metabolites alters the sensitivity of cells of the immune system to glucocorticoids

Table 3 Cytokines and other mediators of inflammation that influence the hypothalamic–pituitary–adrenal axis

- Inflammatory cytokines
  Tumour necrosis factor α
  Interleukin-1α and interleukin-1β
  Interleukin-6
- Other cytokines
  Interferon-α
  Interferon-γ
  Interferon-2
- Growth factors
  Epidermal growth factor
  Transforming growth factor β
- Lipid mediators
  Prostanoids
  Platelet-activating factor

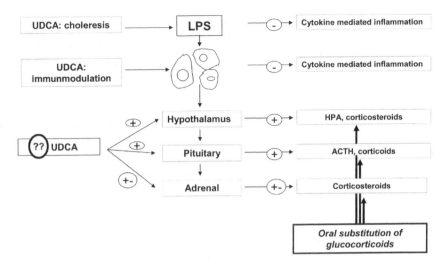

Figure 2 Hypothesis on the mode of action of an UDCA–glucocorticoid combination therapy in primary biliary cirrhosis

Based on these observations we hypothezise on the mode of action of a UDCA–glucocorticoid combination therapy as follows (Figure 2) :

1. UDCA induces choleresis, and in this way

2. decreases endotoxin (LPS) concentrations.

3. UDCA also has immunomodulatory properties[34], reducing the cytokine-mediated inflammation.

4. Oral substitution of glucocorticoids overcomes corticosteroid dysregulations, caused by resistance to glucocorticoids or adaptation of the HPA axis to immune-mediated inflammation.

5. Finally, since UDCA as a steroid molecule probably transgresses the blood–liquor barrier[35], it could influence hypothalamic and pituitary functions (Table 4).

CLINICAL TRIALS WITH UDCA COMBINATION THERAPIES

These physiological and pathophysiological observations raise the question of whether any clinical observations exist suggesting that the combined administration of UDCA with glucocorticoids could improve PBC treatment compared to UDCA monotherapy.

Table 4 Hypotheses: UDCA + corticoids for PBC

- UDCA has choleretic and immunomodulatory properties, it reduces LPS in the blood and the expression of cytokines

- Oral glucocorticoids compensate the amount of less active or inactive corticoid metabolites, seen in chronic liver diseases

- Oral glucocorticoids overcome the resistance of T cells against steroids

- Oral glucocorticoids overcome the adaptation of the HPA axis in chronic cholestatic inflammation

- UDCA 'enters' the brain (blood–brain barrier) and influences the CRH cascade

Indeed, they do exist. For example, in pregnant women with PBC or primary sclerosing cholangitis (PSC) pregnancy significantly further improved liver function tests during or without UDCA therapy[36–38]. That the fetoplacental unit, as described in healthy pregnant women, redirects maternal immunity away from Th1 to Th2, that HLA-G isoforms expressed on the cytotrophoblast inhibit NK and antigen-specific cytolysis, and that placental proteins inhibit T cells at the fetomaternal interface, represents a status of strong temporary immunosuppression (Figure 3)[39–41]. The shift from Th1 to Th2 is caused by a high concentration of oestrogens, by progesterone and testosterone, but not by low oestrogen concentrations or prolactin. Accordingly, in a study in pregnant PBC patients we could show that, during pregnancy and continued UDCA therapy, the anti-inflammatory Th2-mediated cytokines IL-5 and IL-10 increased in the serum of the mother until delivery, and the proinflammatory Th1-mediated cytokines IFN-γ and IL-2 slightly decreased. A short time before delivery the anti-inflammatory cytokines rapidly decreased, IFN-γ and TNF-α increased, PBC-typical liver biochemistries deteriorated but returned to pre-pregnancy data a couple of weeks later[38]. These observations strongly suggest that a powerful immunosuppressive effect, caused by pregnancy, is able to further improve UDCA therapy in pregnant women with PBC.

That a combination of UDCA plus prednisone can be superior to UDCA monotherapy has also been shown in clinical studies with non-pregnant PBC patients[42,43], although the influence on routine liver function tests was not different between the combination (UDCA plus prednisone) and the UDCA plus placebo group. However, it was of great interest that immunoglobulins G, M and A, and liver histology, significantly improved during UDCA–prednisolone combination treatment which was rarely seen in clinical studies with UDCA monotherapy. In two recent randomized controlled trials, in which patients with an autoimmune hepatitis–PBC overlap syndrome had been painstakingly excluded, and where UDCA had been combined with the topical steroid budesonide in a dose of 9 and 6 mg per day respectively, the combination treatment improved laboratory data, liver histology, fibrosis and the histological stage significantly better than UDCA alone or UDCA plus placebo[44,45]. Most patients were in early stages of the disease. These two

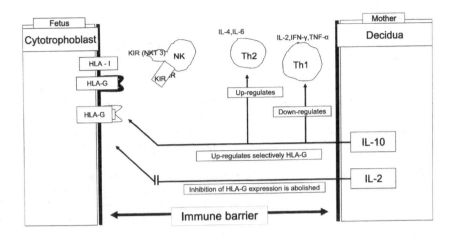

Figure 3 Pregnancy: a model of allograft tolerance. KIR, killing-inhibitory receptor binds to HLA-I; KIR (NKAT 3), killing-inhibitory receptor binds to HLA-G

studies with budesonide are of special interest, since this topical steroid has a 10–20-fold higher relative receptor binding affinity than the conventional glucocorticoids[46], so that budesonide could overcome the above-mentioned T cell resistance against steroids[31,32]. Since the spillover of budesonide into the systemic circulation due to its high first-pass effect in the liver is only 10–15%[47] the risk of side-effects is low.

From these data we conclude that UDCA – in combination with glucocorticoids not only in patients with an autoimmune hepatitis–PBC overlap syndrome but also in patients with stage I and II of a 'regular' PBC – seems to be superior to UCDA monotherapy: UDCA induces choleresis, has immunomodulatory properties, and oral glucocorticoids compensate corticoid dysregulation or overcome adaptation to steroid therapy, as seen in cholestasis. Whether hormones such as oestrogens, progesterone or testosterone additionally given in low doses could further improve treatment results needs to be investigated.

Other UDCA combination therapies have been performed with azathioprine, methotrexate, colchicine, silymarine, bezafibrate, the NSAID sulindac or the combination prednisone plus azathioprine. All these showed some positive results, which suggests that the combination therapy of UDCA with an anti-inflammatory and immunosuppressive compound is superior to UDCA alone. Patients most suitable for combination therapies need to be characterized.

SUMMARY

At present UDCA is the treatment of choice for patients with the autoimmune liver disease PBC. As monotherapy, UDCA, like immunosuppressants, is unable to cure the disease. The combination of UDCA with glucocorticoids seems to be superior to UDCA alone. Best support for the use of a combination therapy is pregnancy in patients with PBC where a strong physiological immunosuppression further improves the positive results obtained with UDCA. However, clinical studies in non-pregnant PBC patients also revealed better results with the combination of UDCA plus prednisone or the topical steroid budesonide than with UDCA monotherapy.

Since interactions between the HPA axis, autoimmune liver diseases such as PBC and humoral and cellular immune responses are poorly understood, future investigations should try to elucidate mechanisms which prevent therapeutic effects of immunosuppressants in biliary liver diseases. In patients with the cholestatic PBC, cholestasis seems to play a crucial role.

References

1. Goulis J, Leandro G, Burrough AK. Randomised controlled trials of ursodeoxycholic acid therapy for primary biliary cirrhosis: a meta-analysis. Lancet. 1999;354:1053–60.
2. Gluud C, Christensen E. Ursodeoxycholic acid for primary biliary cirrhosis. Cochrane Library, Issue 3, 2002.
3. Papatheodoridis GV, Hadziyannis ES, Deutsch M et al. Ursodeoxycholic acid (UDCA) in primary biliary cirrhosis. Trial results of a 12-year prospective randomized, controlled trial. Am J Gastroenterol. 2002;32:561–6.
4. Christensen E, Crowe J, Doniach D et al. Clinical pattern and course of disease in primary biliary cirrhosis based on the analysis of 236 patients. Gastroenterology. 1980;236:46–78.
5. Parés A, Rodés J. Natural history of primary biliary cirrhosis. Clin Liver Dis. 2003;7:779–94.
6. Christensen E. Prognostic modelling. In: J Neuberger, editor. Primary Biliary Cirrhosis. West End Studios, Eastbourne, UK, 1999:93–9.
7. Jensen DM. Cholestasis. In: N Gitlin, editor. Clinics in Liver Disease, vol. 3. Saunders, 1999:529–70.
8. Dickson ER, Grambsch PM, Fleming TR et al. Prognosis in primary biliary cirrhosis: model for decision making. Hepatology. 1989;10:1–7.
9. Angulo P, Lindor KD, Therneau TM et al. Utilization of the Mayo risk score in patients with primary biliary cirrhosis receiving ursodeoxycholic acid. Liver. 1999;19:115–21.
10. Lindor KD, Therneau TM, Jorgensen RA et al. Effects of ursodeoxycholic acid on survival in patients with primary biliary cirrhosis. Gastroenterology. 1996;110:1515–18.
11. Emond M, Carithers RL Jr, Luketi VA et al. Does ursodeoxycholic acid improve survival in patients with primary biliary cirrhosis? Comparison of outcome in the US multicenter trial to expected survival using the Mayo Clinic prognosis model. Hepatology. 1996;24:168A.
12. Markus BH, Dickson ER, Grambsch PM et al. Efficacy of liver transplantation in patients with primary biliary cirrhosis. N Engl J Med. 1989;320:1709–13.
13. Parés A, Caballeria L, Rodés J et al. Long-term effects of ursodeoxycholic acid (UDCA) in primary biliary cirrhosis: results of a double-blind controlled multicentric trial. J Hepatol. 2000;32:561–6.
14. Parés A, Caballeria L, Bruguera M et al. Factors influencing histological progression of early primary biliary cirrhosis. Effect of ursodeoxycholic acid. J Hepatol. 2001;34(Suppl. 1):189–90.
15. Parés A, Caballeria L, Rodés J. Long-term ursodeoxycholic acid treatment delays progression of mild primary biliary cirrhosis. J Hepatol. 2001;34(Suppl. 1):187–8.
16. Yamamoto Y, Sezai S, Sakurabayashi S et al. A study of endotoxemia in patients with primary biliary cirrhosis. J Int Med Res. 1994;22:95–9.

17. Pain JA, Bailey ME. Measurement of operative plasma endotoxin levels in jaundiced and non-jaundiced patients. Eur Surg Res. 1987;19:207–16.
18. Imura H, Fukata J, Mori T. Cytokines and endocrine functions: an interaction between the immune and neuroendocrine systems. Clin Endocrinol. 1991;35:107–15.
19. Bernardini R, Kamilaris TC, Calogero AE et al. Interactions between tumor necrosis factor-α, hypothalamic, corticotropin-releasing hormone, and adrenocorticotropin secretion in the rat. Endocrinology. 1990;126:2876–81.
20. Perlstein RS, Whitnall MH, Abrams JS et al. Synergistic roles of interleukin-6, interleukin-1, and tumor necrosis factor in adrenocorticotropin response to bacterial lipopolysaccharide *in vivo*. Endocrinology. 1993;132:946–52.
21. Zanker B, Walz G, Wieder KJ et al. Evidence that glucocorticosteroids block expression of the human interleukin-6 gene by accessory cells. Transplantation. 1990;49:183–5.
22. Zitnik RJ, Whiting NL, Elias JA. Glucocorticoid inhibition of interleukin-1-induced interleukin-6 production by human lung fibroblasts: evidence for transcriptional and post-transcriptional regulatory mechanisms. Am J Respir Cell Mol Biol. 1994;10:643–50.
23. Swain MC, Maric M, Carter L. Defective interleukin-1 induced ACTH-release in cholestatic rats: impaired hypothalamic PGE 2 release. Am J Physiol. 1995;268:G404–9.
24. Swain MC, Maric M. Impaired stress and interleukin-1 induced hypothalamic expression of the neuronal activation marker FOS in cholestatic rats. Hepatology. 1996;24:914–18.
25. Swain MC, Maric M. Tumor necrosis factor-alpha stimulates adrenal glucocorticoid secretion in cholestatic rats. Am J Physiol. 1996;270:G987–91.
26. Swain MC. Cytokines and endocrine abnormalities in cholestasis. In: Manns MP, Boyer JL, Jansen PLM, Reichen J, editors. Cholestatic Liver Diseases. Dordrecht: Kluwer, 1998: 155–62.
27. Chrousos GP. The hypothalamic–pituitary–adrenal axis and immune mediated inflammation. N Eng J Med. 1995;332:1351–62.
28. Swain MC, Patchev V, Vergalla J et al. Suppression of hypothalamic–pituitary–adrenal axis responsiveness to stress in a rat model of acute cholestasis. J Clin Invest. 1993;91:1903–8.
29. Culman J, Tschope C, Jost N et al. Substance P and neurokinin A induced desensitization to cardiovascular and behavioral effects: evidence of the involvement of different tachykinin receptors. Brain Res. 1993;625:75–83.
30. Jessop DS, Chowdrey HS, Larsen PJ et al. Substance P: multifunctional peptide in the hypothalamic pituitary system? J Endocrinol. 1992;132:331–7.
31. Almawi WY, Lipman ML, Stevens AC et al. Abrogation of glucocortico-mediated inhibition of T cell proliferation by the synergistic action of IL-1, IL-6 and TNF-α. J Immunol. 1991;146:3523–7.
32. Kam JC, Szefler SJ, Surs W et al. Combination IL-2 and IL-4 reduces glucocorticoid receptor-binding affinity and T cell response to glucocorticoids. J Immunol. 1993;151: 3460–6.
33. Klein A, Buskila D, Gladman D et al. Cortisol catabolism by lymphocytes of patients with lupus erythematosus and rheumatoid arthritis. J Rheumatol. 1990;17:30–3.
34. Calmus Y, Podevin P, Correia L et al. Immune modulation by cholestasis and bile acids. In: Leuschner U, Berg PA, Holtmeier J, editors. Bile Acids and Pregnancy. Dordrecht: Kluwer, 2002:57–69.
35. Rodrigues CMP, Steer CS. Tauroursodeoxycholic acid for the treatment of acute and chronic neurodegenerative diseases. In: Paumgartner G, Leuschner U, Keppler D, Stiehl A, editors. Bile Acids: From Genomics to Disease and Therapy. Dordrecht: Kluwer, 2003: 270–86.
36. Landon MB, Soloway RD, Freedman LJ et al. Primary sclerosing cholangitis and pregnancy. Obstet Gynecol. 1987;69:457–60.
37. Rudi J, Schonig T, Stremmel W. Therapy with ursodeoxycholic acid in primary biliary cirrhosis in pregnancy. Z Gastroenterol. 1996;34:188–91.
38. Holtmeier J, Leuschner M, Stiehl A et al. Ursodeoxycholic acid in the treatment of primary biliary cirrhosis and primary sclerosing cholangitis. In: Leuschner U, Berg PA, Holtmeier J, editors. Bile Acids and Pregnancy. Dordrecht: Kluwer, 2002:70–4.
39. Formby B, Wiley TS. Ten ways to suppress the maternal immune system. In: Leuschner U, Berg PA, Holtmeier J, editors. Bile Acids and Pregnancy. Dordrecht: Kluwer, 2002:20–5.

40. Riteau B, Rouas-Freiss N, Menier C et al. HLA-G2, -G3, -G4 isoforms as nonmature cell surface glycoproteins inhibit NK and antigen-specific cytolysis. J Immunol. 2001;166: 5018–26.
41. Rachmilewitz J, Rieley GJ, Tykocinski ML. Placental protein 14 functions as a direct T cell inhibitor. Cell Immunol. 1999;191:26–35.
42. Wolfhagen FHJ, van Buuren HR, Schalm SW. Combined treatment with ursodeoxycholic acid and prednisone in primary biliary cirrhosis. Neth J Med. 1994;44:84–90.
43. Leuschner M, Güldütuna S, You T et al. Ursodeoxycholic acid and prednisolone versus ursodeoxycholic acid and placebo in the treatment of early stage primary biliary cirrhosis. J Hepatol. 1996;25:49–57.
44. Leuschner M, Maier K-P, Schlichting J et al. Oral budesonide and ursodeoxycholic acid for treatment of primary biliary cirrhosis: results of a prospective double-blind trial. Gastroenterology. 1999;117:918–25.
45. Rautiainen HM, Kärkkäinen P, Karvonen A et al. Combination of budesonide and ursodeoxycholic acid (UDCA) compared with UDCA alone in PBC: results of a 3 year randomised trial. Gut. 2003;52(Suppl. VI):A1.
46. Würthwein G, Rehder S, Rodewald P. Lipophilicity and receptor affinity of glucocorticoids. Pharm Ztg Wiss. 1992;137:161–7.
47. Dahlberg E, Thalén A, Brattsand R et al. Correlation between chemical structure, receptor binding, and biological activity of some novel, highly active, 16α,17α-acetal substituted glucocorticoids. Mol Pharmacol. 1983;25:70–8.

22
Prognosis of autoimmune liver diseases

K. M. BOBERG and E. SCHRUMPF

INTRODUCTION

Autoimmune hepatitis (AIH), primary sclerosing cholangitis (PSC), and primary biliary cirrhosis (PBC) are chronic liver disorders that over time progress to cirrhosis. All three disorders can be treated by liver transplantation with favourable results. In the Scandinavian countries PBC and PSC represent main indications for transplantation[1]. For the selection of patients for liver transplantation, and for timing of this event, knowledge of prognosis is of particular importance in these two disorders.

Several prognostic models for PBC and PSC have been published. The models can be useful to describe prognosis in groups of patients. Unfortunately, their ability to predict prognosis in the individual patient is less reliable. Reported prognosis in any group of patients depends upon patient selection. Most prognostic studies are based on cohorts of patients seen at referral centres. These patients are likely to have more serious and advanced disease than those who are not referred. Population-based studies give a better indication of the true disease course, but such studies are more scarce. AIH, PSC, and PBC all have an insidious nature, and it is difficult to define the starting point of the disease. With better diagnostic techniques, chronic liver diseases are currently diagnosed in earlier stages than was the case in initial reports. General care of patients has also improved. Consequently, survival from time of diagnosis will appear to be longer in more recent studies. In this chapter we present data on prognosis in AIH, PSC, and PBC. For therapeutic options in each disorder please refer to other chapters of this book.

AUTOIMMUNE HEPATITIS

AIH is characterized by female preponderance, elevated serum aminotransferase levels, markedly increased immunoglobulin G concentration, seropositivity for autoantibodies, and interphase hepatitis on histological examination[2,3]. Diagnosis is based on assessment of a number of factors, as well as exclusion of

233

other causes of chronic hepatitis that can give similar clinical and laboratory findings, as outlined in the diagnostic criteria proposed by the International Autoimmune Hepatitis Group[4]. Aetiology is unknown, but the risk of developing AIH is associated with the HLA haplotype A1,B8,DR3 and HLA DR4[5]. AIH is commonly divided into subtypes distinguished by autoantibody patterns. AIH type 1 is characterized by the presence of antinuclear antibodies (ANA) and/or smooth muscle antibodies (SMA). AIH type 2 connotes the presence of liver/kidney microsomal antibodies (anti-LKM-1), whereas AIH type 3 is characterized by antibodies to soluble liver antigen/liver pancreas antigen (anti-SLA/LP). AIH appears to be a heterogeneous disorder[2,6], and variation in disease characteristics must be taken into account in assessment of prognosis. Differences in prognosis have been associated with clinical presentation, genetic features, as well as with autoantibody profiles[7].

Survival in AIH

In most cases untreated AIH progresses to cirrhosis[8]. The early studies of AIH, carried out before the widespread use of corticosteroid therapy, demonstrated poor survival rates[6,9–11]. In a study of 26 patients Bearn et al.[10] commented that death from the disease within 10 years of onset was the rule. Among the 12 patients who died during the study, average survival was about 7 years (range 3–20 years). The mean duration of the disease among those still living was 5.5 years. A beneficial effect of corticosteroids in acute exacerbations of the disease was observed[10]. In the study by Read et al.[11], there were 26 fatal cases among the 79 patients with adequate follow-up. Average survival among those who died was 3.4 years. Long-term follow-up revealed that only three among the 81 patients were alive 10 years after disease onset[12].

Prognosis was also poor in untreated patients in controlled studies with corticosteroids published in the early 1970s[13,14]. Fifteen among 27 patients in the control group were dead after follow-up of mean 23 months in the study from the Royal Free Hospital[13]. Long-term follow-up of the patients from this centre revealed that 27% of patients in the control group were alive at 10 years, with median survival 3.3 years[8,15]. On the contrary, 10-year survival among patients treated with prednisolone was 63%. Median survival was 12.2 years in the latter group[15]. In the study from the Mayo Clinic[14], five among 17 patients in the placebo group died within 6 months. Prognosis in untreated patients is influenced by biochemical and histological indicators of severe AIH at presentation[16]. Patients with > 10-fold elevation of aminotransferase (AST) or > 5 times higher AST in conjunction with at least 2-fold elevation of gamma-globulins have a 3-year mortality of 50% and 10-year mortality of 90%[16]. Among patients who present with histological findings of bridging or multilobular necrosis, 82% develop cirrhosis within 5 years, and mortality is 45%[16]. It must, however, be taken into account that patients in the early studies and in the subsequent controlled trials had more severe AIH than those who are diagnosed today[2,6]. The studies were furthermore limited by the failure to exclude cases of viral hepatitis.

The early treatment studies also documented the benefit of corticosteroid treatment in AIH[13–15]. Treatment with prednisolone and/or azathioprine definitely improves survival. With immunosuppressive treatment survival rates similar to those of a normal population can be obtained[17]. For obvious reasons placebo-controlled trials in AIH will not be possible to conduct today. The natural course of untreated AIH diagnosed on the basis of current diagnostic methods and stringent criteria[4] therefore cannot be exactly assessed[2]. AIH onset is often acute, but response to therapy and prognosis in these cases appear to be similar to what is observed in patients who present as chronic hepatitis[18].

Prognostic factors in AIH

Cirrhosis

The presence of cirrhosis is a marker of advanced liver disease and was associated with poor prognosis in initial studies of patients with AIH. Survival rates as low as 40% at 5 years and 18% at 10 years were reported in cirrhotic patients, despite corticosteroid therapy[11,12]. Schalm et al.[19] found that patients within the spectrum of severe chronic active liver disease who presented with cirrhosis had a slower response to treatment, a greater incidence of treatment failure, and a higher mortality rate from liver failure than those without cirrhosis. In the early series from the Mayo Clinic, 5- and 10-year survival in patients with cirrhosis at presentation were 80% and 65%, respectively[20]. In a later report from the Mayo Clinic the impact of cirrhosis on treatment response to corticosteroids and survival in 128 well-characterized type 1 AIH patients was studied[17]. Thirty-seven patients (29%) had histological features of cirrhosis at presentation. Cirrhotic patients entered remission, had a relapse after drug withdrawal, sustained remission after discontinuation of treatment, and failed to respond to therapy as frequently as did patients without cirrhosis[17]. The conclusion of this study thus was that histological cirrhosis does not reduce survival expectations in corticosteroid treated type 1 AIH patients. Overall 5- and 10-year survival was 96% and 93%, respectively, quite similar to survival expectancies of an age- and sex-matched control population. The frequency of remission on treatment was similar in patients with and without cirrhosis (78% and 76%, respectively), as was the likelihood of relapse after drug withdrawal (76% and 74%, respectively). With death and liver transplantation as endpoints, 5- and 10-year survival rates were 97% and 89% for patients with cirrhosis at entry versus 94% and 90% for those without. These results are similar to those of Schvarcz et al.[21] and Kanzler et al.[22]. The majority of patients with cirrhosis in the study by Roberts et al. had well-compensated liver disease. Among seven patients with cirrhosis and ascites, five (71%) entered remission during corticosteroid therapy. The results underscore the importance of evaluating even cirrhotic AIH patients for therapy[17].

Among 91 patients without histological cirrhosis at entry in the study by Roberts et al.[17], 36 (40%) developed cirrhosis during or after corticosteroid therapy after a mean of 39 months. Development of cirrhosis was predicted by higher prothrombin times ($p < 0.001$) and lower serum albumin levels ($p = 0.04$). Histological features at presentation could also distinguish the patients who developed cirrhosis. Patients with confluent necrosis progressed to cirrhosis in 53% of cases, as compared to 32% among those with only periportal hepatitis ($p = 0.05$).

Asymptomatic versus symptomatic AIH

In a retrospective evaluation of 68 AIH patients in Israel, the 23 patients who were classified as asymptomatic at diagnosis responded better to treatment and had a more favourable prognosis than the symptomatic group[23]. However, only seven (30.4%) among the asymptomatic patients remained asymptomatic during follow-up of a median 36.5 months. Interestingly, the histopathological features at presentation were indistinguishable between asymptomatic and symptomatic patients.

HLA type

HLA genes that are associated with susceptibility to AIH also have an impact on clinical features and prognosis[7,24–28]. HLA B8-positive patients have more severe disease at presentation than those who are HLA B8-negative[24]. Patients who carry the A1-B8-DR3 haplotype relapse more frequently and are more often referred for liver transplantation[27,28]. Later studies have linked HLA DR3 to more severe disease at presentation and to a lower likelihood of entering remission during corticosteroid therapy, as well as higher relapse rate[22,25,26]. More specifically, the DRB1*0301 allele and the DRB1*0301-DRB3*0101 haplotype are significantly more common in AIH patients with poor response to corticosteroid therapy[26]. In contrast, the DRB1*0401 allele and the DRB1*0401-DRB4*0103 haplotype (DR4) are associated with a better prognosis, including lower frequency of death from liver failure or need for transplantation[26]. HLA DR3-positive patients generally present at younger age than do DR4-positive cases.

Autoantibody profile

The autoantibody reactivity has been associated to prognosis in AIH, but evidence for a primary role of autoantibodies in determining disease course is lacking[7,29]. Death and liver transplantation occurred more frequently among patients with antibodies to actin than in anti-actin-negative patients with ANA in a study of 99 patients with type 1 AIH[29]. Patients with anti-SLA/LP have been found to be more prone to relapse after treatment withdrawal[30]. Type 2 AIH with anti-LKM1 was originally postulated to have a poorer prognosis than type 1 AIH[31], but this observation has not been confirmed[2,7]. ANA/SMA-positive and LKM-1-positive AIH in childhood appear to have similar severity and long-term outcome[32]. Serum levels of bilirubin and INR are independent predictors of death and/or liver transplantation in childhood AIH[32]. Fulminant LKM-1 positive AIH patients may be a subgroup with less favourable outcome[32].

Hepatocellular carcinoma appears to be an uncommon complication in AIH cirrhosis[2,7].

With early diagnosis and proper medical therapy, AIH patients ideally should not need to be liver-transplanted. However, some patients deteriorate despite treatment, or are diagnosed at an advanced stage and become candidates for transplantation. Five-year survival rate after transplantation is approximately 90%[2,7,28]. Recurrent AIH in the graft occurs in 11% to 35% of cases[2], but is typically mild and easy to treat[3].

PRIMARY SCLEROSING CHOLANGITIS

PSC is a chronic cholestatic liver disease that is characterized by inflammation and fibrosis of the intrahepatic and extrahepatic bile ducts. Diagnosis is confirmed by cholangiographic findings of bile duct irregularities with multi-focal strictures and segmental dilations. Approximately two-thirds of patients are male, and mean age at diagnosis is 30–40 years[33]. Commonly, there is a delay of several years from first biochemical sign or symptom of liver disease until diagnosis of PSC[34]. Aetiology and pathogenesis of PSC is unknown, but there is evidence that the disease occurs in genetically predisposed individuals and that immunological factors are involved. In particular, several studies have confirmed association to genes within the HLA-region[35,36]. PSC is strongly associated with inflammatory bowel disease (IBD), in that approximately 80% of PSC patients have concurrent IBD. In the majority of cases (80%), these patients have ulcerative colitis. The most feared complication of PSC is cholangiocarcinoma (CC) that develops in 6–15% of cases[34,37–39].

The presentation, complications, and clinical course of PSC vary considerably between patients. PSC thus appears to be a heterogeneous disorder[40]. Although up to 50% of patients have been reported to be asymptomatic at time of diagnosis[34], patients often present and are diagnosed during periods of exacerbation, having symptoms of liver disease. Thereafter they often experience prolonged remissions. Some patients remain asymptomatic for decades; others can go through alternating episodes of deterioration and spontaneous improvement. In most cases PSC is, however, progressive and eventually leads to development of cirrhosis and liver failure. There is no effective medical therapy in PSC, and liver transplantation is the only curative option.

Survival in PSC

Survival from diagnosis until death or liver transplantation in PSC varies between studies. Median survival has been reported to be as low as less than 1 year from referral[41] and up to 17[42] and 18[43] years from diagnosis. Several groups have reported median survival to death or liver transplantation in larger cohorts of patients to be approximately 12 years[34,37,44,45]. In a study of 53 patients from Yale University as many as 75% of the patients were alive at 9 years from diagnosis[46].

Development of CC dramatically changes prognosis in PSC. Survival from diagnosis of CC is only 5–6 months[39,47]. In patients who are liver-transplanted and prove to have hepatobiliary malignancy in the explanted liver, survival, however, appears to be fair[48]. Among 31 patients with CC, the 1-, 3-, and 5-

year survival rates after liver transplantation were 65%, 35%, and 35%, respectively[48].

Prognostic factors in PSC

In attempts to predict disease course in PSC, several centres have looked for prognostic variables in their cohorts of patients[34,37,38,41,43-45,49-51]. Prognostic variables have then been combined into prognostic models, using the Cox proportional hazards model[52]. A number of clinical, biochemical and histological variables have been associated with prognosis in univariate analyses. In multivariate analyses, factors with an independent impact on prognosis remain in the model and contribute in proportion to their independent association with survival[52].

Age at diagnosis of PSC appears to be be a strong prognostic variable in most models. Older age at diagnosis is associated with shorter survival. Serum bilirubin concentration is also a very strong indicator of prognosis. The higher the bilirubin value is at diagnosis, the higher is the risk of death or liver transplantation. Low albumin and haemoglobin levels, high alkaline phosphatase, AST and cholesterol levels have been independently associated with prognosis in various studies. Clinical findings or events such as hepatomegaly, splenomegaly, and variceal bleeding have also appeared as prognostic factors in some multivariate analyses[52,53]. Advanced histological stage was a prognostic variable in the first prognostic models. To avoid the necessity of a liver biopsy this parameter has been omitted in several later models, including the revised model from the Mayo Clinic[54]. The Child–Pugh score has also been suggested as a useful predictor of survival in PSC[55]. Comparing the Child–Pugh classification with the Mayo natural-history model, Kim et al.[54] concluded that the Mayo model is superior at least in early stages of PSC. The presence of IBD was independently associated to poorer prognosis in one study[37]. More severe cholangiographic changes, including both intrahepatic and extrahepatic strictures, have also been linked to poorer prognosis[43]. Some studies have suggested that HLA genes influence the rate of disease progression in PSC[56-58].

Prognostic models in PSC

Most prognostic models in PSC have been so-called time-fixed Cox regression models. These models assume that the variables recorded at one single time point for each patient are sufficient to predict survival. They may not be appropriate to update prognosis at later patient visits. Multivariate Cox regression analyses with time-dependent variables are caculated from follow-up data and can be used to update short-term prognosis according to a change in the patient's clinical condition[59]. A time-dependent Cox regression model has been applied to PBC, giving more precise estimates on short-term prognosis compared to a time-fixed procedure[60]. In a time-dependent Cox regression model later developed for PSC, age, bilirubin and albumin were independent predictors of prognosis, as was found in a time-fixed model in the same group of patients[45]. However, the prognostic information of bilirubin and albumin was much stronger in the time-dependent model. The time-dependent model was superior to the time-fixed variant in assigning low 1-year survival probabilities to patients who actually survived less than 1 year. Despite all efforts that have been put into elaboration of prognostic models in PSC as well as in PBC, the models have limited applicability in clinical routine. Hopefully, data that are more essential to disease development can be entered into the models in the future[61].

Cholangiocarcinoma in PSC

CC in PSC is particularly difficult to diagnose, carries a poor prognosis, and is usually considered a contraindication to liver transplantation. CC is a major cause of death in PSC. The risk of getting CC is particularly high in newly diagnosed cases of PSC; up to 30–50% of cases of CC have been diagnosed during the first year from diagnosis of PSC[34,39]. After the first year the incidence rate of CC is 1.5% per year[62]. Although duration of PSC before diagnosis is unknown, it appears that long-standing disease is not a prerequisite for development of malignancy. CC in PSC affects a rather young group of patients, with mean age at diagnosis of CC around 45 years[34,39]. Prognosis is very poor, with median survival of only 5–6 months. Results of liver transplantation in PSC patients with recognized CC are poor, due to a high recurrence rate[63]. However, in cases of incidental finding of CC in the explanted liver, and cases of early CC suspected on the basis of clinical, radiological, or laboratory findings pretransplantation, survival rates may be better than previously believed[48,64].

Prognosis in small-duct PSC

The clinical course of small-duct PSC has recently been described in three studies comprising a total of 83 patients[65–67]. These patients are characterized by having clinical, biochemical and histological features similar to PSC, but presenting with a normal cholangiogram. Both patients with and without IBD have been included in the entity. All three studies indicate that small-duct PSC has a more favourable prognosis than large-duct PSC. Among 33 small-duct PSC patients followed for a mean of 106 months, four patients (12%) underwent liver transplantation or died[66]. Survival was significantly better compared with the 260 large-duct PSC patients with similar follow-up; in this group of patients as many as 122 (47%) required transplantation or died. One of the small-duct PSC patients in these three studies developed hepatocellular carcinoma, but none developed CC. Repeat cholangiograms had been performed in 5/18–27/32 of the small-duct PSC patients, so potential development to large-duct PSC has not been fully investigated. From available data it can be estimated that about 15% of patients progress to large-duct PSC[68].

PRIMARY BILIARY CIRRHOSIS

PBC is a chronic, cholestatic liver disease that is characterized by a non-suppurative destruction of interlobular bile ducts, resulting in progressive ductopenia and fibrosis[69]. Eventually the liver injury progresses to cirrhosis and liver failure. Aetiology and pathogenesis of PBC are unknown, but evidence supports the involvement of immune mechanisms following stimulation by an environmental or infectious agent in genetically predisposed individuals[70-72]. There is an association of PBC to HLA DR8[70,73]. The primary susceptibility allele is DRB1*0801 in northern European white patients and DRB1*0803 in Japanese patients[70,72]. HLA polymorphisms, however, do not seem to be major determinants of susceptibility and clinical expression in PBC[72]. Approximately 95% of PBC patients have elevated serum levels of antimitochondrial antibodies (AMA). The role of AMA in pathogenesis of PBC remains unclear. PBC primarily affects women, with a female/male ratio of 9/1. Median age at diagnosis is approximately 50 years, but it may range from 20 years to 90 years[71,74]. Approximately 60% of patients were asymptomatic at diagnosis in the study by Prince et al.[75]. Currently, a majority of patients diagnosed with PBC have asymptomatic disease[72].

Survival in PBC

Disease progression in PBC varies considerably between patients[76]. In general the disease progresses slowly with gradual development of symptoms in patients who are asymptomatic at diagnosis. Thereafter PBC gradually progresses to cirrhosis and liver failure in a substantial proportion of patients[76,77]. Prognosis appears to differ between asymptomatic and symptomatic patients[75,78,79]. In a study comparing asymptomatic patients ($n = 36$) with symptomatic patients ($n = 243$), median survival from time of diagnosis was 16.0 years and 7.5 years, respectively[79]. Median follow-up was 12.1 years in the asymptomatic patients and 6.4 years in the symptomatic group. Most presymptomatic patients who are followed for a long duration will develop symptomatic disease[76]. Between 36% and 89% of asymptomatic patients become symptomatic during mean/median follow-up of 4.5–17.8 years[76]. Specific features at presentation in asymptomatic PBC that predicted subsequent outcome could not be identified in a study of 91 asymptomatic cases by Springer et al.[78]. The presence of other autoimmune diseases suggested subsequent disease progression in one report[80], but this observation could not be confirmed in another study[78].

Studies of asymptomatic patients have suggested that their survival is reduced compared with an age- and sex-matched population[78,80]. In a prospective study from the Mayo Clinic, 33 (89%) of 37 presymptomatic PBC patients developed one or more symptoms of liver disease after a median follow-up of 7.6 years[80]. On the other hand, patients who remain asymptomatic have been reported to have equivalent survival rates compared with a control population[78] or reduced survival that first becomes apparent after 11 years of follow-up[79]. Once previous asymptomatic patients develop symptoms, survival becomes equal to that of patients who are symptomatic at presentation[79].

Prognostic factors in PBC

As described above for PSC, several groups have investigated potential prognostic factors in their cohorts of PBC patients[60,81–85]. According to their relative impact, prognostic variables have then been combined in prognostic models. Serum bilirubin level is an independent variable predictive of survival in all groups of patients. Bilirubin is also the most heavily weighted variable in each of the prognostic models[86]. Rises in bilirubin levels above 102 μmol/L and 171 μmol/L have been associated with median survival of 25 months and 17 months, respectively[87]. Older age is an independent predictor of adverse outcome in most studies. Other variables appearing in the various prognostic models include hepatomegaly, fibrosis/cirrhosis, cholestasis, and complications of cirrhosis such as ascites, variceal bleeding and oedema[86]. Apparently, overall survival does not depend upon gender[88]. Piecemeal necrosis is a histological parameter predictive for the development of cirrhosis[89].

In a study from the Mayo Clinic, measurements of AMA titres were not useful for predicting the clinical progression of disease in individual PBC patients[90]. Neither did AMA, or particular AMA profiles, predict outcome in PBC in the study by Joshi et al.[91]. These results contrast to those of Klein et al.[92] who in a prospective study of 200 patients followed for 10 years found that anti-AMA profiles discriminated between a benign and a progressive course of PBC already at early stages. Züchner et al.[93] found that the occurrence of the ANA recognizing the Sp100 and promyelocytic leukaemia protein correlated with an unfavourable disease course. Reactivity to gp210, a particular PBC-specific ANA, has been associated with more serious liver disease[94].

An impact of HLA genes on prognosis has also been reported in PBC. In a study comprising 164 PBC patients Donaldson et al.[95] found that there was an increased frequency of the HLA DRB1*0801-DQA1*0401-DQB1*0402 haplotype in patients who had progressed to late-stage disease.

Prognostic models in PBC

As pointed out in the discussion of prognostic models for PSC (above), time-dependent Cox regression survival models also predict short-term survival in PBC better than time-fixed models for this disease[59,60,96]. Such models allow estimates of prognosis to be updated during the course of the disease. The Mayo risk score for prognosis of PBC is also applicable in patients receiving ursodeoxycholic acid[97].

Hepatocellular carcinoma in PBC

Several studies have indicated that hepatocellular carcinoma (HCC) complicates PBC, in particular late-stage PBC and males with PBC[76,98–101]. In a study comprising 667 PBC patients from the UK, 16 (5.9%) among the 273 patients with histological stage III or IV disease had HCC[98]. HCC was not seen in any of the 394 patients with stage I and II PBC followed over the same time period[98]. HCC was a relatively common cause of death in male PBC patients with cirrhosis[98]. Among 1692 PBC patients seen at the Mayo Clinic, 114 patients were identified with a primary cancer[99]. There was a relative risk of 46 for women and 55 for men for hepatobiliary malignancies, compared to expected rates of malignancies for the population at large[99]. In a population-based Swedish study, including 559 cases of PBC, overall cancer incidence was moderately increased[102]. However, the number of HCC did not differ significantly from expected[102]. HCC did not affect patient survival in the study by Shibuya et al.[101].

References

1. Bjøro K, Frimann S, Höckerstedt K et al. Liver transplantation in the Nordic countries, 1982–1998: changes of indications and improving results. Scand J Gastroenterol. 1999;34: 714–22.
2. Manns MP, Strassburg CP. Autoimmune hepatitis: clinical challenges. Gastroenterology. 2001;120:1502–17.
3. Czaja AJ, Freese DK. Diagnosis and treatment of autoimmune hepatitis. Hepatology. 2002; 36:479–97.
4. Alvarez F, Berg PA, Bianchi FB et al. International Autoimmune Hepatitis Group report: Review of criteria for diagnosis of autoimmune hepatitis. J Hepatol. 1999;31:929–38.
5. Manns MP, Krüger M. Genetics in liver diseases. Gastroenterology. 1994;106:1676–97.
6. Johnson PJ, McFarlane IG, Eddleston ALWF. The natural course and heterogeneity of autoimmune-type chronic active hepatitis. Semin Liver Dis. 1991;11:187–96.
7. Gordon SC. Diagnostic criteria, clinical manifestations and natural history of autoimmune hepatitis. In: Krawitt EL, Wiesner RH, Nishioka M, editors. Autoimmune Liver Diseases, 2nd edn. Amsterdam: Elsevier, 1998:343–60.
8. Sherlock S, Dooley J. Chronic hepatitis: general features, and autoimmune chronic disease. In: Sherlock S, Dooley J, editors. Diseases of the Liver and Biliary System, 11th edn. Oxford: Blackwell Publishing, 2002:321–33.
9. Czaja AJ. Natural history of chronic active hepatitis. In: Czaja AJ, Dickson ER, editors. Chronic Active Hepatitis, The Mayo Clinic Experience. New York: Marcel Dekker, 1986.
10. Bearn AG, Kunkel HG, Slater RG. The problem of chronic liver disease in young women. Am J Med. 1956;21:3–15.
11. Read AE, Sherlock S, Harrison CV. Active 'juvenile' cirrhosis considered as part of a systemic disease and the effect of corticosteroid therapy. Gut. 1963;4:378–93.
12. Mistilis SP, Blackburn CRB. The treatment of active chronic hepatitis with 6-mercaptopurine and azathioprine. Australas Ann Med. 1967;16:305–11.
13. Cook GC, Mulligan R, Sherlock S. Controlled prospective trial of corticosteroid therapy in active chronic hepatitis. Q J Med. 1971;40:159–85.
14. Soloway RD, Summerskill WHJ, Baggenstoss AH et al. Clinical, biochemical, and histological remission in severe chronic active liver disease: a controlled study of treatments and early prognosis. Gastroenterology. 1972;63:820–33.
15. Kirk AP, Jain S, Pocock S, Thomas HC, Sherlock S. Late results of Royal Free Hospital prospective controlled trial of prednisolone therapy in hepatitis B surface antigen negative chronic active hepatitis. Gut. 1980;21:78–83.
16. Czaja AJ. Drug therapy in the management of type 1 autoimmune hepatitis. Drugs. 1999; 57:49–68.
17. Roberts SK, Therneau TM, Czaja AJ. Prognosis of histological cirrhosis in type 1 autoimmue hepatitis. Gastroenterology. 1996;110:848–57.
18. Nikias GA, Batta KP, Czaja AJ. The nature and prognostic implications of autoimmune hepatitis with an acute presentation. J Hepatol. 1994;21:866–71.
19. Schalm SW, Korman MG, Summerskill WHJ, Czaja AJ, Baggenstoss AH. Severe chronic active liver disease. Prognostic significance of initial morphologic patterns. Am J Dig Dis. 1977;22:973–80.
20. Czaja AJ. Natural history, clinical features, and treatment of autoimmune hepatitis. Semin Liver Dis. 1984;4:1–12.
21. Schvarcz R, Glaumann H, Weiland O. Survival and histological resolution of fibrosis in patients with autoimmune chronic active hepatitis. J Hepatol. 1993;18:15–23.

22. Kanzler S, Gerken G, Löhr H, Galle PR, Meyer zum Büschenfelde K-H, Lohse AW. Duration of immunosuppressive therapy in autoimmune hepatitis. J Hepatol. 2001;34:354–5.

23. Kogan J, Sadafi R, Ashur Y, Shouval D, Ilan Y. Prognosis of symptomatic versus asymptomatic autoimmune hepatitis: a study of 68 patients. J Clin Gastroenterol. 2002;35:75–81.

24. Czaja AJ, Rakela J, Hay JE, Moore SB. Clinical and prognostic implications of HLA B8 in corticosteroid-treated severe autoimmune chronic active hepatitis. Gastroenterology. 1990;98:1587–93.

25. Doherty DG, Donaldson PT, Underhill JA et al. Allelic sequence variation in the HLA class II genes and proteins in patients with autoimmune hepatitis. Hepatology. 1994;19:609–15.

26. Czaja AJ, Strettell MD, Thomson LJ et al. Associations between alleles of the major histocompatibility complex and type 1 autoimmune hepatitis. Hepatology. 1997;25:317–23.

27. Donaldson PT, Doherty DG, Hayllar KM, McFarlane IG, Johnson PJ, Williams R. Susceptibility to autoimmune chronic active hepatitis: human leukocyte antigens DR4 and A1-B8-DR3 are independent risk factors. Hepatology. 1991;13:701–6.

28. Sanchez-Urdazpal L, Czaja AJ, van Hoek B, Krom RAF, Wiesner RH. Prognostic features and role of liver transplantation in severe corticosteroid-treated autoimmune chronic active hepatitis. Hepatology. 1991;15:215–21.

29. Czaja AJ, Cassani F, Cataleta M, Valentini P, Bianchi FB. Frequency and significance of antibodies to actin in type 1 autoimmune hepatitis. Hepatology. 1996;24:1068–73.

30. Czaja AJ, Donaldson PT, Lohse AW. Antibodies to soluble liver antigen/liver pancreas and HLA risk factors for type 1 autoimmune hepatitis. Am J Gastroenterol. 2002;97:413–19.

31. Homberg J-C, Abuaf N, Bernard O et al. Chronic active hepatitis associated with antiliver/kidney microsome antibody type 1: a second type of 'autoimmune' hepatitis. Hepatology. 1987;7:1333–9.

32. Gregorio GV, Portmann B, Reid F et al. Autoimmune hepatitis in childhood: a 20-year experience. Hepatology. 1997;25:541–7.

33. Boberg KM, Clausen OPF, Schrumpf E. Primary sclerosing cholangitis: diagnosis and differential diagnosis. In: Leuschner U, Broomé U, Stiehl A, editors. Cholestatic Liver Diseases. Therapeutic Options and Perspectives. Dordrecht: Kluwer Academic Publishers, 2004:203–17.

34. Broomé U, Olsson R, Lööf L et al. Natural history and prognostic factors in 305 Swedish patients with primary sclerosing cholangitis. Gut. 1996;38:610–15.

35. Donaldson PT. The HLA system and other genetic markers in primary sclerosing cholangitis. In: Leuschner U, Broomé U, Stiehl A, editors. Cholestatic Liver Diseases. Therapeutic Options and Perspectives. Dordrecht: Kluwer Academic Publishers, 2004:257–72.

36. Wiencke K, Spurkland A, Schrumpf E, Boberg KM. Primary sclerosing cholangitis is associated to an extended B8-DR3 haplotype including particular MICA and MICB alleles. Hepatology. 2001;34:625–30.

37. Wiesner RH, Grambsch PM, Dickson ER et al. Primary sclerosing cholangitis: natural history, prognostic factors and survival analysis. Hepatology. 1989;10:430–6.

38. Schrumpf E, Abdelnoor M, Fausa O, Elgjo K, Jenssen E, Kolmannskog F. Risk factors in primary sclerosing cholangitis. J Hepatol. 1994;21:1061–6.

39. Boberg KM, Bergquist A, Mitchell S et al. Cholangiocarcinoma in primary sclerosing cholangitis: risk factors and clinical presentation. Scand J Gastroenterol. 2002;37:1205–11.

40. Schrumpf E, Boberg KM. Primary sclerosing cholangitis and inflammatory bowel disease: when and how do they relate to each other? In: Leuschner U, Broomé U, Stiehl A, editors. Cholestatic Liver Diseases. Therapeutic Options and Perspectives. Dordrecht: Kluwer Academic Publishers, 2004:283–8.

41. Ismail T, Angrisani L, Powell JE et al. Primary sclerosing cholangitis: surgical options, prognostic variables and outcome. Br J Surg. 1991;78:564–7.

42. Aadland E, Schrumpf E, Fausa O et al. Primary sclerosing cholangitis: a long-term follow-up study. Scand J Gastroenterol. 1987;22:655–64.

43. Ponsioen CY, Vrouenraets SM, Prawirodirdjo W et al. Natural history of primary sclerosing cholangitis and prognostic value of cholangiography in a Dutch population. Gut. 2002;51:562–6.

44. Farrant JM, Hayllar KM, Wilkinson ML et al. Natural history and prognostic variables in primary sclerosing cholangitis. Gastroenterology. 1991;100:1710–17.
45. Boberg KM, Rocca G, Egeland T et al. Time-dependent Cox regression model is superior in prediction of prognosis in primary sclerosing cholangitis. Hepatology. 2002;35:652–7.
46. Helzberg JH, Petersen JM, Boyer JL. Improved survival with primary sclerosing cholangitis. A review of clinicopathologic features and comparison of symptomatic and asymptomatic patients. Gastroenterology. 1987;92:1869–75.
47. Bergquist A, Glaumann H, Persson B, Broomé U. Risk factors and clinical presentation of hepatobiliary carcinoma in patients with primary sclerosing cholangitis: a case–control study. Hepatology. 1998;27:311–6.
48. Brandsæter B, Isoniemi H, Broomé U et al. Liver transplantation for primary sclerosing cholangitis; predictors and consequences of hepatobiliary malignancy. J Hepatol. 2004;40: 815–22.
49. Dickson ER, Murtaugh PA, Wiesner RH et al. Primary sclerosing cholangitis: refinement and validation of survival models. Gastroenterology. 1992;103:1893–901.
50. Okolicsanyi L, Fabris L, Viaggi S et al. Primary sclerosing cholangitis: clinical presentation, natural history and prognostic variables: an Italian multicentre study. Italian PSC Study Group. Eur J Gastroenterol Hepatol. 1996;8;685–91.
51. Kim WR, Therneau TM, Wiesner RH et al. A revised natural history model for primary sclerosing cholangitis. Mayo Clin Proc. 2000;75:688–94.
52. Christensen E. Prognosis of untreated primary sclerosing cholangitis. In: Leuschner U, Broomé U, Stiehl A, editors. Cholestatic Liver Diseases. Therapeutic Options and Perspectives. Dordrecht: Kluwer Academic Publishers, 2004:231–40.
53. Wiesner RH. Liver transplantation for primary biliary cirrhosis and primary sclerosing cholangitis: predicting outcomes with natural history models. Mayo Clin Proc. 1998;73: 575–88.
54. Kim WR, Poterucha JJ, Wiesner RH et al The relative role of the Child–Pugh classification and the Mayo natural history model in assessment of survival in patients with primary sclerosing cholangitis. Hepatology. 1999;29:1643–8.
55. Shetty K, Rybicki L, Carey WD. The Child–Pugh classification as a prognostic indicator for survival in primary sclerosing cholangitis. Hepatology. 1997;25:1049–53.
56. Farrant JM, Doherty DG, Donaldson PT et al. Amino acid substitutions at position 38 of the DRβ polypeptide confer susceptibility to and protection from primary sclerosing cholangitis. Hepatology. 1992;16:390–5.
57. Mehal WZ, Lo Y-MD, Wordsworth BP et al. HLA DR4 is a marker for rapid disease progression in primary sclerosing cholangitis. Gastroenterology. 1994;106:160–7.
58. Boberg KM, Spurkland A, Rocca G et al. The HLA-DR3,DQ2 heterozygous genotype is associated with an accelerated progression of primary sclerosing cholangitis. Scand J Gastroenterol. 2001;36:886–90.
59. Christensen E. Prognostic models in chronic liver disease: validity, usefulness and future role. J Hepatol. 1997;26:1414–24.
60. Christensen E, Altman DG, Neuberger J, de Stavola BL, Tygstrup N, Williams R, the PBC1 and PBC2 trial groups. Updating prognosis in primary biliary cirrhosis using a time-dependent Cox regression model. Gastroenterology. 1993;105:1865–76.
61. Christensen E. Prognostic models including the Child–Pugh, MELD and Mayo risk scores – where are we and where should we go? J Hepatol. 2004;41:344–50.
62. Bergquist A, Ekbom A, Olsson R et al. Hepatic and extrahepatic malignancies in primary sclerosing cholangitis. J Hepatol. 2002;36:321–7.
63. Nashan B, Schlitt H, Tusch G et al. Biliary malignancies in primary sclerosing cholangitis: timing for liver transplantation. Hepatology. 1996;23:1105–11.
64. Goss JA, Shackleton CR, Farmer DG et al. Orthotopic liver transplantation for primary sclerosing cholangitis. A 12-year single center experience. Ann Surg. 1997;225:472–81.
65. Broomé U, Glaumann H, Lindström E et al. Natural history and outcome in 32 Swedish patients with small duct primary sclerosing cholangitis (PSC). J Hepatol. 2002;36:586–9.
66. Björnsson E, Boberg KM, Cullen S et al. Patients with small duct primary sclerosing cholangitis have a favourable long term prognosis. Gut. 2002;51:731–5.
67. Angulo P, Maor-Kendler Y, Lindor K. Small-duct primary sclerosing cholangitis: a long-term follow-up study. Hepatology. 2002;35:1494–500.

68. Olsson R. Small duct primary sclerosing cholangitis: a separate disease? In: Leuschner U, Broomé U, Stiehl A, editors. Cholestatic Liver Diseases. Therapeutic Options and Perspectives. Dordrecht: Kluwer Academic Publishers, 2004:225–30.
69. Kaplan MM. Primary biliary cirrhosis. N Engl J Med. 1996;335:1570–80.
70. Maeda T, Iwasaki S, Onishi S. Immunogenetic studies of primary biliary cirrhosis. In: Krawitt EL, Wiesner RH, Nishioka M, editors. Autoimmune Liver Diseases, 2nd edn. Amsterdam: Elsevier, 1998:167–78.
71. Talwalkar JA, Lindor KD. Primary biliary cirrhosis. Lancet 2003;362:53–61.
72. Bergasa NV, Mason A, Floreani A, et al. Primary biliary cirrhosis: report of a focus study group. Hepatology. 2004;40:1013–20.
73. Gores GJ, Moore SB, Fisher LD, Powell FC, Dickson ER. Primary biliary cirrhosis: associations with class II major histocompatibility complex antigens. Hepatology. 1987;7: 889–92.
74. Heathcote EJ. Management of primary biliary cirrhosis. Hepatology. 2000;31:1005–13.
75. Prince M, Chetwynd A, Newman W, Metcalf JV, James OFW. Survival and symptom progression in a geographically based cohort of patients with primary biliary cirrhosis: follow-up for up to 28 years. Gastroenterology. 2002;123:1044–51.
76. Kurtovic J, Riordan SM. Cinical aspects and prognosis of primary biliary cirrhosis. In: Leuschner U, Broomé U, Stiehl A, editors. Cholestatic Liver Diseases. Therapeutic Options and Perspectives. Kluwer Academic Publishers, Dordrecht, 2004:79–93.
77. Pasha TM, Dickson ER. Diagnostic criteria, clinical manifestations and natural history of primary biliary cirrhosis. In: Krawitt EL, Wiesner RH, Nishioka M, editors. Autoimmune Liver Diseases, 2nd edn. Amsterdam: Elsevier, 1998:361–79.
78. Springer J, Cauch-Dudek K, O'Rourke K, Wanless IR, Heathcote EJ. Asymptomatic primary biliary cirrhosis: a study of its natural history and prognosis. Am J Gastroenterol. 1999;94:47–53.
79. Mahl TC, Shockcor W, Boyer JL. Primary biliary cirrhosis: survival of a large cohort of symptomatic and asymptomatic patients followed for 24 years. J Hepatol. 1994;20:707–13.
80. Balasubramaniam K, Grambsch PM, Wiesner RH. Lindor KD, Dickson ER. Diminished survival in asymptomatic primary biliary cirrhosis: a prospective study. Gastroenterology. 1990;98:1567–71.
81. Roll J, Boyer JL, Barry D, Klatskin G. The prognostic importance of clinical and histologic features in asymptomatic and symptomatic primary biliary cirrhosis. N Engl J Med. 1983; 308:1–7.
82. Christensen E, Neuberger J, Crowe J et al. Beneficial effect of azathioprine and prediction of prognosis in primary biliary cirrhosis. Final result of an international trial. Gastroenterology. 1985;89:1084–91.
83. Dickson ER, Grambsch PM, Fleming TR, Fisher LD, Langworthy A. Prognosis in primary biliary cirrhosis: model for decision making. Hepatology. 1989;10:1–7.
84. Rydning A, Schrumpf E, Abelnoor M, Elgjo K, Jenssen E. Factors of prognostic importance in primary biliary cirrhosis. Scand J Gastroenterol. 1990;25:119–26.
85. Goudie BM, Burt AD, MacFarlane GJ et al. Risk factors and prognosis in primary biliary cirrhosis. Am J Gastroenterol. 1989;84:713–16.
86. Wiesner RH, Porayko MK, Dickson ER et al. Selection and timing of liver transplantation in primary biliary cirrhosis and primary sclerosing cholangitis. Hepatology. 1992;16:1290–9.
87. Shapiro JM, Smith H, Schaffner F. Serum bilirubin: a prognostic factor in primary biliary cirrhosis. Gut. 1979;20:137–40.
88. Christensen E, Crowe J, Doniach D et al. Clinical pattern and course of disease in primary biliary cirrhosis based on an analysis of 236 patients Gastroenterology. 1980;78:236–46.
89. Corpechot C, Carrat F, Poupon R, Poupon R-E. Primary biliary cirrhosis: incidence and predictive factors of cirrhosis development in ursodiol-treated patients. Gastroenterology. 2002;122;652–8.
90. Van Norstrand MD, Malinchoc M, Lindor KD et al. Quantitative measurement of autoantibodies to recombinant mitochondrial antigens in patients with primary biliary cirrhosis: relationship of levels of autoantibodies to disease progression. Hepatology. 1997; 25:6–11.

91. Joshi S, Cauch-Dudek K, Heathcote J, Lindor K, Jorgensen R, Klein R. Antimitochondrial antibody profiles: are they valid prognostic indicators in primary biliary cirrhosis? Am J Gastroenterol. 2002;97:999–1002.

92. Klein R, Pointner H, Zilly W et al. Antimitochondrial antibody profiles in primary biliary cirrhosis distinguish at early stages between a benign and a progressive course: a prospective study on 200 patients followed for 10 years. Liver. 1997;17:119–28.

93. Züchner D, Sternsdorf T, Szostecki C, Heathcote EJ, Cauch-Dudek K, Will H. Prevalence, kinetics and therapeutic modulation of autoantibodies against Sp100 and promyelocytic leukemia protein in a large cohort of patients with primary biliary cirrhosis. Hepatology. 1997;26:1123–30.

94. Muratori P, Muratori L, Ferrari R et al. Characterization and clinical impact of antinuclear antibodies in primary biliary cirrhosis. Am J Gastroenterol. 2003;98:431–7.

95. Donaldson P, Agarwal K, Craggs A, Craig W, James O, Jones D. HLA and interleukin 1 gene polymorphisms in primary biliary cirrhosis: associations with disease progression and disease susceptibility. Gut. 2001;48:397–402.

96. Murtaugh PA, Dickson ER, Van Dam GM et al. Primary biliary cirrhosis: prediction of short-term survival based on repeated patient visits. Hepatology. 1994;20:126–34.

97. Angulo P, Lindor KD, Therneau TM et al. Utilization of the Mayo risk score in patients with primary biliary cirrhosis receiving ursodeoxycholic acid. Liver. 1999;19:115–21.

98. Jones DEJ, Metcalf JV, Collier JD, Bassendine MF, James OFW. Hepatocellular carcinoma in primary biliary cirrhosis and its impact on outcomes. Hepatology. 1997;26:1138–42.

99. Nijhawan PK, Therneau TM, Dickson ER, Boynton J, Lindor KD. Incidence of cancer in primary biliary cirrhosis: the Mayo experience. Hepatology. 1999;29:1396–8.

100. Caballería L, Parés A, Castells A, Ginés A, Bru C, Rodés J. Hepatocellular carcinoma in primary biliary cirrhosis: similar incidence to that in hepatitis C virus-related cirrhosis. Am J Gastroenterol. 2001;96:1160–3.

101. Shibuya A, Tanaka K, Miyakawa H et al. Hepatocellular carcinoma and survival in patients with primary biliary cirrhosis. Hepatology. 2002;35:1172–8.

102. Lööf L, Adami H-O, Sparen P et al. Cancer risk in primary biliary cirrhosis: a population-based study from Sweden. Hepatology. 1994;20:101–4.

23
Transplantation for autoimmune liver diseases

J. NEUBERGER

INTRODUCTION

The three major autoimmune liver diseases are primary biliary cirrhosis (PBC), primary sclerosing cholangitis (PSC) and autoimmune hepatitis (AIH). Despite the widespread use of ursodeoxycholic acid (UDCA) for the treatment of PBC and PSC, and corticosteroids and azathioprine for AIH, some patients progress to end-stage liver disease or develop symptoms making life intolerable. For these patients liver transplantation may be indicated. In this chapter the indications and contra-indications for liver transplantation will be discussed, together with the outcomes, postoperative complications and evidence for disease recurrence.

INDICATIONS FOR TRANSPLANTATION

Liver transplantation is indicated for patients with liver disease either because of a quality of life, because of the liver disease, that is unacceptable to the patient, or because the anticipated survival in the absence of transplantation is estimated to be less than 1 year[1].

Quality of life

Assessing the quality of life for an individual patient is difficult, and the clinician has to distinguish poor quality of life as a consequence of liver disease from other causes of poor quality, such as depression, side-effects of treatment and other treatable conditions. Since autoimmune liver diseases are associated with other autoimmune diseases, it is important to exclude diseases such as thyroid disease, Addison's disease and gluten-sensitive enteropathy. Antidepressants are usually ineffective and ondansetron may be of benefit[2].

Length of life

Although most transplant centres consider transplantation when the antici-pated length of life is less than 1 year, this figure is arbitrary and more detailed analysis of the risks and benefits of transplantation and surgical non-interven-tion is required to refine the decision-making process. In practice, however, the indications for transplantation are relatively well defined and these are summarized in Table 1. Once hepatic decompensation occurs the short-term mortality starts to increase. The decision to place a patient on the transplant list will depend on many factors, summarized in a recent consensus document[3]. The development of prognostic models to give a quantitative estimate of short term survival has allowed a more objective assessment to be made; in particular, the adoption of the MELD (Model for End-Stage Liver Disease) has dramatically altered the distribution of liver allografts in North America (see below), with a reduction in the mortality of patients on the waiting list but no significant effect on outcome.

Table 1   General indications for transplantation for chronic parenchymal disease

Serum bilirubin
  > 180 µmol/L for > 6 months in cholestatic liver disease
  > 50 µmol/L for non-cholestatic liver disease
Serum albumin 30g/L
Ascites resistant to treatment
Spontaneous bacterial peritonitis
Hepatic encephalopathy
Progressive muscle wasting*
Recurrent variceal bleeding
Development of primary liver cell cancer
Intractable pruritus/lethargy
Progressive osteopenia*
Developing hepatopulmonary syndrome*
Early pulmonary hypertension
MELD score > 15

*These may also, paradoxically, be contraindications to liver replacement.

CONTRAINDICATIONS TO TRANSPLANTATION

In general, these contraindications can be divided into those conditions which preclude good survival after surgery or those conditions which make it unlikely the patient will survive surgery, and for these the patients are referred to recent reviews. A summary is given in Table 2. Contraindications need regular review: for example, the advent of effective antiretroviral agents means that transplan-tation of some HIV-infected patients is now possible[1].

Table 2 Contraindications for liver transplantation (although these will vary to some extent between centres, and few are absolute)

Extensive portal and mesenteric vein thrombosis
Active bacterial/fungal/viral infection
Extrahepatic primary malignancy
Primary liver cell cancer extending outside the liver
Cholangiocarcinoma
Severe extrahepatic disease (cardiac, respiratory, neurological) preventing successful transplantation or rehabilitation

TIMING OF TRANSPLANTATION AND USE OF PROGNOSTIC MODELS

The timing of transplantation is one of the more difficult areas in transplantation practice: transplantation too late in the course of the disease may increase the risk of the patient dying from the procedure, whereas transplantation too early may reduce the patient's life because of the small but finite risk of the patient dying from the surgery. Thus choosing the optimal time of transplantation depends on understanding the risks and benefits of surgery and no surgery at any time during the course of the patient's illness.

The development of prognostic models has given the clinician some guidance for selecting the optimal time for surgery. However, these models must be used with caution: estimated survival probabilities have a relatively wide confidence interval so extrapolation to the individual should be done in the light of the clinical situation. Furthermore, introduction of new therapies may invalidate the models.

Disease non-specific models

The MELD score was introduced to predict short-term survival in patients undergoing transjugular intrahepatic shunts; one of the strengths of the model is that its components are laboratory analytes (serum creatinine, prothrombin time and bilirubin) and, in contrast to the Child–Turcotte–Pugh model, not dependent on clinical inputs such as ascites or encephalopathy. The MELD model has been well validated but does not take into account complications of liver disease, such as the onset of hepatocellular carcinoma or hepatopulmonary syndrome, that may affect survival independently of liver function; furthermore, different laboratory methods of measuring the component analytes can affect the estimated survival[4]. Nonetheless, the use of MELD has proven to be robust and reliable[5].

The Child–Turcotte–Pugh model is an effective guide to prognosis but does not offer a quantitative estimate of survival probabilities and is dependent on clinical observations.

Disease-specific models

These models have been well developed for PBC[6-12] and less so for PSC[13-15]. There are very few prognostic models for patients with AIH, and indications for transplantation follow those for general parenchymal liver disease. These models compare well with each other[16]. In these models the risks of surviving in the absence of transplantation and the probability of surviving after transplantation can be estimated for an individual patient at any time during the course of the illness.

Prognostic models for PBC

Shapiro and colleagues[17] first emphasized the importance of serum bilirubin as a prognostic marker for PBC; there are now several models that have been described (Table 3). While serum bilirubin remains the major prognostic factor, the presence of cirrhosis, age and serum albumin are also important prognostic factors. The Mayo Clinic model, when applied to patients with PBC treated with UDCA for 6 months, still accurately predicts outcome[18].

Table 3 Prognostic factors in different models for prognosis PBC

	Age	Variceal bleeding	Ascites	Bilirubin	Pro-thrombin time	Hepato-megaly	Fibrosis/cirrhosis	Cholestasis	Mallory's hyaline	Albumin
European	+			+			+	+		+
Mayo	+		(+)	+	+					+
Yale	+			+		+	+			
Australia	+			+						+
Glasgow	+	+	+	+			+	+	+	
Oslo		+		+						
Royal Free	+		+	+						+

(+), oedema.

Other prognostic variables have been described: these include serum hyaluronic acid[19], procollagen peptide type III[20] and dynamic liver function tests such as the lignocaine metabolism (MEGX test), indocyanine green clearance, caffeine clearance and galactose elimination capacity[21,22]. Although many of these markers and tests have been shown to correlate well with survival, none has yet achieved widespread use in clinical practice.

Prognostic models for PSC

Some of the models for PSC are defined in Table 4[23]. In contrast to PBC the natural history of PSC is more variable, with a more fluctuating course which may in part depend on the presence of coincidental cholangitis. Selection and timing of transplantation in patients with PSC is associated with two additional problems: the presence of inflammatory bowel disease and the probability of cholangiocarcinoma.

Table 4 Prognostic factors in different models for prognosis in PSC

	Age	Bilirubin	IBD	Hepato-megaly	Spleno-megaly	Alkaline phosphatase	Histo-logical stage	Haemo-globin
Mayo	+	+	+				+	+
King's	+			+	+	+	+	
Multicentre	+	+			+		+	

IBD, inflammatory bowel disease.

In many patients with PSC there is coexisting inflammatory bowel disease, usually ulcerative colitis. The colitis is often mild and may be subclinical. There is an increased risk of colonic neoplasia, usually in the right side of the colon[24]; the risk is reduced by treatment with UDCA. It is inappropriate to transplant patients with active colitis although, once this is treated, transplantation can be undertaken. In general, non-transplant surgical intervention on the biliary tree in patients with PSC is unhelpful, and may indeed increase the morbidity of transplantation[25]; where there is a single dominant extrahepatic biliary stricture, stenting and/or dilation may be beneficial.

The major diagnostic problem in patients with PSC is the increased risk of cholangiocarcinoma. Many studies have shown convincingly that the presence of cholangiocarcinoma has a significant adverse effect on survival after transplantation[24]. Smoking seems to be a risk factor for hepatobiliary malignancy[26]. In most cases the cholangiocarcinoma develops insidiously and it is often very difficult to diagnose this condition pretransplant. Serum markers for cholangiocarcinoma, such as CEA and CA 19-9, lack the necessary sensitivity and specificity required for a certain diagnosis of cholangiocarcinoma[27]. These tumours are often difficult to visualize on imaging whether by ERCP, PTC, ultrasound, CT scanning or MRI scanning. We have found that biopsy of any possible stricture is highly specific but not very sensitive. Although brushings or bile cytology may be of value in this instance, it is often difficult to obtain satisfactory brushings of hilar lesions in those with extensive extrahepatic sclerosing cholangitis. Furthermore, there is a significant risk of inducing biliary sepsis, which may be fatal. A high clinical suspicion and, where appropriate, frozen section histology during the laparotomy preceding hepatectomy, will reduce the number of liver replacements in this condition. The Mayo Clinic has recently described excellent results in a very selected group of patients, where chemotherapy and brachytherapy will allow successful transplantation[28].

Survival after transplantation

Survival rates after liver transplantation have improved over time, with those grafted for PBC having the best outcomes[29] (Figure 1); those with PBC have the best outcome but the reasons for this are not clear and persist even when allowance is made for case mix.

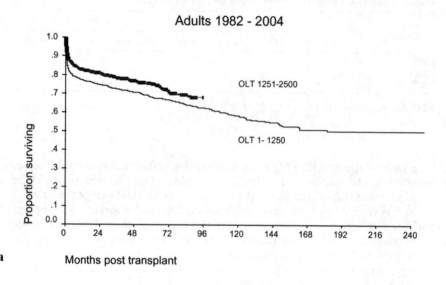

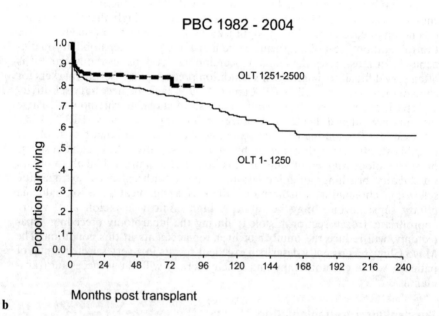

Figure 1 (and opposite) Overall survival of patients transplanted in Birmingham, UK. **(a)** = overall survival; **(b)** = patients grafted for PBC; **(c)** = patients grafted for AIH; **(d)** = patients grafted for PSC (courtesy of Bridget Gunson)

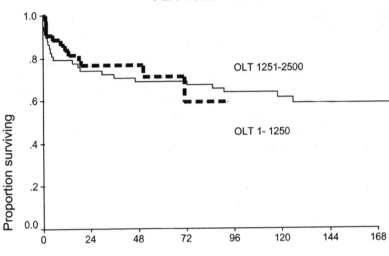

AIH 1982 - 2004

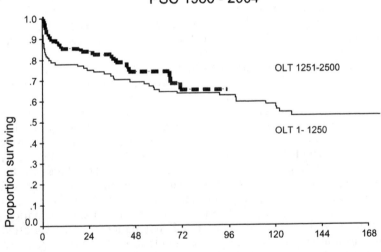

PSC 1986 - 2004

Prediction of survival after transplantation

To predict survival after transplantation is of value for several reasons: not only does this allow for identification of the risk factors predicting survival but also, by comparing prognosis with and without transplantation, it allows the optimal timing of transplantation to be determined. Those factors which predict survival after transplantation are not necessarily those that predict survival without transplantation. Wiesner et al.[30] compared the predictive accuracy of the Child–Pugh score with the Mayo risk score for PSC (developed to predict survival in the absence of transplantation) for survival and resource utilization after transplantation. They concluded that the Child–Pugh was a better overall predictor of outcome and resource utilization after transplantation.

PSC. Univariate analysis of 118 patients grafted at the Liver Unit, Birmingham, and followed for up to 9 years, identified seven variables associated with a poor outcome after surgery[31]: high serum creatinine, high serum bilirubin, biliary tree malignancy, previous upper abdominal surgery, hepatic encephalopathy, ascites and coexisting Crohn's disease. In contrast, coexisting ulcerative colitis had a beneficial association with outcome. The final Cox model included inflammatory bowel disease, ascites, previous upper abdominal surgery, high serum creatinine and biliary tree malignancy. The model, validated in another group of patients with PSC, grafted in another centre, may help in understanding the timing of the procedure, and suggests that upper abdominal surgery pretransplant should, if possible, be avoided. The reasons for the discordant association of ulcerative colitis and Crohn's disease are uncertain.

PBC: Assessment of those variables associated with a poor, short-term survival after transplantation for PBC identified earlier year of transplant, ascites (with or without diuretic treatment), low serum albumin, high serum bilirubin, high serum urea and high serum creatinine, low serum sodium and rhesus negativity as significant factors[32]. In the final Cox regression model, year of transplant, albumin/age ratio, plasma bilirubin and ascites or diuretic treatment were the only factors included. Several of these factors also predict outcome in the absence of transplantation, although the prognostic value differs in the two models. By assessing the 6-month survival probability with and without transplantation, it can be clearly shown that there is a benefit of transplantation when the 6-month survival probability in the absence of transplantation falls below 0.85. However, it must be noted that patients may be grafted for conditions other than end-stage disease, and that patients should be listed at a time which will also reflect the likely length of wait for a liver to become available.

Survival after transplantation for PBC

Several studies have confirmed that the actual, observed survival after transplantation is superior to the length of life predicted from prognostic models[33] (Figure 2). As will be seen, survival after transplantation has

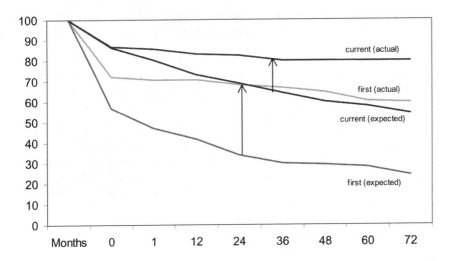

Figure 2 Survival of the first 400 patients following liver transplantation for PBC in Birmingham: the actual survival of each of the two cohorts (first and current) after transplantation (actual) and expected in the absence of transplantation (expected) is shown

improved over time: however, comparison of actual survival after transplantation with anticipated survival in the absence of transplantation suggests that part of this improvement is because less sick patients are grafted.

Survival after transplantation for PSC

The survival of patients transplanted for PSC is lower than for PBC[29] (see Figure 1). The reasons for this difference are not clear, and do not relate solely to the associated cholangiocarcinoma.

Survival after transplantation for AIH

Survival after transplantation for patients with AIH is broadly similar to that seen after transplantation for other autoimmune liver diseases[29] (Figure 1).

Rejection post-transplant

Several studies have shown that patients with autoimmune liver disease are at greater risk of both acute and chronic allograft rejection[34]. For example, in our own series, the incidence of severe acute allograft rejection at 7 days in patients grafted for PBC was 63%, for AIH 61% and for PSC 50%. In contrast, for patients grafted for liver disease where immune mechanisms were not implicated in the pathophysiology, acute rejection was found in 42% of those with alcoholic liver disease and in 37% of those with alcohol-related liver disease.

However, there is no difference in graft survival in patients with and without early acute graft rejection. Early acute rejection does not adversely impact on outcome.

In contrast, ductopenic rejection is associated with graft loss: although there are many factors which are associated with ductopenic rejection, it does seem that all three autoimmune diseases are risk factors for graft loss from chronic rejection[34]. It remains unclear whether the immunosuppressive protocol used should be different according to the indication for transplant.

It remains unclear why there is a higher incidence of acute and chronic rejection in patients grafted for autoimmune diseases. Farges et al.[35] suggested this may be due to pretransplant immunoglobulin protecting the patient from rejection, but this could not be confirmed in our series and does not explain why those with a toxic cause of acute liver failure have a lower risk than those with autoimmune disease.

Quality of life after transplantation

Now that many of the technical problems of transplantation have been overcome, clinicians are starting to assess the quality of life after transplantation. While it can never be stated to be normal, it is better than no life at all. Quality of life improves[36] but the quality of life after transplantation does not correlate with pretransplant state of health.

Inflammatory bowel disease (IBD) after transplantation

PSC is associated with IBD: the course of IBD in patients with PSC after transplantation is variable. Riley et al.[37] described the development of IBD in 14 of 6800 liver and kidney allograft recipients; all but two had received a liver graft, but of these 12 only two were grafted for PSC, the remainder for a variety of indications including four with autoimmune hepatitis. In a study of 30 patients, Papatheodoridis et al.[38] found that in half the colitis remained the same after transplantation, and in the other half the colitis deteriorated. Colitis developed *de novo* in three patients and none had developed colonic cancer. It is unknown why IBD will develop in patients already receiving immunosuppressive drugs which, in the ungrafted patient, are effective therapies. While Haagsma et al.[39] found that immunosuppression affected the prevalence of IBD after transplantation: tacrolimus had a promoting effect and azathioprine a protective effect. Our own studies (Vera, unpublished) have not confirmed this, with no observed effect of immunosuppression on the pattern of bowel disease post-transplantation.

RECURRENCE OF AUTOIMMUNE LIVER DISEASE AFTER TRANSPLANTATION

If autoimmune disease is defined in the most simplistic way as an immune response against self-antigens, then in patients following transplantation the immune response will be against the donor alloantigen, on bile duct cells or hepatocytes which retain the donor phenotype. Thus, a more appropriate term would be 'alloimmune disease'.

Recurrent autoimmune disease has been suggested after transplantation of other organs, such as recurrent sarcoidosis in the lung or liver[40] and auto-immune diabetes after pancreas transplantation[41]. The diagnosis of recurrent autoimmune disease in the allograft is often difficult since there will be several confounding factors which may render invalid those diagnostic criteria which apply to the native liver.

Immunosuppression

After transplantation, patients are maintained on immunosuppressive therapy using drugs which may also be used for the treatment of autoimmune disease and may mask or modify the features of recurrent disease.

Rejection

The graft usually remains at risk of rejection and it is rare for an allograft biopsy to be devoid of infiltrating leucocytes. There may be overlap between the histological features of rejection from recurrent disease.

Other causes of graft damage

The allograft may be affected by factors which are not found in the native liver and may mimic features of recurrent disease. Such factors include infections (bacterial, viral or protozoan), ischaemia, preservation damage, reperfusion damage and drug toxicity.

De-novo autoimmunity

De-novo autoimmune features has been described after transplantation[42]. This has since been defined by the following:

1. Elevation of serum transaminases.

2. Interface hepatitis.

3. Elevated serum IgG.

4. Autoantibodies: anti-smooth muscle, antinuclear, typical or atypical anti-LKM and antimitochondrial.

5. Response to treatment with prednisolone 60 mg/day.

De-novo autoimmune hepatitis was reported by Kerkar and colleagues[43], who showed that seven of 200 paediatric liver allograft recipients have developed, after transplantation, autoimmune liver disease associated with elevated titres of autoantibodies, elevated serum immunoglobulin G and histological features of chronic active hepatitis. The indications for transplantation were not those in which autoimmunity is believed to be implicated (such as

257

extrahepatic biliary atresia or Alagille's syndrome). These autoimmune features, which became apparent at a median of 24 months after transplantation, responded well to increased immunosuppression. Heneghan and colleagues identified three adult liver allograft recipients who developed features of *de-novo* autoimmune hepatitis after transplantation for ecstasy overdose, HCV cirrhosis and PSC: in all three cases there was a significant transaminitis, liver/kidney microsomal antibodies at a titre greater than 1 in 640 and portal and periportal hepatitis on liver histology. Markers of inflammation resolved on treatment with high-dose corticosteroids[44]. Many others have described de novo AIH but not in all cases did the inflammation resolve with increased corticosteroids[45].

The implications of *de-novo* AIH are unclear: whilst it is clear that transplantation can trigger autoimmunity, it may be that the immune response is towards a donor antigen: Aguilera and colleagues have shown that these patients have antibodies to glutathione *S*-transferase T1, suggesting that the immune response is directed against the donor GSTT1 that was absent in the host; this supports the concept that this is in reality a form of rejection[46].

Primary sclerosing cholangitis

In the native liver the diagnosis of PSC is made on a combination of clinical, serological, pathological and radiological features. The clinical, biochemical and immunological (including the antineutrophil cytoplasmic antibody (ANCA)) are not specific for PSC. Although the histological features of fibrous cholangitis and fibro-obliterative lesions are characteristic of the condition, these are found in less than half the cases of PSC on needle biopsies and may be found in other conditions affecting the biliary tree. The definitive diagnosis of sclerosing cholangitis is made by imaging the biliary tree showing the characteristic dilation and strictures affecting intrahepatic and/or extrahepatic biliary tree. Imaging will not, however, differentiate primary from secondary biliary cirrhosis, so the confirmation of PSC is made by excluding the secondary causes. Many of the causes of secondary biliary cirrhosis will be present in the allograft. For example, most liver allograft recipients have Roux-en-Y loops fashioned at the time of surgery, which is associated with cholangitis and thus sclerosing cholangitis.

Sebagh and colleagues[47] studied 24 patients who developed non-anastomotic strictures after transplantation. The histological diagnosis of sclerosing cholangitis was made on the basis of periductal fibrosis and large-duct obstruction, which were present in 23 of the 24 cases. In all instances there were histological features present in the first postoperative year; cholangiographic features of sclerosing cholangitis were detected in all but two in the first year (median 3.5 months). Of these, 10 patients had ABO-incompatible grafts; 12 patients having hepatic artery thrombosis, three focal fibro-intimal hyperplasia, four preservation-related ischaemia and three arteriopathic chronic ductopenia rejection (some patients had more than one possible risk factor). Thus, after transplantation, cholangitis may be attributed to several causes. Patel et al.[48] analysed 10 patients who underwent regrafting for non-anastomotic biliary strictures. Recurrence of biliary strictures requiring intervention was significantly more common in those regrafted for biliary strictures than those re-grafted for other indications (50% compared with 11% respectively). Similar findings were recently reported from Oslo[49], where the prevalence of biliary strictures was investigated in 50 patients transplanted for PSC and 45 controls, using magnetic resonance cholangiography: biliary strictures were reported in 38% of PSC-transplanted patients and 9% of controls. While other possible causes for biliary strictures were seen in nine (18%) in those grafted for PSC, acute rejection was associated with recurrent PSC.

It remains uncertain whether ANCA are markers of PSC or are implicated in the pathogenesis of the disease. Haagsma and colleagues[50] showed that titres of p-ANCA fall in the first few months after transplantation but thereafter rise so that, after 1 year, titres are similar to those found prior to surgery. In the nine patients studied between 3 and 48 months after transplantation there was no histological evidence of PSC recurrence. Our own studies have reached broadly similar conclusions, although in two ANCA became negative after transplantation.

We therefore assessed the histological features of PSC in 22 patients grafted for PSC who had survived more than 1 year with 22 patients who had been grafted for other indications but had a Roux-loop[51] (Figure 3). None of the patients had an ABO-incompatible graft, hepatic artery thrombosis or arteriopathic or ductopenic rejection. Features of biliary obstruction were seen in 10–32% of patients. Periductal fibrosis was observed more commonly in those grafted for PSC than either the Roux-control group or the general allograft population. Fibro-obliterative lesions were found in only three of the patients grafted for PSC but in none of the others. The Mayo Clinic[52] found features

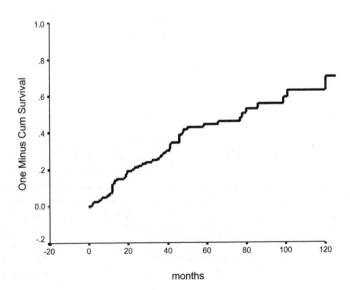

Figure 3 Rate of recurrence of PSC

suggestive of recurrent PSC in 24 (20%) of 120. In 11 patients histological features compatible with recurrent PSC were reported. The median time to the diagnosis of recurrent disease was 421 days if the diagnosis was based on cholangiography and 42 months if based on histology. Five patients with recurrent PSC died and two underwent regrafting. In the control group only one of 464 had histological features of recurrent PSC. More recently, in an analysis of 380 patients grafted for PSC in Pittsburgh and followed up for a median of 6 years, Abu-Elmagd and colleagues[53] reported recurrence of PSC in 52 grafts (13.7%) at a mean onset of 4 years after transplantation. The diagnosis of recurrence was made either on imaging the biliary tree or on histology. The probability of recurrence increased with time: 3.6% at 1 year, 12% at 5 years, 25% at 10 years. Factors associated with recurrence included older recipient age and the presence of inflammatory bowel disease.

Not all studies have confirmed these observations. Haagsma found no histological features of recurrent PSC in the allograft in nine patients[37]. However, in a larger study, Sheng et al.[54] analysed the cholangiograms of 643 liver allograft recipients with a choledochojejunostomy. Overall, in those transplanted for conditions other than PSC, 13% had intrahepatic biliary strictures, whereas 27% of those grafted for PSC had such strictures. When those with hepatic artery thrombosis were excluded, intrahepatic strictures were found in 18% of those grafted for PSC compared with 9% in the comparison group. That there was no statistically significant difference between the two groups (20 of 112 and 53 of 575 respectively) may be a consequence of small numbers rather than a true difference. Others have reported a greater incidence of biliary complications in patients grafted for PSC than for other indications[55], but in both series the numbers of patients were relatively small.

Thus, in conclusion, it is likely that PSC recurs in the allograft but uncertainty remains because of the difficulties in differentiating primary recurrent sclerosing cholangitis from secondary sclerosing cholangitis.

Primary biliary cirrhosis (PBC)

As with PSC, the diagnosis of PBC in the native liver is made on the basis of the clinical, biochemical, immunological and histological features. The presence of antimitochondrial antibodies reacting with the components of the pyruvate dehydrogenase complex is virtually diagnostic of the condition. However, in the allograft, demonstration of these antibodies may not imply disease recurrence, but merely persistence of the immune or other abnormalities which allow development of the disease. After transplantation there is a small and transient fall in the titre of antimitochondrial antibody which is followed by a persistence of the AMA with similar but not always identical antigen recognition; other PBC-specific autoantibodies, Sp100 and gp210, also persist after transplantation but do not correlate with histological evidence of disease recurrence[56]. The persistence or *de-novo* development of the autoimmune conditions associated with PBC (such as thyroid disease or sicca syndrome) do not necessarily imply disease recurrence.

The possibility of disease recurrence has generated much controversy since the initial report of three patients with clinical and histological features of PBC in the allograft[57]. There is, however, consensus that the diagnosis of PBC recurrence can be made only on the basis of histology, which in the native liver is best described as a non-suppurative destructive cholangitis or a granulomatous cholangitis. While granulomas around the bile ducts are strongly suggestive of recurrent PBC, granulomatous infiltration may be associated with a variety of causes including infection (viral, especially recurrent HCV, microbial, especially TB, foreign-body reaction) but granulomas occurring after 1 year are most commonly associated with recurrent PBC[58].

The initial report[57] suggesting recurrence of PBC was followed by a number of other studies which have confirmed these observations. Not all studies, however, have confirmed these findings[59]. Features of early recurrence may have been masked by features of rejection or other complications. The study of Gouw and co-workers[60] involved careful prospective histological evaluation of 19 patients grafted for PBC for a period of up to 11 years (median 5). Although the AMA were positive, the authors were unable to find any convincing evidence of recurrent PBC and could find no other serological or histological differences between patients grafted for PBC and those grafted for other indications.

Some centres (including our own in Birmingham) which reported in preliminary studies that there was no evidence for recurrent PBC have later reached different conclusions. There are several explanations which may account for these revisions:

Numbers and duration of follow-up

Our preliminary study from Birmingham on 19 patients grafted for PBC and who had survived for more than 1 year led us to conclude that there was no evidence for PBC recurrence[61]; however, when more patients were evaluated over a longer period these conclusions were reversed[62].

Immunosuppression

Some patients grafted for PBC were included in a prospective, randomized study comparing cyclosporin-based immunosuppression with tacrolimus-based immunosuppression. When the 1- and 2-year post-transplant biopsies were studied, it was found that those receiving tacrolimus-based immunosuppression had histological evidence of PBC recurrence not only earlier but more severely (Figure 4). These findings have been confirmed by a larger retrospective study and by others[62]. The median time to the diagnosis of recurrence was 62 months for those on tacrolimus and 123 months for those on cyclosporin. A study from Pittsburgh[63], contrary to previous reports, found evidence of recurrent PBC; histological features of PBC were found in 7.9% of patients at 5 years after liver transplantation and in 21.6% at 10 years. Risk factors for recurrence included recipient age, cold ischaemia age and immunosuppression.

Survival Functions

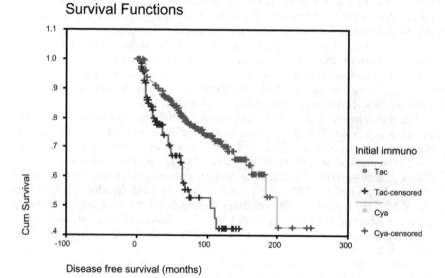

Figure 4 Effect of immunosuppression on PBC recurrence

Implications of PBC recurrence

The possibility of disease recurrence has important implications for the pathogenesis of the disease: AMA and other PBC-specific autoantibodies persist after transplantation but have not been shown to be pathogenic. Biliary epithelial cells from the native livers of patients with PBC, but not with other conditions, have plasma membrane expression of the antigen recognized by AMA. We studied the grafts of patients with and without histological evidence of recurrent PBC with an affinity-purified antibody to the E2 component of pyruvate dehydrogenase[64] and were unable to find any evidence of aberrant antigen expression in any of the groups studied. However, Van der Water et al.[65] used a different antibody (a monoclonal antibody which reacts with the inner lipoyl domain of PDC-E2) and found that all sections from patients grafted for PBC showed a pattern of staining identical to that seen in the native PBC liver, whether or not there was evidence of disease recurrence.

These results raise the possibility that there is some factor (possibly infective or toxic) which is associated with the development of aberrant antigen expression on biliary epithelial cells and sets the scene for cell damage. The other implications are that immunosuppressive therapy may be of limited value in prevention or possibly progression of the disease process.

Autoimmune hepatitis

The diagnosis of autoimmune hepatitis (AIH) is made in the patient with a native liver on the basis of internationally agreed criteria which include clinical, biochemical and histological features. The combination of elevated serum immunoglobulins (especially IgG), high titres of autoantibodies, interface hepatitis and a prompt response to corticosteroids, in the absence of known risk factors, such as drug ingestion or HCV infection, is virtually pathognomonic of the disease.

Following liver transplantation, lymphocytic infiltration around the portal tracts is found in allograft rejection and autoantibodies may be present in low titre. In children following liver transplantation, significant tires of autoantibodies are seen frequently, and are associated with graft dysfunction[66].

The first report of recurrence AIH after transplantation was a single-case study in which the diagnosis was suggested by the development of spider naevi in a woman 18 months after transplantation[67]. The diagnosis was suggested by the development of raised IgG and antibodies to actin and liver-specific lipoprotein. The biopsy showed typical histological features of AIH. The reintroduction of corticosteroids was associated with a prompt resolution of all these abnormalities. The recipient was HLA B8 DR3-positive whereas the donor had neither of these antigens.

The role of HLA remains controversial[68]. In 1992 Wright and colleagues from Pittsburgh published a retrospective analysis of 43 patients grafted for putative AIH[69]. Disease recurrence occurred in 11 (26%). All patients had hyperglobulinaemia and elevated levels of autoantibodies. The authors suggested that recurrence was most likely in HLA DR3-positive recipients of DR3-negative grafts. These findings must be interpreted with caution since molecular confirmation of HLA assignments was not available at the time. The Mayo Clinic[70] described recurrence of AIH in four (11%) of their series at a mean interval of 72 months. Serum anti-SMA or anti-ANA were present in two. Two of the four patients were mismatched at the DR3 locus. Remission, induced by increased immunosuppression, occurred in all but one. Devlin et al.[71] described severe AIH occurring 10 years after transplantation in a woman; there was a prompt response to the introduction of corticosteroids. The recipient was HLA A11 B8 DR3 and the donor was negative for all three antigens. Prados and colleagues[72] estimated the risk of AIH recurrence at 8% in the first year and 68% at 5 years. Interestingly, in that series from Madrid, none of those with liver–kidney microsomal antibodies developed recurrence.

Birnbaum and colleagues[73] reported recurrence of AIH in five of six children at a mean time of 11 months after transplantation. They suggest that the recurrence is more severe in children, since in three patients retransplantation was required. The recurrence was both in cyclosporin- and tacrolimus-treated patients. Others[74] have also described graft loss from recurrent AIH. Of note, the type of AIH does not affect the rate of recurrence[75]. Histological evidence of disease recurrence may precede the biochemical and immunological abnormalities by several years[76].

Overall, the diagnosis of recurrent AIH is readily made on serological and histological evidence. There is usually a prompt response to increased immunosuppression, and recurrence has little adverse impact on the graft. The data do suggest that HLA mismatching may play a role in the development of recurrent AIH, although the mechanism is unclear. The recurrence of AIH suggests that the defect leading to AIH lies in the lymphocytes rather than the liver, an observation supported by the recent case report of remission of AlH in a patient following successful bone-marrow transplantation[77].

CONCLUSIONS

For the three major autoimmune liver conditions treated by transplantation there is good evidence that in both PBC and AIH there is evidence for recurrent disease affecting the graft. Recurrent AIH responds well to increased immuno-suppression and donor/recipient HLA matching may be implicated, although the factors which determine which patients have a recurrence have not been identified. The type or degree of immunosuppression may affect the rate and severity of recurrence. PSC recurrence is difficult to define because of the inherent problems in distinguishing primary and secondary sclerosing cholangitis in the graft. Histological and radiological studies do suggest that the condition recurs, although further and longer-term studies are required.

References

1. Neuberger J. Developments in liver transplantation. Gut. 2004;53:759–68.
2. Ter Borg PC, van Os E, van der Broek WW, Hansen BE, van Buuren W. Fluvoxamine for fatigue in primary biliary cirrhosis and primary sclerosing cholangitis: a randomised controlled trial. BMC Gastroenterol. 2004;4:13.
3. Lucey M, Brown KA, Everson GT et al Minimal listing criteria for placement of adults on the liver transplant waiting list. Transplantation. 1998;66:956–62.
4. Trotter JF, Brimhall B, Arjal R, Phillips C. Specific laboratory methodologies achieve higher model for end-stage liver disease (MELD) scores for patients listed for liver transplantation. Liver Transplant. 2004;10:995–1000.
5. Neuberger J. Allocation of donor livers – is MELD enough? Liver Transplant. 2004;10:908–10.
6. Roll J, Boyer JL, Barry D, Klatskin G. The prognostic importance of clinical and histologic features in asymptomatic and symptomatic primary biliary cirrhosis. N Engl J Med. 1983;308:1–8.
7. Christensen E, Neuberger J, Crowe J et al. Beneficial effect of azathioprine and prediction of prognosis in primary biliary cirrhosis: final result of an international trial. Gastroenterology. 1985;89:1084–91.
8. Dickson E, Grambsch P, Fleming T, Fisher L, Langworthy A. Prognosis in primary biliary cirrhosis: model for decision making. Hepatology. 1989;10:1–7.
9. Rydning A, Schrumpf E, Abdel-noor M, Elgio K, Jenssen E. Factors of prognostic importance in primary biliary cirrhosis. Scand J Gastroenterol. 1990;25:119–26.
10. Goudie B, Burt AD, Macfarlane G et al. Risk factors and prognosis in primary biliary cirrhosis. Am J Gastroenterol. 1989;84:713–16.
11. Jeffrey G, Reed W, Shilkin K. Natural history and prognostic variables in primary biliary cirrhosis. Hepatology. 1990;12:955.
12. Hughes M, Raskino C, Pocock S, Biagini M, Burroughs A. Prediction of short-term survival with an application in primary biliary cirrhosis. Stat Med. 1992;11:1731–45
13. Wiesner R, Grambsch P, Dickson ER et al. Primary sclerosing cholangitis: natural history, prognostic factors and survival analysis. Hepatology. 1989;10:430–6.
14. Farrant M, Hayllar K, Wilkinson M et al Natural history and prognostic variables in primary sclerosing cholangitis. Gastroenterology. 191;100:1710–17.

15. Dickson ER, Murtaugh P, Wiesner R et al. Primary sclerosing cholangitis: refinement and validation of survival models. Gastroenterology. 1992;103:1893–901.
16. Klion F, Fabry T, Palmer M, Schaffner F. Prediction of survival of patients with primary biliary cirrhosis: examination of the Mayo Clini model on a group of patients with known end-point. Gastroenterology. 1992;102:310–13.
17. Shapiro J, Smith H, Schaffner F. Serum bilirubin: a prognostic factor in primary biliary cirrhosis. Gut. 1979;20:137–40.
18. Poupon R, Balkau B, Guechot J, Heintzman F. Predictive factors in ursodeoxycholic acid treated patients with primary biliary cirrhosis. Hepatology. 1994;19:635–40.
19. Nyberg A, Engstrom-Laurent A, Loof L. Serum hyaluronate in primary biliary cirrhosis: a biochemical marker for progressive liver damage. Hepatology. 1988;8:142–6.
20. Eriksson E, Zettervall O. The N-terminal propeptide of collagen type III in serum as a prognostic indicator in primary biliary cirrhosis. J Hepatol. 1982;2:370–8.
21. Clements D, Elias E, McMaster P. Preliminary study of indocyanine green clearance in primary biliary cirrhosis. Scand J Gastroenterol. 1991;26:119–23.
22. Reichen J, Widmer T, Cutting J. Accurate prediction of death by serial determination of galactose elimination capacity in primary biliary cirrhosis: a comparison with the Mayo model. Hepatology. 1991;14:504–10.
23. Wiesner R, Grambsch P, Dickson E et al. Primary sclerosing cholangitis: natural history, prognostic factors and survival analysis. Hepatology. 1989;10:430–6.
24. Bjoro K, Schrumpf E. Liver Transplantation for primary sclerosing cholangitis. J Hepatol. 2004;40:570–7.
25. Ismail T, Angrisani L, Powell J et al. Primary sclerosing cholangitis: surgical options, prognostic variables and outcome. Br J Surg. 1991;78:564–7.
26. Bergquist A, Glaumann H, Persson B, Broome U. Risk factors and clinical presentation of hepatobiliary carcinoma in patients with primary sclerosing cholangitis: a case–control study. Hepatology. 1998;27:311–16.
27. Fisher A, Thiese N, Min A et al. CA19-9 does not predict cholangiocarcinoma in patients with primary sclerosing cholangitis undergoing liver transplantation. Liver Transplant Surg. 1995;1:94–8.
28. Heimbach JK, Gores GJ, Haddock MG, Alberts SR, Nyberg SL, Ishi , Rosen CB. Liver transplantation for unresectable perihilar cholangiocarcinoma. Semin Liver Dis. 2204;24:201–7.
29. Roberts MS, Angus DC, Bryce CL, Valenta Z, Weissfeld L. Survival after liver transplantation in the United States: a disease-specific analysis of the UNOS database. Liver Transplant. 2004;10:886–97.
30. Wiesner RH, Seaberg EC, Kim WR et al Predicting outcomes of liver transplantation in patients with primary sclerosing cholangitis (PSC) using the Mayo PSC model and the Child–Pugh score. Transplantation. 1999;67:S244 (abstract).
31. Neuberger J, Gunson B, Komolit P, Davies MH, Christensen E. Pretransplant prediction of prognosis after liver transplantation in primary sclerosing cholangitis using a Cox regression model. Hepatology. 1999;29:1375–9.
32. Christensen E, Gunson B, Neuberger J. Optimal timing of liver transplantation for patients with primary biliary cirrhosis: use of prognostic models. J Hepatol. 1999;30:285–92.
33. Liermann Garcia RF, Garcia CE, McMaster P, Neuberger J. Transplantation for primary biliary cirrhosis: retrospective analysis of 400 patients from a single center. Hepatology. 2001;33:22–7.
34. Neuberger J. Incidence, timing and risk factors for acute and chronic rejection. Liver Transplant Surg. 2005 (In press)
35. Farges O, Saliba F, Fahramant H et al. Incidence of rejection and infection after liver transplantation as a factor of the primary disease: possible influence of alcohol and polyclonal immunoglobulins. Hepatology. 1996;23:240–8.
36. Gross CR, Malinchoc M, Kim WR et al. Quality of life before and after transplantation for cholestatic liver disease. Hepatology. 1999;29:356–64.
37. Riley TR, Schoen RE, Lee RG, Rakela J. A case series of transplant recipients who despite immunosuppression developed inflammatory bowel disease. Am J Gastroenterol. 1997;92:279–82.

38. Papatheodoridis GV, Hamilton M, Mistry PK, Davidson B, Rolles K, Burroughs AK. Ulcerative colitis has an aggressive course after orthotopic liver transplantation for primary sclerosing cholangitis. Gut. 1998;43:639–44.
39. Haagsma EB, van der Berg AP, Kleibeuker JH, Slooff MJ, Dijkstra W. Inflammatory bowel disease after liver transplantation: the effect of different immunosuppressive regimes. Aliment Pharmacol Ther. 2003;18:33–44.
40. Muller C, Briegel J, Haller M et al. Sarcoidosis recurrence following lung transplantation. Transplantation. 1996;61:1117–19.
41. Stegall M, Lafferty K, Kam I, Gill R. Evidence of recurrent autoimmunity in human allogeneic islet transplantation. Transplantation. 1996;61:1272–4.
42. Mieli Vergani G, Vergani D. *De novo* autoimmune hepatitis after liver transplantation. J Hepatol. 2004;40:3–7.
43. Kerkar N, Hadzic N, Davies E et al. Graft dysfunction associated with autoimmunity in paediatric liver transplantation. Hepatology. 1996;24:235A (abstract).
44. Heneghan MA, Portmann B, Rela M, Heaton ND, O'Grady JG. Autoimmune hepatitis occurring *de-novo* following orthotopic liver transplantation in adults. Transplantation. 1999;67:S199 (abstract).
45. Neuberger J. Transplantation for autoimmune hepatitis. Semin Liver Dis. 2002;22:379–86.
46. Aguilera I, Sousa JM, Gavilan F, Bernardos A, Wichmann I, Nunez-Roldan A. Glutathione S-transferase T1 mismatch constitutes a risk factor for *de novo* immune hepatitis after liver transplantation. Liver Transplant. 2004;10:1166–72.
47. Sebagh M, Farges O, Kalil A et al. Sclerosing cholangitis following human orthotopic liver transplantation. Am Surg Pathol. 1995;19:81–90.
48. Patel T, Reddy S, Gores G, Wiesner R, Krom R. Recurrence following retransplantation for biliary strictures. Hepatology. 1996;24:243A (abstract).
49. Brandsaeter B, Scrumpf E, Bentdal O, Braband K, Smith HJ, Abilgaard A, Clausen OP, Bjoro K. Bile duct strictures after liver transplantation in patients with primary sclerosing cholangitis. Transplantation. 2004;78:P534 (abstract).
50. Haagsma E, Mulder A, Gouw A et al. Neutrophil cytoplasmic autoantibodies after liver transplantation in patients with primary sclerosing cholangitis. J Hepatol. 1993;19:8–14.
51. Harrison RF, Davies MH, Neuberger JM et al. Fibrous and obliterative cholangitis in liver allografts: evidence of recurrent primary sclerosing cholangitis. Hepatology. 1994;20:356–61.
52. Graziadei I, Wiesner R, Marotta P et al Strong evidence for recurrence of primary sclerosing cholangitis after liver transplantation. Hepatology. 1997;26: 76A (abstract).
53. Abu-Elmagd K, Demetris AJ, Rakela J et al Recurrence of primary sclerosing cholangitis (PSC) after hepatic transplantation: single center experience with 380 grafts. Transplantation. 1999;67:S236 (abstract).
54. Sheng R, Zajko AB, Campbell WL et al. Biliary structures in hepatic transplants:prevalence and types in patients with primary sclerosing cholangitis vs those with other liver diseases. Am J Radiol. 1993;161:297–300.
55. McEntee G, Wiesner RH, Rosen C et al. A comparative study of patients undergoing liver Transplantation for primary sclerosing cholangitis and primary biliary cirrhosis. Transplant Proc. 1991;23:1563–4.
56. Luettig B, Boeker KH, Schoessler W et al.The antinuclear autoantibodies Sp100 and gp210 persist after orthotopic liver transplantation in patients with primary biliary cirrhosis. J Hepatol. 1998;28:824–8.
57. Neuberger J, Portmann B, MacDougall B, Calne RY, Williams R. Recurrence of primary biliary cirrhosis after liver transplantation. N Engl J Med. 1982;306:1–4.
58. Ferrell L, Lee R, Brixko C, et al Hepatic granulomas following liver transplantation. Hepatology. 1995;60:926–44.
59. Demetris AJ, Markus BH, Esquivel C et al. Pathologic analysis of liver transplantation for primary biliary cirrhosis. Hepatology. 1988;6:937–47.
60. Gouw A, Haagsma E, Manns M et al. Is there recurrence of primary biliary cirrhosis after liver transplantation. J Hepatol. 1994;20:500–6.
61. Buist L, Hubscher S, Vickers C, Neuberger J, McMaster P. Does transplantation cure primary biliary cirrhosis? Transplant Proc. 1989;21:2402.

62. Neuberger J, Gunson B, Hubscher SG, Nightingale P. Immunosuppression affects the rate of recurrent primary biliary cirrhosis after liver transplantation. Liver Transplant. 2004;10: 488–91.
63. Abu-Elamgd K, Demetris J, Rakela J et alTransplantation for primary biliary cirrhosis: recurrence and outcome in 421 patients. Hepatology. 1997;26:176A (abstract).
64. Neuberger J, Wallace L, Joplin R, Babbs C, Hubscher S. Hepatic distribution of E2 component of pyruvate dehydrogenase complex after transplantation. Hepatology. 1995; 22:798–801.
65. Van de Water J, Gerson LB, Ferrell L et al. Immunohistochemical evidence of disease recurrence after liver transplantation for primary biliary cirrhosis. Hepatology. 1996;74: 1079–84.
66. Richter A, Ganschow R, Rogiers X, Burdelski M, Manns M. Autoantibodies after paediatric liver transplantation. Transplantation. 2004;78:P537 (abstract).
67. Neuberger J, Portmann B, Calne R, Williams R. Recurrence of autoimmune chronic active hepatitis following orthotopic liver grafting. Transplantation. 1984;37:363–6.
68. Neumann UP, Guckelberger O, Langrehr JM et al. Impact of human leucocyte antigen matching in liver transplantation. Transplantation. 2003;75:132–7.
69. Wright HL, Ben-Abboud C, Hussapein T et al. Disease recurrence and rejection following liver transplantation for autoimmune chronic active liver disease. Transplantation. 1992;53: 136–9.
70. Roberts S, Czaja A, Chariton M et al. Recurrent autoimmune hepatitis following orthotopic liver transplantation. Hepatology. 1995;72:215A (abstract).
71. Devlin J, Donaldson P, Portmann B, Heaton N, Tan K-C, Williams R. Recurrence of autoimmune hepatitis following liver transplantation. Liver Transplant Surg. 1995;1:162–5.
72. Prados E, Cuervas-Mons V, de la Mata M et al. Outcome of autoimmune hepatitis after liver transplantation. Transplantation. 1998;66:1645–50.
73. Birnbaum AH, Benkov KJ, Pittman NS, McFarlane-Ferreira Y, Rosh JR, LeLeiko NS. Recurrence of autoimmune hepatitis in children after liver transplantation. J Pediatr Gastroenterol Nutr. 1997;25:20–5.
74. Ratziu V, Samuel D, Sebagh M et al. Long-term follow-up after liver transplantation for autoimmune hepatitis: evidence of recurrent disease. J Hepatol. 1999;30:131–41.
75. Vogel A, Heinrich E, Bahr MJ et al. Long-term outcome of liver transplantation for autoimmune hepatitis. Clin Transplant. 2004;18:62–9.
76. Duclos-Vallee JC, Sebagh M, Rifai K et al. A 10 year follow up study of patients transplanted for autoimmune hepatitis: histological recurrence precedes clinical and biochemical recurrence. Gut. 2003;52:983–97.
77. Vento S, Cainelli S, Renzini C, Ghironzi G, Concia E. Resolution of autoimmune hepatitis after bone marrow transplantation. Lancet. 1996;348:544–5.

Section VIII
Paediatric aspects of autoimmune liver disease

Chair: M. BURDELSKI and X. ROGIERS

24
Autoimmune hepatitis in children: clinical and diagnostic aspects

F. ALVAREZ

Autoimmune hepatitis (AIH) is a disease of unknown aetiology that progresses spontaneously to cirrhosis in most cases. Clinical and laboratory observations have led to the hypothesis that it is a multifactorial disease. Genetic and environmental factors play an important role in the pathogenesis of AIH.

Two types of AIH have been recognized from the nature of the autoantibody detected in children at the time of diagnosis[1-3]. Patients with type 1 AIH display anti-smooth muscle (SMA) and/or anti-nuclear autoantibodies (ANA) in their sera. Type 2 AIH is characterized by anti-liver–kidney microsome (LKM1) and/or anti-cytosol (LC1) autoantibodies. Some differences between the two types of AIH have been described in the genetic background and clinical manifestations. The choice of the immunosuppressor treatment, as well as follow-up and prognosis, appear not to be greatly influenced by the type of AIH. However, type 2 AIH presents more frequently with cirrhosis at an early age[2,3].

Susceptibility to AIH is probably determined by a complex combination of genetic factors, including major histocompatibility complex (MHC) and non-MHC genes. HLA-DRB1*1301 and DQB1*0201 haplotypes have been associated with susceptibility to type 1 and type 2 AIH, respectively[4]. Polymorphisms of the CTLA-4 gene[5] and mutations of the AIRE gene[6], two non-MHC genes, have been linked with the development of AIH. Mutations of the AIRE gene are responsible for autoimmune polyendocrinopathy-candidiasis-ectodermal dystrophy (APECED) syndrome. The AIRE gene codes for a transcription factor that regulates the ectopic expression of peripheral tissue-restricted antigens in medullary epithelial cells of the thymus. Central tolerance in the thymus is important in controlling autoimmunity. Mutations of the gene coding for complement factor 4 have also been found in AIH patients[7]. The C4A null allele is in linkage disequilibrium with the HLA A1, B8 and DR haplotypes. Selective IgA deficiency, the most frequent primary immunodeficiency, is detected in children with AIH.

CLINICAL FEATURES

AIH is primarily a paediatric disease, with a female preponderance (Table 1). AIH has been diagnosed as early as 6 months after birth, but its peak incidence occurs at a pre-pubertal age. Type 1 AIH is diagnosed in 40% of cases before the age of 18 years, with a mean age at onset of 10 years for this group[2,3]. In 80% of cases type 2 AIH is found before the age of 18 years with a mean age of 6.5 years[2,3].

Table 1 Clinical features in children with AIH

Clinical features	Type 1 AIH	Type 2 AIH
Mean age at onset	10 years	6.5 years
Females (%)	~75%	90%
Form of presentation[a]		
Acute hepatitis[b]	~45%	~50%
Chronic hepatitis	~35%	~30%
Others	~20%	~20%
Extrahepatic autoimmune diseases		
Patients	50%	40%
Family	5-10%	~20%

[a]These percentages are obtained from previously published series.

[b]Including fulminant and subfulminant liver failure.

Acute hepatitis syndrome, rarely fulminant or subfulminant liver failure[8,9], is the most frequent form of presentation[2,3]. Non-specific symptoms, such as fatigue, anorexia and weight loss, lead to the diagnosis of AIH. In 10–15% of cases hepatomegaly or an unexplained increase of transaminases is the only sign[2,3]. In most of these cases some manifestations of chronic liver disease are already present[10]. These signs, such as spider naevi, palmar erythema, clubbing or ascites, orient the physician. Less frequently, extrahepatic symptoms or the presence of other autoimmune diseases lead to the discovery of an autoimmune process in the liver[2,3]. Unfortunately, in a considerable number of patients, a long time interval between the appearance of symptoms and confirmation of the diagnosis[2,3] delays the institution of appropriate treatment.

Most patients present a firm or hard hepatomegaly at diagnosis, and splenomegaly due to portal hypertension is seen in 50% of them[2,3]. Gastrointestinal bleeding secondary to portal hypertension is an infrequent event at onset in children with AIH as well as after the initiation of therapy. Jaundice occurs at onset, or an episode of jaundice is recorded in the history of more than 50% of these children[2,3].

Extrahepatic autoimmune diseases are seen in more than one-third of children at onset or show up after the beginning of treatment. Some extrahepatic diseases are particular to type 1 or type 2 AIH (Table 2). The association of AIH with coeliac disease in children is still a controversial matter. The liver disorder in most patients with coeliac disease is corrected by

Table 2 Extrahepatic autoimmune diseases in children with AIH

Extrahepatic autoimmune disease	Type 1 AIH	Type 2 AIH
Ulcerative colitis	+	−
Crohn's disease	+	−
Vasculitis	+	−
Arthritis	+	−
Thrombocytopenia	+	−
Fibrosing alveolitis	+	−
Haemolytic anemia	+	+
Glomerulonephritis	+	+
Autoimmune enteropathy	+	++
Thyroiditis	+	++
Diabetes	+	++
APECED	−	+
Vitiligo	−	+
Alopecia	−	+
Autoimmune lymphoproliferative syndrome	−	+

the complete elimination of gluten from the diet. However, some rare cases of AIH have been described in patients compliant with the restricted diet, and after correction of the enteropathy[11].

A history of autoimmune diseases is found in 20–40% of first-degree relatives: the type of autoimmune disease observed does not differ from that described in patients with AIH[2,3]. Rarely is AIH already diagnosed in another family member.

LABORATORY DATA

AIH shows a fluctuating course, which explains why serum aminotransferase levels vary between 1.5 and 50 times the normal values at the time of diagnosis[2,3,12]. Gammaglutamyl transferase (GGT) and alkaline phosphatase (AP) are frequently normal or increased slightly[12]. When plasma levels of these two enzymes, related to cholestasis, are more than 4 or 5 times the normal values for GGT or more than 2 or 3 times for AP, bile duct injury should be suspected and cholangiography proposed.

One of the main characteristics of AIH is the non-specific proliferation of B lymphocytes and, as a consequence, hypergammaglobulinaemia. A marked increase of IgG is observed[12]. The highest IgG levels are found in patients with type 1 AIH. Deficiencies in C4 and IgA should be investigated. In more than 50% of patients low levels of serum albumin and clotting factors are detected at onset[2,3]. These findings indicate liver failure, and denote the severity of the disease.

The analysis of circulating autoantibodies is of diagnostic value and is useful for follow-up (Table 3). SMA titres of 1:100 to 1:500 000, associated or not with ANA titres between 1:10 to 1:100 000, are characteristic of patients with type 1

Table 3 Serological markers in AIH

Autoantibody	Antigen	Type 1 AIH	Type 2 AIH
SMA	Actin filaments	+ (90–100%)	–
ANA	Various	+ (0–10%)	±
SMA/ANA	–	+ (40–60%)	–
LKM1	Cytochrome P450 2D6	–	+ (40–50%)
LC1	Formiminotransferase cyclodeaminase	–	+ (5–10%)
LKM1/LC1	–	–	+ (35–45%)
SLA	tRNP(Ser)See	+ (50%)	+ (45)
ASGP-R	Asialoglycoprotein receptor	+ (75%)	+ (40%)

AIH (Table 3). Titres of SMA and mainly ANA must be carefully interpreted for patients in different regions of the world. In countries with poor hygienic conditions repeated stimulation of the immune system leads to the production of low titres of these autoantibodies. LKM1 titres between 1:10 and 1:100 000 are found in serum from type 2 AIH patients. LC1 is frequently associated with LKM1, but is the only serological marker in some children with type 2 AIH (Table 3). Currently, indirect immunofluorescence in cryostat sections or cells in culture is the most commonly used technique for the detection of these autoantibodies. The recent discovery of autoantigens recognized by LKM1 and LC1 autoantibodies should facilitate the development of new and more performant detection tests (Table 3)[13–15]. Some of these are already available, but their specifity and sensitivity have not yet been validated in a large series of patients.

Other circulating autoantibodies that can be detected in sera from children with AIH are: anti-soluble liver antigen[16] and anti-asialoglycoprotein receptor (ASGP-R)[17,18]. These autoantibodies are found in both types of AIH, generally at lower titres than those mentioned above, and tests for their detection are available in a few specialized laboratories. Exceptionally, antimitochondrial autoantibody (AMA) is the only serological marker in children responding to immunosuppressive treatment[19]. AMA positivity characterizes primary biliary cirrhosis, an autoimmune liver disease observed in adults.

None of these autoantibodies (SMA, ANA, LKM1, LC1, SLA, or ASGP-R) is specific of AIH; they can be also found in sera from some patients with chronic hepatitis C infection.

HISTOPATHOLOGICAL DATA

An inflammatory lymphoplasmocytic infiltrate constituted of T and B lymphocytes is observed in the portal tract, invading the lobule and producing interface hepatitis[12]. The fluctuating course of the disease is reflected in the degree of portal, periportal and lobular inflammation. Inflammatory potential and aggressivity are mild or even absent in 10–40% of liver biopsies in different series[2,3]. Low levels of inflammation, or its absence, have been described in adolescents with type 2 AIH.

In 10–20% of patients multinucleated giant cells are discerned in the liver, without any viral particles on electron microscopy[3]. This possibility should be remembered in the differential diagnosis of other forms of hepatitis. Recently a new histological picture associated with AIH has been characterized by centrilobular necrosis and inflammation that almost completely spares the portal tracts[20]. Toxic or hypoxaemic liver injury must be excluded before starting immunosuppressive treatment. This type of therapy has been successful in the restoration of hepatic tissue.

The bile duct epithelium can be injured in AIH, without signs of sclerosing cholangitis, in up to 30% of adult patients[21]. Similar studies in children are required.

Liver biopsy is of great help in the diagnosis of AIH. However, frequently, clotting abnormalities due to hepatic failure preclude this practice. In addition, liver biopsy underestimates the level of cirrhosis. With laparoscopic examination up to 90% of patients show liver nodules at onset[10]; however, this procedure is not recommended as standard practice. Fibrosis is reversible in AIH patients who respond well to immunosuppressive treatment.

DIAGNOSIS

Criteria helpful for the diagnosis of AIH have been proposed as a scoring system by the International Autoimmune Hepatitis Group[12]. Female gender, hypergammaglobulinaemia, circulating autoantibodies, low levels of complement factor 4, the presence of extrahepatic autoimmune diseases in patients or first-degree relatives, and chronic active hepatitis on liver biopsy are basic features in the diagnosis of AIH. This scoring system showed a high degree of sensitivity when it was applied retrospectively to series of adult and paediatric patients[22].

In cases with an acute presentation, viral and toxic causes of hepatitis should be investigated. Minocycline, an antibiotic administered to adolescents with severe acne, has been reported to induce hepatitis in several cases in association with a lupus-like syndrome[23,24]. Circulating autoantibodies (mainly SMA and ANA) and extrahepatic autoimmune-like signs could made the differential diagnosis of type 1 AIH quite difficult. Caeruloplasmin, urinary copper, and eventually liver copper should be measured to eliminate Wilson's disease. Both hepatitis A and Wilson's disease can display high gammaglobulins and circulating autoantibodies. Low titres of autoantibodies are also found in sera from patients with acute or chronic hepatitis B or C infection. The differential diagnosis of AIH and sclerosing cholangitis is not easy, and some cases of overlap syndrome have improved with immunosuppressive treatment.

CONCLUSION

The diagnosis of AIH must be undertaken in patients with acute or chronic hepatitis. This also applies to children with liver test abnormalities and non-specific symptoms, such as fatigue or amenorrhoea. The presence of an extrahepatic autoimmune disorder or other signs of autoimmunity must lead to the search for AIH. The fluctuating course of the disease can be responsible for alternative periods of remission and relapse. A low inflammatory syndrome at onset does not preclude the beginning of immunosuppressive treatment. Rapid and complete control of the liver inflammation improves the short- and long-term outcome since it has been demonstrated that liver fibrosis can regress in patients responding to treatment, justifying an aggressive approach to the diagnosis of AIH as early as possible.

References

1. Odievre M, Maggiore G, Homberg JC et al. Seroimmunologic classification of chronic hepatitis in 57 children. Hepatology. 1983;3:407–9.
2. Maggiore G, Veber F, Bernard O et al. Autoimmune hepatitis associated with anti-actin antibodies in children and adolescents. J Pediatr Gastroenterol Nutr. 1993;17:376–81.
3. Gregorio GV, Portmann B, Reid F et al. Autoimmune hepatitis in childhood: a 20-year experience. Hepatology. 1997;25:541–7.
4. Djilali-Saiah I, Renous R, Caillat-Zucman S, Debray D, Alvarez F. Linkage disequilibrium between HLA class II region and autoimmune hepatitis in pediatric patients. J Hepatol. 2004;40:904–9.
5. Djilali-Saiah I, Ouellette P, Caillat-Zucman S, Debray D, Kohn JI, Alvarez F. CTLA-4/CD 28 region polymorphisms in children from families with autoimmune hepatitis. Hum Immunol. 2001;62:1356–2.
6. Vogel A, Liermann H, Harms A, Strassburg CP, Manns MP, Obermayer-Straub P. Autoimmune regulator AIRE: evidence for genetic differences between autoimmune hepatitis and hepatitis as part of the autoimmune polyglandular syndrome type 1. Hepatology. 2001;33:1047–52.
7. Vergani D, Wells L, Larcher VF et al. Genetically determined low C4: a predisposing factor to autoimmune chronic active hepatitis. Lancet. 1985;2:294–8.
8. Porta G, Gayotto LC, Alvarez F. Anti-liver–kidney microsome antibody-positive auto-immune hepatitis presenting as fulminant liver failure. J Pediatr Gastroenterol Nutr. 1990;11:138–40.
9. Herzog D, Rasquin-Weber AM, Debray D, Alvarez F. Subfulminant hepatic failure in autoimmune hepatitis type 1: an unusual form of presentation. J Hepatol. 1997;27:578–82.
10. Vajro P, Hadchouel P, Hadchouel M, Bernard O, Alagille D. Incidence of cirrhosis in children with chronic hepatitis. J Pediatr. 1990;117:392–6.
11. Iorio R, Sepe A, Giannattasio A, Spagnuolo MI, Vecchione R, Vegnente A. Lack of benefit of gluten-free diet on autoimmune hepatitis in a boy with celiac disease. J Pediatr Gastroenterol Nutr. 2004;39:207–10.
12. Johnson PJ, McFarlane IG. Meeting report: International Autoimmune Hepatitis Group. Hepatology. 1993;18:998–1005.
13. Manns MP, Johnson EF, Griffin KJ, Tan EM, Sullivan KF. Major antigen of liver kidney microsomal autoantibodies in idiopathic autoimmune hepatitis is cytochrome P450db1. J Clin Invest. 1989;83:1066–72.
14. Gueguen M, Meunier-Rotival M, Bernard O, Alvarez F. Anti-liver kidney microsome antibody recognizes a cytochrome P450 from the IID subfamily. J Exp Med. 1988;168:801–6.
15. Lapierre P, Hajoui O, Homberg JC, Alvarez F. Formiminotransferase cyclodeaminase is an organ-specific autoantigen recognized by sera of patients with autoimmune hepatitis. Gastroenterology. 1999;116:643–9.

16. Vitozzi S, Djilali-Saiah I, Lapierre P, Alvarez F. Anti-soluble liver antigen/liver-pancreas (SLA/LP) antibodies in pediatric patients with autoimmune hepatitis. Autoimmunity. 2002;35:485–92.
17. McFarlane BM, McSorley CG, Vergani D, McFarlane IG, Williams R. Serum autoantibodies reacting with the hepatic asialoglycoprotein receptor protein (hepatic lectin) in acute and chronic liver disorders. J Hepatol. 1986;3:196–205.
18. Hajoui O, Debray D, Martin S, Alvarez F. Auto-antibodies to the asialoglycoprotein receptor in sera of children with auto-immune hepatitis. Eur J Pediatr. 2000;159:310–13.
19. Gregorio GV, Portmann B, Mowat AP, Vergani D, Mieli-Vergani G. A 12-year-old girl with antimitochondrial antibody-positive autoimmune hepatitis. J Hepatol. 1997;27:751–4.
20. Misdraji J, Thiim M, Graeme-Cook FM. Autoimmune hepatitis with centrilobular necrosis. Am J Surg Pathol. 2004;28:471–8.
21. Czaja AJ, Carpenter HA. Autoimmune hepatitis with incidental histologic features of bile duct injury. Hepatology. 2001;34:659–65.
22. Czaja A, Carpenter HA. Validation of scoring system for diagnosis of autoimmune hepatitis. Dig Dis Sci. 1996;41:305–14.
23. Teitelbaum JE, Perez-Atayde AR, Cohen M, Bousvaros A, Jonas MM. Minocycline-related autoimmune hepatitis: case series and literature review. Arch Pediatr Adolesc Med. 1998;152:1132–6.
24. Schrodt BJ, Callen JP. Polyarteritis nodosa attributable to minocycline treatment for acne vulgaris. Pediatrics. 1999;103:503–4.

25
Therapeutic aspects of autoimmune liver disease in children

G. MIELI-VERGANI, K. BARGIOTA, M. SAMYN and D. VERGANI

INTRODUCTION

Autoimmune hepatitis (AIH) in children is exquisitely responsive to immuno-suppression. The rapidity and degree of response depends on the disease severity at presentation. AIH can present as prolonged 'acute' hepatitis, compensated chronic liver disease, acute decompensation of chronic liver disease with severe hepatic synthetic failure, or, rarely, as fulminant hepatitis with encephalopathy. The latter responds only exceptionally to immunosup-pression, and transplantation is usually the only effective management. All other presentations generally respond to standard treatment with prednisolone, to which azathioprine may be added.

AUTOIMMUNE HEPATITIS

In our unit, treatment for AIH consists of prednisolone 2 mg/kg per day (maximum 60 mg/day), which is gradually decreased over a period of 4–8 weeks if there is progressive normalization of the transaminases, and then the patient is maintained on the minimal dose able to sustain normal transaminase levels, usually 5 mg/day[1]. During the first 6–8 weeks of treatment, liver function tests are checked weekly to allow a constant and frequent fine-tuning of the treatment, avoiding severe steroid side-effects. If progressive normal-ization of the liver function tests is not obtained over this period of time, or if too high a dose of prednisolone is required to maintain normal transaminases, azathioprine is added at a starting dose of 0.5 mg/kg per day which, in the absence of signs of toxicity, is increased up to a maximum of 2 mg/kg per day until biochemical control is achieved. Azathioprine is not recommended as first-line treatment because of its hepatotoxicity, particularly in severely jaundiced patients. A preliminary report in a cohort of 30 children with AIH suggests that the measurements of the azathioprine metabolites 6-thioguanine and 6-methylmercaptopurine are useful in identifying drug toxicity and non-adherence, and in achieving a level of 6-thioguanine considered therapeutic for inflammatory bowel disease[2]. In our experience, although an 80% decrease of initial transaminase levels is obtained within 6 weeks from starting treatment in most patients, complete normalization of liver function may take several months. In our own series normalization of transaminase levels occurred at medians of 0.5 years (range 0.2–7 years) in ANA/SMA-positive children and 0.8 years (range 0.02–3.2 years) in LKM-1-positive children[1]. Relapse while on treatment is common, affecting about 40% of the patients and requiring a temporary increase of the steroid dose. The risk of relapse is higher if steroids are administered on an alternate-day schedule, often instituted in the unsubstantiated belief that it has a less negative effect on the child's growth. Small daily doses should be used because they are more effective in maintaining disease control and minimize the need for high-dose steroid pulses during relapses (with attendant more severe side-effects). If a liver biopsy shows minimal or no inflammatory changes after 1 year of normal liver function tests, cessation of treatment should be considered, but not during or immediately before puberty, when relapses are more common. In 13 children (4 LKM-1-positive), the only ones fulfilling these criteria in our series, discontinuation of treatment was attempted[1]. This was successful in six children, all ANA/SMA-positive, after a median duration of 3.2 (range 1–11) years of treatment. All six have remained in remission for a period of 9–13 years. The remaining children (three ANA/SMA-positive and all four LKM-1-positive) relapsed between 1 and 15 months (median 2 months) after immunosuppression was discontinued. They all responded to the reintroduction of treatment. These data indicate that most children with AIH, particularly those who are LKM-1-positive, are likely to require lifelong immunosuppressive treatment.

In paediatrics an important role in monitoring the response to treatment is the measurement of autoantibody titres and immunoglobulin G (IgG) levels, the fluctuation of which is correlated with disease activity[3]. Despite the efficacy of current treatment, severe hepatic decompensation may develop even after many years of apparently good biochemical control. Thus, four of our patients who responded satisfactorily to immunosuppression ultimately required transplant 8–14 years after diagnosis. Overall, in our series, 46 of the 47 patients treated with immunosuppression were alive between 0.3 and 19 years (median 5 years) after diagnosis, including four patients after liver transplant. Side-effects of steroid treatment were mild, the only serious complication being psychosis during induction of remission in two patients, which resolved after prednisolone withdrawal. All patients developed a transient increase in appetite and mild cushingoid features during the first few weeks of treatment. After 5 years of treatment 56% of the patients maintained the baseline centile for height or went up across a centile line, while only 6% dropped across two centile lines.

Sustained remission of AIH has been reported in adult patients maintained on azathioprine alone[4]. Following this observation we have attempted to stop prednisolone, maintaining azathioprine, in five children, two who were ANA/SMA-positive and three who were LKM-1-positive. Although the attempt was successful in the ANA/SMA-positive cases, all LKM-1-positive children relapsed and required reinstitution of steroid treatment.

Alvarez et al. report remission in 25 of 32 children with AIH treated with cyclosporin A alone for 6 months followed by combined low-dose prednisone and azathioprine for 1 month, after which cyclosporin A was stopped and the other two drugs were continued[5]. The side-effects of cyclosporin A were mild, though patients did experience transient physical alteration due to hairiness and gum hypertrophy. High-dose steroid side-effects were avoided. A disadvantage of this schedule was that all patients were eventually treated with the prednisone/azathioprine combination, whereas by using the conventional treatment schedule about a third of the children can maintain remission with very-low-dose steroids alone. Moreover, longer follow-up of the patients is necessary to establish possible long-term toxicity of cyclosporin A, a nephrotoxic drug that in experimental animal models has been shown to induce a paradoxical autoaggressive syndrome by increasing the number of self-reactive T cells escaping thymic selection and by interfering with regulatory T cell function[6–8].

Mycophenolate mofetil has been successfully used in adult patients with type 1 AIH who have been either intolerant of, or not responsive to, azathioprine[9]. Mycophenolate mofetil is an inhibitor of purine nucleotide synthesis and has a mechanism of action similar to that of azathioprine. It is not hepatotoxic or nephrotoxic, and its main side-effects are diarrhoea, vomiting, and bone-marrow suppression. In our experience the drug was able to resolve laboratory abnormalities in 21 of 25 children who did not tolerate or respond to azathioprine. In one other child, it reduced serum aminotransferase levels to a degree that allowed a decrease in the dose of prednisolone. Only three patients did not respond. Side-effects of mycophenolate mofetil included mild hair loss, headache, diarrhoea, nausea, dizziness and neutropenia, the latter being the cause of permanent withdrawal of the drug in three patients.

Children who present with acute hepatic failure pose a particularly difficult therapeutic problem. If not encephalopathic, they may benefit from conventional immunosuppressive therapy[1,10,11], but only one of the six children with acute liver failure and encepaholpathy in our own series responded to immunosuppression and survived without transplant[1]. Of the four LKM-1-positive patients, one died before a donor organ could be found and two died soon after transplant. Encouraging results have been reported using cyclosporin A in LKM-1-positive patients presenting with fulminant hepatitis[11,12]. These results should be evaluated on a larger number of patients because our own experience has not confirmed the value of this therapeutic approach.

AUTOIMMUNE SCLEROSING CHOLANGITIS (ASC)

In paediatrics, half of the children who present to a tertiary hepatology service with serological (i.e. autoantibodies, high IgG levels) and histological (i.e. interface hepatitis) features of autoimmune liver disease have bile duct abnormalities if a cholangiogram is performed[13]. We have proposed to call this AIH/sclerosing cholangitis overlap syndrome autoimmune sclerosing cholangitis (ASC).

Children with ASC respond to the same immunosuppressive treatment described above for typical AIH[13]. In our series liver test abnormalities resolved in almost 90% of our patients within a median of 2 months after starting treatment. This good response is in contrast to the outcome in adults with primary sclerosing cholangitis (PSC) who appear to have no beneficial effects from corticosteroid treatment[14,15]. The PSC of adults, however, is usually diagnosed at an advanced stage, and may be the result of various aetiologies. Disappointing results with immunosuppressive agents have been reported in a small number of children with sclerosing cholangitis associated with autoimmune features, but these children appear to have had more advanced disease at the start of treatment than those recruited into our prospective study[16]. Ursodeoxycholic acid (UDCA) was added to our treatment schedule in 1992 following preliminary reports of its value in the treatment of adult PSC[17,18]. The small number of patients, the lack of a control group, and the relatively short follow-up period do not allow us to determine whether treatment with UDCA from onset is successful in arresting the progression of ASC. In adults with well-established PSC, UDCA treatment has been disappointing, possibly because of the advanced stage of the disease at the time of diagnosis[19]. Measurement of autoantibody titres and IgG levels is useful in monitoring disease activity and response to treatment also in ASC[3]. Follow-up liver biopsies in our series have shown no progression to cirrhosis, although one patient did develop vanishing bile duct syndrome. Follow-up endoscopic retrograde cholangiograms have shown static bile duct disease in half of our patients with ASC and progression of the bile duct abnormalities in the other half. Interestingly, one of the children with AIH who was followed prospectively developed sclerosing cholangitis 8 years after presentation despite treatment with corticosteroids and no biliary changes on several follow-up liver biopsies. This observation suggests that AIH and ASC are part of the same pathogenic process and that prednisolone and azathioprine may be more effective in controlling the liver parenchyma inflammatory changes than the bile duct disease. The medium-term prognosis of ASC is good[13]. All patients in our series were alive after a median follow-up of 7 years. Four patients with ASC, however, required liver transplant after 2–11 years of observation (median interval of follow-up 7 years). In contrast, liver transplant was not required by any of the 28 children with typical AIH who had been followed for this same time.

References

1. Gregorio GV, Portmann B, Reid F et al. Autoimmune hepatitis in childhood: a 20-year experience. Hepatology. 1997;25:541–7.
2. Rumbo C, Emerick KM, Emre S, Shneider BL. Azathioprine metabolite measurements in the treatment of autoimmune hepatitis in pediatric patients: a preliminary report. J Pediatr Gastroenterol Nutr. 2002;35:391–8.
3. Gregorio GV, McFarlane B, Bracken P, Vergani D, Mieli-Vergani G. Organ and nonorgan specific autoantibody titres and IgG levels as markers of disease activity: a longitudinal study in childhood autoimmune liver disease. Autoimmunity. 2002;35:515–19.
4. Johnson PJ, McFarlane IG, Williams R. Azathioprine for longterm maintenance of remission in autoimmune hepatitis. N Engl J Med. 1995;333:958–63.
5. Alvarez F, Ciocca M, Canero-Velasco C et al. Short-term cyclosporine induces a remission of autoimmune hepatitis in children. J Hepatol. 1999;30:222–7.
6. Bucy PB, Yan Xu X, Li J, Huang GQ. Cyclosporin A-induced autoimmune disease in mice. J Immunol. 1993;151:1039–50.

7. Sakaguchi S, Sakaguchi N. Role of genetic factors in organ-specific autoimmune disease induced by manipulating the thymus or T cells, and non-self antigens. Rev Immunogenet. 2000;2:147–53.
8. Demoiseaux JG, van Breda Vriesman PJ. Cyclosporin A-induced autoimmunity: the result of defective *de novo* T-cell development. Folia Biol. 1998;44:1–9.
9. Richardson PD, James PD, Ryder SD. Mycophenolate mofetil for maintenance of remission in autoimmune hepatitis in patients resistant to or intolerant of azathioprine. J Hepatol. 2000;33:371–5.
10. Maggiore G, Hadchouel M, Alagille D. Life-saving immunosuppressive treatment in severe autoimmune chronic active hepatitis. J Pediatr Gastroenterol Nutr. 1985;4:655–8.
11. Debray D, Maggiore G, Girardet JP, Mallet E, Bernard O. Efficacy of cyclosporine A in children with type 2 autoimmune hepatitis. J Pediatr. 1999;135:111–14.
12. Debray D, Bernard O. Autoimmune hepatitis (AIH) in children. Cyclosporin treatment of autoimmune hepatitis. J Pediatr Gastroenterol Nutr. 1995;20:470.
13. Gregorio GV, Portmann B, Karani J et al. Autoimmune hepatitis/sclerosing cholangitis overlap syndrome in childhood: a 16-year prospective study. Hepatology. 2001;33:544–53.
14. Czaja AJ. Frequency and nature of the variant syndromes of autoimmune liver disease. Hepatology. 1998;28:360–5.
15. Lee YM, Kaplan MM. Primary sclerosing cholangitis. N Engl J Med. 1995;332:924–33.
16. Wilschanski M, Chait P, Wade JA et al. Primary sclerosing cholangitis in 32 children: clinical, laboratory, and radiographic features, with survival analysis. Hepatology. 1995;22: 1415–22.
17. Beuers U, Spengler U, Kruis W et al. Ursodeoxycholic acid for treatment of primary sclerosing cholangitis: a placebo controlled trial. Hepatology. 1992;16:707–14.
18. Lebovics E, Salama M, Elhosseiny A. Resolution of radiographic abnormalities with ursodeoxycholic acid therapy of primary sclerosing cholangitis. Gastroenterology. 1992; 102:2143–7.
19. Lindor KD. Ursodiol for primary sclerosing cholangitis. Mayo Primary Sclerosing Cholangitis-Ursodeoxycholic Acid Study Group. N Engl J Med. 1997;336:691–5.

26
Indications and results of liver transplantation for autoimmune liver disease in children

D. KELLY

INTRODUCTION

The rapid improvement in liver transplantation has led to long-term survival and good quality life for many children and adults[1]. The outcome for autoimmune liver disease, including autoimmune hepatitis and primary sclerosing cholangitis following transplantation, has important differences from other indications.

AUTOIMMUNE LIVER DISEASE TYPES I AND II

The majority of children with autoimmune hepatitis (AIH) types I or II respond to immunosuppression with prednisolone or azathioprine but between 25% and 50% require liver transplantation[2-4]. Liver transplantation is indicated for:

1. Failure of medical treatment. Despite an apparent initial response to immunosuppression, gradual histological progression over years may occur. Failure of medical treatment is more likely when established cirrhosis is present at diagnosis, when there is a long history prior to commencing treatment, and in early-onset AIH. Lack of response to immunosuppression despite second-line drugs such as cyclosporin A, tacrolimus, or mycophenolate mofetil may develop early in the disease, or after relapse due to non-adherence in the teenage years.

2. Intolerable side-effects of immunosuppression. In children with resistant autoimmune hepatitis, steroid side-effects such as obesity and cushingoid facies may add to their psychological distress, while osteoporosis and diabetes may be more serious medical problems. Growth suppression as a result of steroids may compound malnutrition related to liver failure.

3. The development of endstage liver failure with jaundice, malnutrition, ascites, encephalopathy, and coagulopathy which is not controlled by medical therapy. The timing of transplantation for chronic liver failure secondary to autoimmune disease may be difficult, as many children have well-compensated liver function but poor quality of life. Use of the Pediatric End-Stage Liver Disease Score (PELD) to predict death[5] or evidence of a persistent rise in total bilirubin >150 mol/L, prolongation of prothrombin ratio (INR >1.4) and a fall in serum albumin >35 g/L and growth failure may be helpful[6].

4. Advanced portal hypertension or intractable variceal bleeding which is not controlled by oesophageal ligation, sclerotherapy or the placement of a trans-jugular intrahepatic portosystemic shunt is an important reason, particularly in adolescents.

5. Poor quality of life is often quoted as an indication for transplantation, but can be difficult to judge in an adolescent. Young people may complain of fatigue or lethargy, and of being unable to attend school, and it may be difficult to differentiate between adolescence, the disease, or the side-effects of treatment.

6. Fulminant hepatic failure. Fulminant autoimmune hepatitis is a relatively rare indication for transplantation in children, accounting for approximately 25% of all cases of fulminant hepatitis. Children with autoimmune hepatitis type II are more likely to present in fulminant hepatic failure and have an increased requirement for liver transplantation[2,3]. The indications for transplantation are similar to other causes of fulminant hepatitis and include:

- rapid onset of coma with progression to grade III or IV hepatic coma;
- diminishing liver size;
- falling transaminases;
- increasing bilirubin (>300 mol/L);
- persistent coagulopathy (>50 s/control; INR >4).

Unlike adults, children with fulminant hepatitis may have more severe coagulopathy than encephalopathy[7]. High doses of steroids pretransplant may exacerbate encephalopathy.

OVERLAP SYNDROME/AUTOIMMUNE SCLEROSING CHOLANGITIS

The overlap syndrome of sclerosing cholangitis (SC) and AIH may be more common in children[3,8,9], and may be associated with inflammatory bowel disease in up to 81%. Treatment with a combination of ursodeoxycholic acid (20 mg/kg), prednisolone and azathioprine may control the liver disease, while olsalazine is helpful to maintain remission if inflammatory bowel disease is present. Approximately 25% of children require transplantation for the development of biliary cirrhosis, with increasing cholestasis, intractable pruritus, fat-soluble vitamin deficiency, malnutrition and growth failure[9].

TRANSPLANT TECHNIQUE

There are now many different liver transplant grafts available which include: orthotopic (whole graft), reduction hepatectomy (a reduced graft) which is mostly used for small infants, a split-liver graft (used for two recipients), living-related graft (usually from a parent), or auxiliary graft (in which only part of the liver is replaced). All grafts have similar outcomes, but auxiliary grafts are not suitable for autoimmune disease because of the risk of persistent disease and/or the development of carcinoma in the native liver[10].

POST-TRANSPLANT MANAGEMENT

Immunosuppression

The choice of immune suppression is critical in patients transplanted for autoimmune disease because of the risk of recurrence. There have been many recent advances in immunosuppressive drugs, but current protocols consist of the calcineurin inhibitors:

1. cyclosporin microemulsion (neoral), prednisolone and azathioprine;

2. tacrolimus combined with prednisolone;

3. interleukin-2 antibodies with either tacrolimus or cyclosporin, and steroids.

Some European units reduce steroids over the first 2 weeks and either withdraw or reduce to alternate-day therapy after 3 months to improve growth. Complete withdrawal of steroids is contraindicated in autoimmune disease because of the risk of recurrence. Azathioprine is usually discontinued after 1 year in normal circumstances, but may be restarted or continued to prevent recurrent disease. Cyclosporin or tacrolimus are continued for life. There is some evidence that recurrent autoimmune disease is more likely with calcineurin inhibitors, and more so with tacrolimus than with cyclosporin, but immediate post-transplant immunosuppression is dependent on these agents[11]. A recent European multicentre study, which directly compared tacrolimus to neoral post-transplant, demonstrated a significant reduction in the incidence of acute and steroid-resistant rejection in the tacrolimus group compared to the cyclosporin group without significant difference in adverse side-effects[12], although the effect on recurrent autoimmune disease was not evaluated.

Mycophenolate mofetil (MMF) (10–40 mg/kg) has recently been developed as an adjuvant immunosuppressive agent. It is an antiproliferative agent which is similar in action to azathioprine and may depress the bone marrow. The long-term safety and efficacy is undetermined but it has no cosmetic side-effects, is renal sparing and does not require drug monitoring. It was initially used as rescue therapy and was found to be effective and safe but with significant gastrointestinal and haematological side-effects in adults. More recently MMF has been used with neoral and prednisolone as primary immunosuppression and to treat recurrent or *de-novo* autoimmune disease[13].

Anti-interleukin-2 receptor antibodies (IL-2 antibodies) are monoclonal antibodies, which selectively target the IL-2 receptors on activated T cells, which is a key step in the development of cell-mediated immunity. Two antibodies are available, basiliximab and daclizumab, both of which are renal-sparing and provide effective induction immunosuppression post-transplant in combination with a calcineurin inhibitor in adults. To date there is little experience with children with autoimmune disease[14,15].

Sirolimus is a macrocyclic triene antibiotic which prevents T cell proliferation by inhibiting cytokine production and does not inhibit calcinueurin. In adults, sirolimus has been evaluated as both primary and rescue immunosuppression for liver transplant recipients and has the advantage of being both renal-sparing and reducing the need for high-dose steroids[16]. Significant side-effects include delayed wound healing, hyperlipidaemia and an increase in the rate of hepatic artery thrombosis. Sirolimus should not be used immediately post-transplant but may be useful for chronic rejection or for recurrent disease.

Future studies with the combination of interleukin 2 antibodies with drugs such as MMF and/or sirolimus may prove beneficial in autoimmune disease, but it is unlikely that plans to use a steroid-free protocol[17] will have a role in transplantation for this indication.

POST-TRANSPLANT PROBLEMS

As all patients transplanted for autoimmune disease will have had long-term or high-dose steroids pretransplant, delays in wound healing or increased infection might be anticipated, but have not been reported in the literature.

Several studies have indicated that there is an increase in both acute and chronic rejection post-transplant in patients transplanted either for AIH or SC. Incidences vary from 64% to 88%, but the mechanism for the increased incidence is unknown, and may relate to relative resistance to steroids because of prolonged use[18,19]. Surprisingly, one study of transplantation for fulminant AIH found a lower than expected incidence of rejection (33%) compared to other indications[20]

Post-transplant lymphoproliferative disease (PTLD), which occurs in 5–10% of paediatric recipients, may be less common in autoimmune disease[21], perhaps because more of the recipients are EBV-positive pretransplant and are therefore less likely to develop a primary EBV infection leading to PTLD.

RECURRENT DISEASE

The recurrence of both AIH and SC is well documented in both adults and children. Several studies have indicated that AIH may recur in approximately 25% of children both immunologically and histologically, and may be more severe than the original disease[22]. A recent study from Hanover identified recurrence in 32% of recipients, while a study from Atlanta noted that children had a similar risk of recurrence as adults[18,19]. Our own data from Birmingham support this with a 28% recurrence rate. Another study reviewing outcome post-transplant in children with both AIH and cryptogenic cirrhosis, noted a higher recurrence rate in Hispanic children[4]. Although the mechanism for recurrence is unknown, it may share similarities with the development of *de-novo* autimmune disease (see below), and is thought to be more common in those with the major histocompatibility antigens (HLA DR 3/4)[23] (Tables 1 and 2).

Table 1 Outcome post-transplant for AIH in children

	5 year survival	Rejection	Recurrence
Hannover (n = 28)	78%	88%	32%
UCLA (n = 30)	87%	–	33%
Birmingham (n = 14)	78%	64%	28%
Spain* (n = 18)	78%	33%	5%

*Fulminant Hep in adults

Table 2 Outcome post-transplant for PSc in children

	% year survival	Rejection	Recurrence
Birmingham (n = 7)	85%	71%	14%
Rochester (n = 11)	64%	–	27%
Atlanta (n = 13)	85%	50%	10%*

*6-fold increase in colitis post Tx compared to adults

The diagnosis is based on the development of clinical symptoms, characteristic histology, abnormal biochemistry with raised hepatic transaminases, elevated immunoglobulin G and positive autoantibodies (ANA, SMA, LKM)[23]. A French study of adult women transplanted for AIH[24] noted a 41% recurrence rate, but also that histological recurrence predated abnormal biochemistry or clinical symptoms, suggesting that protocol biopsies are important. Treatment with steroids and azathioprine and/or MMF usually controls the disease[23] but graft loss and the necessity for retransplantation has been reported[24] in adults and led to retransplantation in 21% of our AIH recipients in Birmingham.

The recurrence rate in adults transplanted for SC is 6–14%[25]. Studies in children indicate that the recurrence rate may be higher, ranging from 14% in our unit in Birmingham to 27% in Rochester[9]. In contrast, a French study reported excellent survival with 11/15 (73%) children alive without recurrence 6 months to 6 years after transplantation[8]. It is even more difficult to detect recurrence following transplantation for SC, as the clinical, radiological and histological appearances may mimic hepatic ischaemia, biliary obstruction or cholangitis, and careful assessment is required[23].

Ulcerative or indeterminate colitis may develop for the first time after liver transplantation[9] and the incidence may be higher in children than in adults[19]. Treatment is as above with ursodeoxcholate, increased immunosuppression and supportive management. Retransplantation for resistant disease has been reported in 1/11[9] and in 1/7 of our group in Birmingham, but a second recurrence is likely and more effective immunosuppression may be required to prevent it.

DE-NOVO AIH

A number of recent studies have documented the development of autoantibodies (ANA, SMA and rarely LKM) post-transplant in both children and adults in recipients who did not have autoimmune disease pretransplant[26,27]. The incidence varies from 2–3% to 50% with time, and is associated with a graft hepatitis and progressive fibrosis and/or cirrhosis[28] (Figure 1). The incidence does not appear to be related to the choice of calcineurin inhibitor or pre-transplant diagnosis. The development of autoantibodies predates, in some cases, the biochemical or histological changes[28]. The clinical symptoms, serology and histology are similar to recurrent AIH, suggesting that the underlying mechanism is the same. The aetiology is unknown; it may represent a form of chronic rejection, but it is more likely to be related to a different immune reaction to the graft, as de-novo AIH has been reported in an auxiliary graft but not the native liver[29]. It has been suggested that the mechanism is related to molecular mimicry with viral antigens from common post-transplant infections such as cytomegalovirus or EBV[30], or prevention of T cell maturation by calcineurin inhibitors allowing the emergence of autoaggressive T cell clones[31]. The hepatitis resolves with steroid therapy, or with azathioprine[27,32] or MMF.

SURVIVAL FOLLOWING LIVER TRANSPLANTATION

Current results from international units indicate that 1-year survival after paediatric liver transplantation may be as high as 90% with long-term survival (5–8 years) from 60% to 80%[1,33,34]. Patients receiving elective living-related liver transplantation may have a higher 1-year survival (94%) compared to those receiving cadaveric grafts (78%)[35]. Survival following transplantation for autoimmune liver disease is similar, with 5-year survival figures of 78% to 87%[4,18] (Figure 2). A Spanish study which evaluated the outcome of transplan-

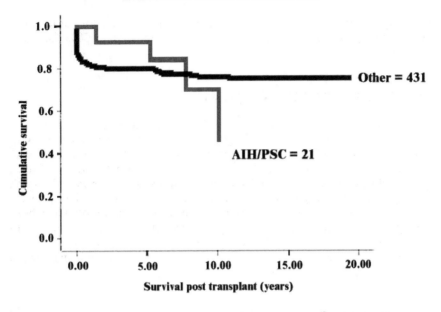

Figure 1 Liver transplantation for autoimmune liver disease compared to other indications

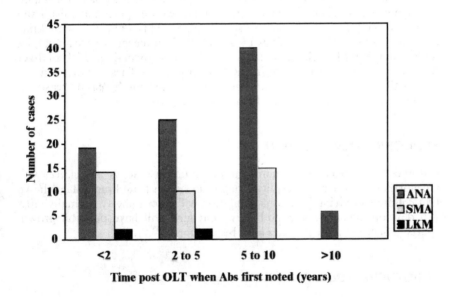

Figure 2 Time post OLT when positive antibodies were first noted in 103 children

289

tation for fulminant hepatitis found that patients transplanted for autoimmune disease had improved 5-year survival of 78% compared to 50% for other indications[20].

QUALITY OF LIFE POST-TRANSPLANT

As more children survive long-term post-transplant, attention has turned to long-term quality of life. There is good evidence that nutrition returns to normal, with 80% of survivors achieving normal growth patterns and body habitus[36,37]. Linear growth may be delayed for between 6 and 24 months, which is directly related to steroid dosage and preoperative stunting[38,39], which may be a particular problem in autoimmune liver disease. The growth-suppressant effects of corticosteroids in liver transplantation have been documented by many units, leading to the use of alternate-day steroids to promote catch-up growth[40]. As steroids cannot be discontinued in patients transplanted for autoimmune liver disease, alternate-day steroids may be the best way to encourage growth.

PSYCHOSOCIAL DEVELOPMENT

There is an initial deterioration in psychosocial development post-transplant, as noted by deterioration in social skills, language development, and eye/hand coordination for up to 1 year post-transplant. This improves with time, returning to pretransplant values by 2 years[41]. Two recent American studies found that Health Related Quality of Life was normal in 75% of liver transplant recipients who were in full-time education. These studies did not specifically evaluate outcome in autoimmune patients, but there is no reason to assume that they would respond differently[42,43].

ENDOCRINE DEVELOPMENT

Chronic liver disease, particularly autoimmune disease, is associated with amenorrhoea and reduced fertility. Long-term studies from France have shown that children surviving liver transplantation will enter puberty normally; girls will develop menarche and both boys and girls will have pubertal growth spurts[38]. Successful pregnancies have been reported[44].

LONG-TERM RENAL FUNCTION

The development of nephrotoxicity with both cyclosporin and tacrolimus is inevitable, although only 4–5% of patients develop severe chronic renal failure long-term requiring renal transplantation. The use of low-dose calcineurin inhibitors or renal-sparing drugs such as MMF or sirolimus for maintenance immunosuppression prevents significant renal dysfunction[45,46]. Acute postoperative hypertension is seen in 65% of children, but only persists in 28%[47]. It is more likely to be a problem in children on long-term steroids pre- and post-transplant.

NON-COMPLIANCE WITH THERAPY

Non-compliance with immunosuppressive therapy is a problem in all adolescents and may be a particular issue for young people transplanted with autoimmune liver disease in the teenage years[48,49] because of the cosmetic side-effects of steroids. Non-compliance is the main cause of late graft loss in adolescence post-liver transplant[50]. The management of non-adherence is difficult, and relies on a non-judgemental approach and efforts to improve education, social functioning and behavioural strategies to encourage self-motivation[51].

TRANSITION TO ADULT CARE

As most children with autoimmune liver disease are transplanted in adolescence, it is important that their post-transplant management includes a plan to transfer them to an adult unit. The ideal situation includes having a clearly defined transition policy which allows plenty of time for preparation for transfer, ideally in an adolescent clinic, giving control of time for transfer to the young persons to choose when they feel emotionally and psychologically ready for transfer, providing adequate information about the organization of the adult clinic and finally the facility to hold joint adult and paediatric clinics at the adult hospital[49,51].

SUMMARY

Liver transplantation is effective therapy for liver failure, leading to >80% survival long-term with excellent quality of life. The prognosis for survival is similar for patients with autoimmune liver disease, but because of the risk of recurrence, morbidity and the need for retransplantation may be increased.

References

1. Kelly DA. Current results and evolving indications for liver transplantation in children. J Pediatr Gastroenterol Nutr. 1998;27:214–21.
2. Gregorio GV, Portmann B, Reid F et al. Autoimmune hepatitis in childhood: a 20-year experience. Hepatology. 1997;25:541–7.
3. Gregorio GV, Portmann B, Karani J et al. Autoimmune hepatitis/sclerosing cholangitis overlap syndrome in childhood: a 16-year prospective study. Hepatology. 2001;33:544–53.
4. Bahar RJ, Yanni GS, Martin MG et al. Orthotopic liver transplantation for autoimmune hepatitis and cryptogenic hepatitis in children. Transplantation. 2001;72:829–33.
5. McDiarmid SV, Anand R, Lindblad AS. The Principal Investigators and Institutions of the Studies of Pediatric Liver Transplantation (SPLIT) Research Group. Development of a

pediatric end-stage liver disease score to predict poor outcome in children awaiting liver transplantation.Transplantation. 2002;74:173–81.

6. Malatack JJ, Schald DJ, Urbach AH et al. Choosing a paediatric recipient for orthotopic liver transplantation. J Pediatr. 1987;112:479–89.

7. Bonatti H, Muiesan P, Connolly S et al. Liver transplantation for acute liver failure in children under 1 year of age. Transplant Proc. 1997;29:434–5.

8. Debray D, Pariente D, Urvoas E, Hadchouel M, Bernard O. Sclerosing cholangitis in children. J Pediatr. 1994;124:49–56.

9. Feldstein AE, Perrault J, El-Youssif M, Lindor KD, Freese DK, Angulo P. Primary sclerosing cholangitis in children: a long-term follow-up study. Hepatology. 2003;38:210–17.

10. Rela M, Muiesan P, Andreani P et al. Auxiliary liver transplantation for metabolic diseases. Transplant Proc. 1997;29:444–5.

11. Neuberger J, Portmann B, Macdougal BR, Calne RY, Williams R. Recurrence of primary biliary cirrhosis after liver transplantation. N Engl J Med. 1982;306:1–4.

12. Kelly DA, Jara P, Rodeck B et al. Tacrolimus and steroids versus cyclosporin microemulsion, steroids, and azathioprine in children undergoing liver transplantation: randomised European multicentre trial. Lancet. 2004;364:1054–61.

13. Chardot C, Nicoluzzi JE, Janssen M, Sokal E, Lerut J, Otte JB, Reding R. Use of mycophenolate mofetil as rescue therapy after paediatric liver transplantation. Transplantation. 2001;71:224–9.

14. Ganschow R, Broering DC, Stuerenburg I, Rogiers X, Hellwege HH, Burdelski M. First experience with basiliximab in pediatric liver graft recipients. Pediatr Transplant. 2001;5:353–8.

15. Arora N, McKiernan PJ, Beath SV, deVille de Goyet J, Kelly DA. Concomitant basiliximab with low-dose calcineurin inhibitors in children post-liver transplantation. Pediatr Transplant. 2002;6:214–18.

16. McAlister VC, Peltekian KM, Malatjalian DA et al. Orthotopic liver transplantation using low-dose tacrolimus and sirolimus. Liver Transplant. 2001;7:401–8.

17. Reding R, Gras J, Sokal E, Otte JB, Davies HF. Steroid free liver transplantation in children. Lancet. 2003;362:2068–70.

18. Vogel A, Heinrich E, Bahr MJ et al. Long-term outcome of liver transplantation for autoimmune hepatitis. Clin Transplant. 2004;18:62–9.

19. Heffron TG, Smallwood GA, Oakley B et al. Adult and pediatric liver transplantation for autoimmune hepatitis. Transplant Proc. 2003; 5;1435–6

20. Nunez-Martinez O, De la Cruz G, Salcedo M et al. Liver transplantation for autoimmune hepatitis: fulminant versus chronic hepatitis presentation. Transplant Proc. 2003;35:1855–6.

21. Khettry U, Keaveny A, Goldar-Najafi A et al. Liver transplantation for primary sclerosing cholangitis: a long-term clinicopathologic study. Hum Pathol. 2003;34:1127–36.

22. Birnbaum AH, Benkov KJ, Pittman NS, McFarlane-Ferrerira Y, Rosh JR, LeLeiko NS. Recurrence of autoimmune hepatitis in children after liver transplantation. J Pediatr Gastroenterol Nutr. 1997;25:20–5.

23. Diego Vergani, Mieli-Vergani G. Autoimmunity after liver transplantation. Hepatology. 2002;36:271–6.

24. Duclos-Vallee JC, Sebagh M, Rifai K et al. A 10-year follow-up study of patients transplanted for autoimmune hepatitis: histological recurrence precedes clinical and biochemical recurrence. Gut. 2003;52:893–7.

25. Harrison RF, Davies MH, Neuberger JM, Hubscher SG. Fibrous and obliterative cholangitis in liver allografts; evidence of recurrent primary sclerosing cholangitis? Hepatology. 1994;20:356–61.

26. Kerkar N, Hadzic N, Davies ET et al. De-novo autoimmune hepatitis after liver transplantation. Lancet. 1998;351:409–13.

27. Andries S, Casamayou L, Sempoux C et al. Post-transplant immune hepatitis in pediatric liver transplant recipients: incidence and maintenance therapy with azathioprine. Transplantation. 2001;72:267–72.

28. Evans HM, McKiernan PJ, Beath SV, de Ville de Goyet J, Kelly DA. Histology of liver allografts following paediatric liver transplantation. J Pediatr Gastroenterol Nutr. 2001;32:383.

29. Miyagawa-Hayashino A, Haga H, Sakurai T, Shirase T, Manage T, Egawa H. *De-novo* autoimmune hepatitis affecting allograft but not the native: auxiliary partial orthotopic liver transplantation. Transplantation. 2003;76271–2.

30. Bogdanos DP, Choudhuri K, Vergani D. Molecular mimicry and autoimmune liver disease: virtuous intentions, malign consequences. Liver. 2001;21:225–32.

31. Bucy PB, Yan Xu X, Li J, Huang GQ. Cyclosporin A-induced autoimmune disease in mice. J Immunol. 1993;151:1039–50.

32. Salcedo M, Vaquero J, Banares R et al. Response to steroids in *de-novo* autoimmune hepatitis after liver transplantation. Hepatology. 2002;35:349–56.

33. Noujaim HM, Gunson B, Mayer D et al. *Ex-situ* split liver transplantation. Impact of a new protocol. Transplantation. 2002;74:1386–90.

34. European Liver Transplant Registry. 2002. http://www.elt.org

35. Hashikura Y, Kawasaki S, Terada M et al. Long-term results of living-related donor liver graft transplantation: a single-center analysis of 110 transplants. Transplantation. 2001;72:95–9.

36. Holt RI, Broide E, Buchanan CR et al. Orthotopic liver transplantation reverses the adverse nutritional changes of end-stage liver disease in children. Am J Clin Nutr. 1997;65:534–42

37. Viner RM, Forton JTM, Cole TJ, Clark IH, Nobel-Jamieson G, Barnes ND. Growth of long term survivors of liver transplantation. Arch Dis Child. 1999;80:235–40.

38. Codoner-Franch P, Bernard O, Alvarez F. Long-term follow-up of growth in height after successful liver transplantation. J Pediatr. 1994;124:368–73.

39. Rodeck B, Melter M, Hoyer PF, Ringe B, Brodehi J. Growth in long-term survivors after orthotopic liver transplantation in childhood. Transplant Proc. 1994;26:165–6.

40. Dunn SP, Falkenstein K, Lawrence JP et al. Monotherapy with cyclosporine for chronic immunosuppression in pediatric liver transplant recipients. Transplantation. 1994;57:544–7.

41. Van Mourik, IDM, Beath SV, Kelly D. Long term nutrition and neurodevelopmental outcome of liver transplantation in infants aged less than 12 months. J Pediatr Gastroenterol Nutr. 200;30:269–76.

42. Alonso EM, Neighbors K, Mattson C et al. Functional outcomes of pediatric liver transplantation. J Pediatr Gastroenterol Nutr. 2003;37:155–60.

43. Bucuvalas JC, Britto M, Krug S et al. Health-related quality of life in pediatric liver transplant recipients: a single-center study. Liver Transplant. 2003;9:62–71.

44. Jain A, Venkataramanan R, Fung JJ et al. Pregnancy after liver transplantation under tacrolimus. Transplantation. 1997;64:559–65.

45. Arora-Gupta N, Davies P, McKiernan P, Kelly DA. The effect of long-term calcineurin inhibitor therapy on renal function in children after liver transplantation. Pediatr Transplant. 2004;8:145–50.

46. Evans HE, McKiernan P, Kelly DA. Mycophenolate mofetil for renal dysfunction following paediatric liver transplantation. Transplantation. 2005 (Submitted).

47. Bartosh SM, Alonso EM, Whitington PF. Renal outcomes in pediatric liver transplantation. Clin Transplant. 1997;11:354–60.

48. Falkstein K, Flynn L, Kirkpatric B, Casa-Melley A, Dunn S. Non-compliance in children post liver transplant. Who are the culprits? Pediatr Transplant. 2004;8:233–6.

49. McDonagh JE, Kelly DA. Transitioning care of the paediatric recipient to adult caregivers. Pediatr Clin North Am. 2003;50:1561–83.

50. Sudan DI, Shaw Jr, BW, Langnas AN. Causes of late mortality in pediatric liver transplant recipients Ann Surg. 1998;227.289 95.

51. Kelly DA. Transition of the liver transplant recipient to adult care. Update Gastroenterology 2004 – New developments in the management of benign gastrointestinal disorders. Post Grad Course, Prague. 2004:181–91.

27
Cyclosporin indication in 'autoimmune hepatitis'

F. ALVAREZ

INTRODUCTION

The importance of immunosuppressant administration depends on laboratory and histological data at autoimmune hepatitis (AIH) onset[1]. The availability of various immunosuppressive medications and the identification of predictive features of response to treatment should guide the choice of a particular drug or combination of drugs. Recent experience in the management of liver transplantation with immunosuppressive drugs should help to reach a final decision. All immunosuppressants have significant side-effects, and extrahepatic autoimmune and non-autoimmune disorders in patients must be assessed carefully to avoid possible disease aggravation.

CHOICE OF THERAPY

Some factors should be considered in the choice of therapy for children with AIH:

1. the liver inflammatory process should be rapidly and completely controlled to allow hepatic regeneration;
2. cirrhosis is reversible[2] if the inflammation regresses;
3. treatment side-effects must be minimized;
4. age and sex of the patient can influence coping with side-effects;
5. some medications can aggravate particular extrahepatic autoimmune diseases and are contraindicated.

Standard treatment of AIH consists of high-dose corticosteroids alone or in association with azathioprine. Patients under such therapy show more than 80% complete response, frequently with deleterious side-effects[3,4]. Introduced in the 1980s as a powerful immunosuppressor, cyclosporin dramatically improved the outcome of liver transplantation. Some years ago it was also used successfully as rescue therapy in AIH patients not responding to corticosteroids and azathioprine[5,6]. Recently cyclosporin was administered successfully for a relatively short time period to control the liver inflammatory process, and was replaced by low doses of corticosteroids and azathioprine 6 months after the beginning of treatment. This protocol avoids the main side-effects of both therapies (corticosteroids and cyclosporin), and achieves more than 90% complete response.

TREATMENT PROTOCOL

Cyclosporin was first used as immunosuppressive medication to prevent or treat cellular rejection. New pharmacological knowledge and experience acquired by clinicians extended its application to several autoimmune diseases. The drug was given successfully to AIH patients when they failed to respond to prednisone/azathioprine therapy, and when serious side-effects of this combination occurred. Recently, we undertook a pilot study that showed the value of cyclosporin as initial therapy in AIH patients[7]. The main problem of cyclosporin administration is its long-term side-effects. To avoid them this powerful immunosuppressor can be proposed as *induction therapy* in patients with severe inflammatory liver disease, followed by low doses of prednisone and azathioprine once serum aminotransferases are less than twice the normal values (Figure 1). Rapid control of the inflammation plays an important role in the prognosis.

Cyclosporin is a member of the calcineurin inhibitor group of immunosuppressors. They reduce the production and release of interleukin-2, an important cytokine in the activation of resting T cells. In AIH patients cyclosporin is administered orally, and its absorption depends on the presence of food, bile

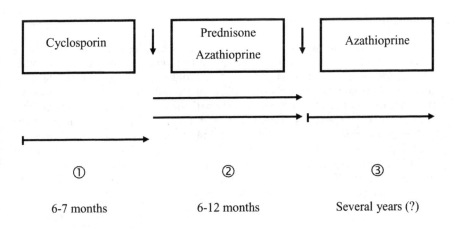

Figure 1   Induction of remission by short-term cyclosporin use

acids, and gastrointestinal motility. The starting oral dose of 4 mg/kg per day must be adjusted until cyclosporin blood levels reach between 250 and 300 ng/ml. Thereafter, these levels are reduced to around 200 ng/ml. When serum aminotransferase levels are less than 2 times the normal values the treatment is switched to low prednisone doses and azathioprine. In a previous study the required dosages were between 2.8 and 6.5 mg/kg per day[1]. Cyclosporin is given in three daily doses, every 8 h, to decrease the blood peak (toxicity) and to reduce the total amount of drug administered (cost). Blood levels should be controlled every 2–3 days during the first week, and once a month when the desired target is reached.

TREATMENT OUTCOME

Almost 95% of patients show a normalization of amintotransferases levels[6]. Cyclosporin allows an increase of growth velocity[6]. Some extrahepatic auto-immune diseases are improved or not affected by this treatment.

The adverse reactions observed are: hypertrichosis in 50% of patients, moderate or mild gingival hypertrophy in one-third of them, and headaches in 3% of children. All these symptoms are transient, and disappear when cyclosporin is replaced by prednisone and azathioprine. No long-term adverse effects have been reported. Three daily doses and close control of cyclosporin blood levels may explain the favourable outcome.

Cyclosporin induces thymus-dependent autoimmunity in laboratory animals[8]. However, this complication has not been described in humans. In our patients no aggravation or onset of autoimmunity has been observed during or after short-term cyclosporin treatment .

No relapse has occurred in any patient during the switch from cyclosporin to prednisone and azathioprine.

CONCLUSION

AIH is an uncommon liver disease affecting children and adults. Early diagnosis and treatment improve the initial response and long-term outcome. The choice of treatment must take into account patient age; clinical, labora-tory, and histological features that allow prediction of the response; and the presence or absence of associated extrahepatic disorders. In specialized centres short-term cyclosporin is used safely and successfully to control the liver inflammatory process. Low doses of prednisone in combination with azathioprine are sufficient to sustain the response. Maintenance treatment must be administered for several years, and withdrawal can be attempted after at least 4 years of complete and sustained response. Future research should focus on the recovery of immune homeostasis in these patients.

References

1. Alvarez F. Treatment of autoimmune hepatitis: current and future therapies. Curr Treat Options Gastroenterol. 2004;7:413–20.
2. Malekzadeh R, Mohamadnejad M, Nasseri-Moghaddam S et al. Reversibility of cirrhosis in autoimmune hepatitis. Am J Med. 2004;117:125–9.
3. Maggiore G, Veber F, Bernard O, et al. Autoimmune hepatitis associated with anti-actin antibodies in children and adolescents. J Pediatr Gastroenterol Nutr. 1993;17:376–81.
4. Gregorio GV, Portmann B, Reid F et al. Autoimmune hepatitis in childhood: a 20-year experience. Hepatology. 1997;25:541–7.
5. Hyams JS, Ballow M, Leichtner AM. Cyclosporine treatment of autoimmune chronic active hepatitis. Gastroenterology. 1987;93:890–3.
6. Jackson LD, Song E. Cyclosporin in the treatment of corticosteroid resistant autoimmune chronic active hepatitis. Gut. 1995;36:459–61.
7. Alvarez F, Ciocca M, Canero-Velasco C et al. Short-term cyclosporine induces a remission of autoimmune hepatitis in children. J Hepatol. 1999;30:222–7.
8. Wu DY, Goldschneider I. Cyclosporin A-induced autologous graft-versus-host disease: a prototypical model of autoimmunity and active (dominant) tolerance coordinately induced by recent thymic emigrants. J Immunol. 1999;162:6926–33.

Index

Falk Symposium Series

Falk Symposium Series